Fluorescence-Guided Surgery

Takeaki Ishizawa
Editor

Fluorescence-Guided Surgery

From Lab to Operation Room

JSFGS

Springer

Editor
Takeaki Ishizawa
Department of Hepatobiliary-Pancreatic Surgery
Graduate School of Medicine
Osaka Metropolitan University
Osaka, Japan

Contribution by
Japanese Society for Fluorescence Guided Surgery (JSFGS)
Tokyo, Japan

ISBN 978-981-19-7371-0 ISBN 978-981-19-7372-7 (eBook)
https://doi.org/10.1007/978-981-19-7372-7

The translation was done with the help of artificial intelligence (machine translation by the service DeepL.com). A subsequent human revision was done primarily in terms of content.

This Springer imprint is published by the registered company Springer Nature Singapore Pte Ltd.
The registered company address is: 152 Beach Road, #21-01/04 Gateway East, Singapore 189721, Singapore

Foreword 1

Fluorescence imaging, especially the use of indocyanine green (ICG), is a technology that originated in Japan, and many innovations and clinical applications of fluorescence-guided surgery using ICG have been made in Japan. Today we commonly use these techniques and methods in a wide range of medical fields and disciplines, and the standardization and internationalization of these methods is progressing year by year. Surgical field physicians have applied this technique to vascular imaging, lymphatic and lymph node imaging, and imaging to identify anatomical structures. The scope of application is very wide, and the application procedure is also diverse.

In this book, the authors provide state-of-the-art information by disease area, application methodology, and historical background. The authors' enthusiasm for fluorescence imaging and fluorescence-guided surgery is evident in their excellent results, ideas, and perspectives. The authors provide essential practical information on the technique, such as agent injection timing and localization, amount of administration, details of the equipment, novel projection systems, new fluorescent probes, and combinations with other imaging techniques. On the other hand, there are examples of increased integration with therapy that has resulted in significant improvements in treatment outcomes. Fluorescence-based techniques and surgery, along with other therapeutic advances, have revolutionized in various areas. In addition, we will discuss not only the benefits of this method, but also its limitations and challenges. This information is also extremely useful. Fluorescent imaging, integrated into endoscopic surgery and robotic surgery, plays a key role in the evolution of smart operating rooms. Artificial intelligence will detect fluorescent signals in near future. Accumulation of efforts to maximize the advantages and overcome the challenges will contribute to the overall progress of surgery.

Training is essential for learning new techniques. This book contains information on education and management of fluorescence-guided surgery, such as recording and video recording methods, links to medical records, and information systems. In addition, as you read this book through, there are organ-specific differences in the details of applications and methodologies that will help you better understand fluorescent-guided surgery.

The book contains a variety of future perspectives, including new fluorescent agents, next-generation equipment, and operating rooms. This book is full of beautiful illustrations, easy to understand, and has a simple yet detailed structure to help you find the essence of the subject.

Finally, I believe that the further development and evolution of fluorescence imaging technique and surgery and the creation of new concepts will continue, which will benefit patients.

Masakazu Toi

Professor, Breast Surgery
Graduate School of Medicine, Kyoto University
Representative Director Japanese Society
for Fluorescence Guided Surgery (JSFGS)
Kyoto, Japan
July 2022

Foreword 2

I still remember how impressed I was when I saw the ICG fluorescence imaging technique applied for the first time in the spring of 2007 by Dr. Takeaki Ishizawa, the editor of this book, in the OR at the University of Tokyo Hospital. I immediately felt that this technique would definitely be useful. It may be a cliché, but "seeing is believing," and all the members of the medical staff who saw the beautiful, clear fluorescent images were captivated by them. The Nobel Prize-winning GFP by Dr. Osamu Shimomura brought epoch-making progress in life science, and I believe that fluorescent imaging using ICG, which is almost non-toxic to the human body, excreted in bile, and available at low cost, is a Nobel Prize-class technology in terms of clinical impact.

The ICG fluorescence imaging technique has been applied to many areas of hepatobiliary surgery, including fluorescence cholangiography (especially in endoscopic surgery), hepatocellular carcinoma imaging, hepatic segmental staining, and visualization of cholestatic areas in the liver. The ICG fluorescence method has expanded to many areas, including blood flow evaluation of the gastrointestinal tract and transplanted liver, lymphangiography, and others. Collaboration with Dr. Yasuteru Urano, a basic medical scientist and an expert in the synthesis of fluorescent materials, was also a great fortunate encounter. New ideas such as cancer imaging and pancreatic fluid imaging using new fluorescent probes developed by Dr. Urano were born one after another.

The ICG Fluorescence Navigation Study Group, organized by Dr. Mitsuo Kusano, one of the pioneers of ICG fluorescence imaging, started in 2008 as a gathering of surgeons fascinated by this new technology. We invited Dr. Hisataka Kobayashi at NIH, who was an expert in photoimmunotherapy, a research area of great clinical significance as a treatment beyond imaging. In May 2020, the Ministry of Health, Labour and Welfare (MHLW) of Japan approved the use of ASP-1929 photoimmunotherapy for head and neck cancer. We look forward to further progress in this clinical trial.

The international expansion of ICG fluorescence imaging in the field of hepatobiliary and pancreatic surgery is largely due to the energetic networking efforts of Dr. Takeaki Ishizawa. As described in "COLUMN 2" of this report, Dr. Ishizawa met with Argentine surgeon Fernando Dip at the International Hepatobiliary and Pancreatic Surgery Association (IHPBA) meeting held in Buenos Aires in 2010, and Dr. Dip introduced this technique to Professor Rosenthal at the Department of Surgery, Cleveland Clinic in Florida, where Dr. Dip moved to. Together with Professor Michael Bouvet of the Department of Surgery at UC San Diego, we founded the International Society for Fluorescence Guided Surgery (ISFGS) in 2014. Back in Japan, the ICG Fluorescence Navigation Study Group was transformed to the Japan Society for Fluorescence Guided Surgery (JSFGS) in 2018 to promote international collaboration in this field.

The publication of this guidebook covering the latest knowledge and techniques of intraoperative fluorescence imaging at this time is truly timely, and we would like to express our sincere appreciation and respect to Dr. Takeaki Ishizawa, the editor, Dr. Masakazu Toi, the supervisor, and all others involved.

President, National Center for Global Health and Medicine Norihiro Kokudo
Tokyo, Japan

Executive Director, Japanese Society
for Fluorescence Guided Surgery (JSFGS)
Tokyo, Japan
July 2022

Preface

In April 2018, the first annual meeting of the "Japanese Society for Fluorescence Guided Surgery (JSFGS)" was held in Tokyo. More than 200 physicians, researchers, and business people gathered for a lively exchange of opinions that transcended the boundaries of their fields of expertise. This book is a guidebook on the development and clinical application of intraoperative fluorescence imaging, based on the program of the first meeting. We are very honored to have the opportunity to publish the English version, which will be of practical use to more and more people working on "fluorescence-guided surgery" in the world.

"Fluorescence-guided surgery" is a surgical procedure that aims to improve outcomes by using intraoperative fluorescence imaging to identify target biological structures and evaluate organ function and blood perfusion. Today, intraoperative fluorescence imaging is beginning to be implemented in almost every surgical department, literally from head to toe, for one application or another. The primary users of this technology are definitely surgeons in the operating room. However, the understanding of nurses, medical engineers, and manufacturers is also essential for the safe and effective use of fluorescent imaging. On the other hand, new fluorescent probes and next-generation imaging devices are being developed each day in research laboratories in university and company. In these circumstances, the main purpose of this book is to share basic knowledge about intraoperative fluorescence imaging and to serve as a guidebook to facilitate the exchange of information for the next step. To achieve this goal, the editorial process was guided by the following principles

1. *To serve as a guide for the introduction of fluorescent probes and imaging devices into the operating room:* The book covers all fluorescent probes and imaging devices currently available for clinical use in Japan and lists the features of each product (Introduction).
2. *To be a practical guide across the medical departments:* The "Practice" part is organized by application of fluorescence imaging (i.e., perfusion assessment, cancer identification, lymph node mapping/lymphography, anatomy visualization) rather than by organ. The book is richly illustrated with intraoperative photographs and videos to show the effectiveness of the techniques, and at the same time, knacks and pitfalls in their application are also clearly stated.
3. *To be a guide for research and development bridging the operating room and the laboratory*: At the beginning of each chapter of the practice parts, we have presented the challenges of conventional surgical methods. In contrast, the development part introduces the latest technologies that are being applied to medical treatment, including clinical veterinary medicine. We hope to match the seeds of researchers with the needs of clinicians.
4. *To serve as a guide for presentations at academic conferences and papers*: In the explanation of the development history and current status of each technology, we have tried to provide examples of first reports and major papers with a high level of evidence that should be cited in future presentations.

We hope that this guidebook will help intraoperative fluorescence imaging to develop into a truly useful technology for patient care. Finally, I would like to express my sincere gratitude to all those who have guided and supported the development of this guidebook and the activities of the JSFGS.

Osaka, Japan
September 2020

Takeaki Ishizawa

Contents

Editors and Contributors

Supervised

Japanese Society for Fluorescence Guided Surgery, Tokyo, Japan

Editor

Takeaki Ishizawa Department of Hepatobiliary-Pancreatic Surgery, Graduate School of Medicine, Osaka Metropolitan University, Osaka, Japan

Japanese Society for Fluorescence Guided Surgery (JSFGS), Tokyo, Japan

Contributors

Daisuke Aoki Department of Obstetrics & Gynecology, Keio University School of Medicine, Shinjuku-ku, Tokyo, Japan

Takeshi Aoki Division of Gastroenterological and General Surgery, Department of Surgery, School of Medicine, Showa University, Tokyo, Japan

Tohru Asai Department of Cardiovascular Surgery, Juntendo University, Bunkyo-ku, Tokyo, Japan

Hironobu Baba Department of Gastrointestinal Surgery, Tokyo Medical and Dental University, Bunkyo-ku, Tokyo, Japan

Toyofumi Fengshi Chen-Yoshikawa Department of Thoracic Surgery, Nagoya University Graduate School of Medicine, Nagoya, Aichi, Japan

Yuichiro Doki Department of Gastroenterological Surgery, Osaka University Graduate School of Medicine, Suita, Osaka, Japan

Hidetoshi Eguchi Department of Gastroenterological Surgery, Osaka University Graduate School of Medicine, Suita, Osaka, Japan

Hideo Fukuhara Department of Urology, Kochi Medical School, Nankoku-shi, Kochi, Japan

Center for Photodynamic Medicine, Kochi Medical School, Nankoku-shi, Kochi, Japan

Takeomi Hamada Division of Hepato-biliary-pancreas Surgery, Department of Surgery, Faculty of Medicine, University of Miyazaki, Miyazaki, Japan

Kazuhiro Hanazaki Department of Surgery, Kochi Medical School, Kochi University, Nankoku, Kochi, Japan

Center for Photodynamic Medicine, Kochi Medical School, Nankoku, Kochi, Japan

Hiro Hasegawa Department of Colorectal Surgery, National Cancer Center Hospital East, Kashiwa, Chiba, Japan

Kiyoshi Hasegawa Department of Hepatobiliary and Pancreatic Surgery, Artificial Organ and Transplantation Surgery, The University of Tokyo, Bunkyo-ku, Tokyo, Japan

Etsuro Hatano Department of Surgery, Graduate School of Medicine, Kyoto University, Sakyo-ku, Kyoto, Japan

Itaru Higuchi Department of Gastroenterological Surgery, Tokai University School of Medicine, Isehara, Kanagawa, Japan

Keiji Inoue Department of Urology, Kochi Medical School, Nankoku-shi, Kochi, Japan

Center for Photodynamic Medicine, Kochi Medical School, Nankoku-shi, Kochi, Japan

Masaaki Ito Department of Colorectal Surgery, National Cancer Center Hospital East, Kashiwa, Chiba, Japan

Masaki Kaibori Department of Surgery, Kansai Medical University, Hirakata, Osaka, Japan

Shinichi Kinami Department of Surgical Oncology, Kanazawa Medical University, Kahoku, Ishikawa, Japan

Yusuke Kinugasa Department of Gastrointestinal Surgery, Tokyo Medical and Dental University, Bunkyo-ku, Tokyo, Japan

Kazuo Kishi Department of Plastic and Reconstructive Surgery, Keio University School of Medicine, Shinjuku, Tokyo, Japan

Norihiro Kokudo National Center for Global Health and Medicine, Shinjuku, Tokyo, Japan

Japanese Society for Fluorescence Guided Surgery (JSFGS), Tokyo, Japan

Yoshiharu Kono Department of Hepatobiliary and Pancreatic Surgery, Artificial Organ and Transplantation Surgery, The University of Tokyo, Bunkyo-ku, Tokyo, Japan

Kazuo Koyanagi Department of Gastroenterological Surgery, Tokai University School of Medicine, Isehara, Kanagawa, Japan

Koshi Kumagai Division of Gastric Surgery, Department of Gastroenterological Surgery, The Cancer Institute Hospital of JFCR, Koto, Tokyo, Japan

Toshihiko Kuroiwa Tesseikai Neurosurgical Hospital, Shijonawate, Osaka, Japan

Yukinori Kurokawa Department of Gastroenterological Surgery, Osaka University Graduate School of Medicine, Suita, Osaka, Japan

Mitsuo Kusano Yoichi Hospital, Hokkaido Social Work Association, Yoichi, Japan

Kushiro Rosai Hospital, Kushiro, Hokkaido, Japan

Tomokazu Kusano Division of Gastroenterological and General Surgery, Department of Surgery, School of Medicine, Showa University, Tokyo, Japan

Ken Masamune Faculty of Advanced Techno Surgery (FATS), Institute of Advanced Biomedical Engineering and Science, Tokyo Women's Medical University, Tokyo, Japan

Fumihiro Matano Department of Neurological Surgery, Nippon Medical School, Bunkyo, Tokyo, Japan

Kazuhiro Matsuda Division of Gastroenterological and General Surgery, Department of Surgery, School of Medicine, Showa University, Tokyo, Japan

Kosuke Matsui Department of Surgery, Kansai Medical University, Hirakata, Osaka, Japan

Akio Morita Department of Neurological Surgery, Nippon Medical School, Bunkyo, Tokyo, Japan

Yoshihiro Muragaki Faculty of Advanced Techno Surgery (FATS), Institute of Advanced Biomedical Engineering and Science, Tokyo Women's Medical University, Tokyo, Japan

Yasuo Murai Department of Neurological Surgery, Nippon Medical School, Bunkyo, Tokyo, Japan

Akihiro Nakajo Department of Digestive Surgery, Breast and Thyroid Surgery, Kagoshima University Graduate School of Medical and Dental Sciences, Kagoshima, Japan

Tsutomu Namikawa Department of Surgery, Kochi Medical School, Kochi University, Nankoku, Kochi, Japan

Center for Photodynamic Medicine, Kochi Medical School, Nankoku, Kochi, Japan

Atsushi Nanashima Division of Hepato-biliary-pancreas Surgery, Department of Surgery, Faculty of Medicine, University of Miyazaki, Miyazaki, Japan

Yamato Ninomiya Department of Gastroenterological Surgery, Tokai University Hachioji Hospital, Hachioji, Tokyo, Japan

Toshihiko Nishidate Department of Surgery and Gastroenterological surgery, Muroran City General Hospital, Muroran, Hokkaido, Japan

Hironori Odaira Department of Gastroenterological Surgery, International University of Health and Welfare Hospital, Nasushiobara-Shi, Tochigi-Ken, Japan

Toshiro Ogura Department of Gastrointestinal Surgery, Saitama Cancer Center, Ina, Kitaadachi District, Saitama, Japan

Keisuke Okabe Department of Plastic and Reconstructive Surgery, Keio University School of Medicine, Shinjuku, Tokyo, Japan

Jun Okamoto Faculty of Advanced Techno Surgery (FATS), Institute of Advanced Biomedical Engineering and Science, Tokyo Women's Medical University, Tokyo, Japan

Yoshiharu Okamoto Department of Veterinary Clinical Medicine, School of Veterinary Medicine, Tottori University, Tottori, Japan

Kenji Okita Department of Surgical Oncology and Science, Sapporo Medical University, Sapporo, Hokkaido, Japan

Koichi Okuya Department of Surgical Oncology and Science, Sapporo Medical University, Sapporo, Hokkaido, Japan

Soji Ozawa Department of Gastroenterological Surgery, Tokai University School of Medicine, Isehara, Kanagawa, Japan

Kensuke Sakai Department of Obstetrics & Gynecology, Keio University School of Medicine, Shinjuku-ku, Tokyo, Japan

Yasuo Sekine Department of General Thoracic Surgery, Tokyo Women's Medical University Yachiyo Medical Center, Chiba, Japan

Satoru Seo Department of Surgery, Kochi Medical School, Nankoku, Kochi, Japan

Yoshiaki Shinden Department of Digestive Surgery, Breast and Thyroid Surgery, Kagoshima University Graduate School of Medical and Dental Sciences, Kagoshima, Japan

Akiko Suganami Department of Bioinformatics, Graduate School of Medicine, Chiba University, Chiba, Japan

Tomoharu Sugie Department of Breast Surgery, Kansai Medical University Hospital, Hirakata, Osaka, Japan

Nobuyuki Susumu Department of Obstetrics and Gynecology, International University of Health and Welfare, Narita, Chiba, Japan

Yutaka Suzuki Department of Gastroenterological Surgery, International University of Health and Welfare Hospital, Nasushiobara-Shi, Tochigi-Ken, Japan

Manami Tada Department of Breast Surgery, Kansai Medical University Hospital, Hirakata, Osaka, Japan

Tsuyoshi Takahashi Medical Education Center, Osaka University School of Medicine, Suita, Osaka, Japan

Department of Gastroenterological Surgery, Osaka University Graduate School of Medicine, Suita, Osaka, Japan

Ichiro Takemasa Department of Surgical Oncology and Science, Sapporo Medical University, Sapporo, Hokkaido, Japan

Nobuyuki Takemura Hepato-Biliary Pancreatic Surgery Division, Department of Surgery, National Center for Global Health and Medicine, Shinjuku, Tokyo, Japan

Manabu Tamura Faculty of Advanced Techno Surgery (FATS), Institute of Advanced Biomedical Engineering and Science, Tokyo Women's Medical University, Tokyo, Japan

Yutaka Tamura Department of Bioinformatics, Graduate School of Medicine, Chiba University, Chiba, Japan

Minoru Tanabe Department of Surgery, Kochi Medical School, Nankoku, Kochi, Japan

Center for Photodynamic Medicine, Kochi Medical School, Nankoku, Kochi, Japan

Masakazu Toi Professor, Breast Surgery, Graduate School of Medicine, Kyoto University, Sakyo-ku, Kyoto, Japan

Japanese Society for Fluorescence Guided Surgery (JSFGS), Tokyo, Japan

Yuichiro Tsukada Department of Colorectal Surgery, National Cancer Center Hospital East, Kashiwa, Chiba, Japan

Shunsuke Tsuzuki Department of Neurosurgery, Tokyo Women's Medical University, Tokyo, Japan

Faculty of Advanced Techno Surgery (FATS), Institute of Advanced Biomedical Engineering and Science, Tokyo Women's Medical University, Tokyo, Japan

Sunao Uemura Department of Surgery, Kochi Medical School, Kochi University, Nankoku, Kochi, Japan

Yasuteru Urano Graduate School of Pharmaceutical Sciences, Faculty of Pharmaceutical Sciences, The University of Tokyo, Tokyo, Japan

Shuichi Watanabe Department of Hepatobiliary and Pancreatic Surgery, Tokyo Medical and Dental University, Bunkyo-ku, Tokyo, Japan

Wataru Yamagami Department of Obstetrics & Gynecology, Keio University School of Medicine, Shinjuku-ku, Tokyo, Japan

Miho Yamamoto Department of Gastroenterological Surgery, Tokai University School of Medicine, Isehara, Kanagawa, Japan

Shinkuro Yamamoto Department of Urology, Kochi Medical School, Nankoku-shi, Kochi, Japan

Takumi Yamamoto Department of Plastic and Reconstructive Surgery, National Center for Global Health and Medicine, Shinjuku, Tokyo, Japan

Makoto Yamasaki Department of Gastroenterological Surgery, Osaka University Graduate School of Medicine, Suita, Osaka, Japan

Kentaro Yatabe Department of Gastroenterological Surgery, Tokai University School of Medicine, Isehara, Kanagawa, Japan

Masashi Yoshida Department of Surgery, International University of Health and Welfare Hospital, Nasushiobara, Tochigi, Japan

Part I

Basics of Intraoperative Fluorescence Imaging [Introduction]

Takeaki Ishizawa

Intraoperative fluorescence imaging requires a "fluorescent reagent" and an "imaging device (light source and camera)" that matches the wavelength of the reagent. In the current clinical settings, indocyanine green (ICG) and 5-aminolevulinic acid (5-ALA) are mainly used as the fluorescent reagents for various applications of intraoperative fluorescence imaging. Imaging devices are broadly classified into those for open surgery, those for endoscopic surgery (laparoscopy and thoracoscopy), and those for microscopic surgery, and the method of displaying fluorescent images and image resolution differ depending on the product. In order to utilize fluorescence imaging in the operating room, it is necessary to fully understand the features of fluorescent reagents and imaging devices. In addition, it will be increasingly important to create appropriate operative records and to store them with the videos in order to evaluate and report the efficacy of fluorescence imaging after surgery.

Clinically Available Fluorescent Reagents

1

Kosuke Matsui and Masaki Kaibori

Summary

- Indocyanine green (ICG) and 5-aminolevulinic acid (5-ALA) are typical fluorescent reagents approved by the pharmaceutical affairs bodies in Japan.
- Fluorescence imaging can be used for (1) angiography and evaluation of blood flow, (2) lymph node mapping and lymphography, (3) visualization of anatomical structures, and (4) cancer localization.

1 Introduction

In recent years, intraoperative fluorescence imaging has been clinically applied to identify the extent of cancer spread and anatomical structures in real time using fluorescent probes. In Japan, indocyanine green (ICG), 5-aminolevulinic acid (5-ALA, a precursor of porphyrin), and fluorescein have been approved by the pharmaceutical affairs bodies for clinical use, although their indications are still limited (Table 1.1). In this chapter, we will focus on the characteristics of ICG and 5-ALA for practical use in clinical situations.

Table 1.1 Characteristics of fluorescent agents approved in Japan

Product	Indocyanine green(ICG)	5-aminolevulinic acid (5-ALA)	Fluorescein
Dosage form	Lyophilized products	Lyophilized products or granule	Aqueous injection
Maximum (maximal) Absorption wavelength, excitation wavelength	805 nm	375 ~ 445 nm (blue light)	494 nm
Maximum (maximal) wavelength of fluorescence	835 nm (near-infrared fluorescence)	600 ~ 740 nm (red fluorescence)	521 nm (visible light region: Green)
Characteristics of fluorescent	Excited by infrared light (maximum absorption wavelength is around 805 nm) and emits fluorescence (maximum fluorescence wavelength is around 835 nm)	Exogenously administered 5-ALA is synthesized from glycine and succinyl CoA in vivo, metabolized to protoporphyrin IX (PP IX) in the process of biosynthesis from 5-ALA to heme, and selectively accumulated in tumor cells. PP IX is a photosensitizer and emits red fluorescence when excited by blue light (400–410 nm).	Represents strong green fluorescence under alkaline conditions

K. Matsui (✉) · M. Kaibori
Department of Surgery, Kansai Medical University, Hirakata, Osaka, Japan
e-mail: matsuik@hirakata.kmu.ac.jp

T. Ishizawa (ed.), *Fluorescence-Guided Surgery*, https://doi.org/10.1007/978-981-19-7372-7_1

2 Indocyanine Green

2.1 Indications (Insurance Coverage)

The insurance coverage of ICG in Japan has been limited to (1) liver function tests (plasma loss rate, blood stagnation rate, and hepatic blood flow measurement); (2) circulatory function tests (cardiac output, mean circulatory time, or abnormal blood flow measurement); (3) cerebral angiography during neurosurgery (by fluorescence measurement during infrared irradiation); and (4) identification of sentinel lymph nodes in breast cancer and malignant melanoma. In the 2018 revision of the medical fee system, intraoperative monitoring of blood flow (by measuring fluorescence during infrared irradiation) was approved. In the revision of medical fees in 2020, an additional fee can be calculated when intraoperative fluorescence imaging is used for the identification of tumors, blood vessels, and gastrointestinal perfusion with ICG or aminolevulinic acid hydrochloride.

2.2 Dosage and Administration

2.2.1 Liver Function Test

- For plasma disappearance rate measurement and blood stagnation rate measurement

A dose equivalent to 0.5 mg of ICG per kilogram of body weight is diluted to about 5 mg/mL with water for injection and gradually administered intravenously through the elbow vein within 30 seconds while paying attention to symptoms.

- For hepatic blood flow measurement

Dissolve 25 mg of ICG in as little water as possible for injection, dilute to a concentration of 2.5–5 mg/mL with saline solution, and administer the above solution equivalent to 3 mg of ICG intravenously. Thereafter, the drug is administered intravenously at a constant rate of 0.27–0.49 mg/min for about 50 minutes until blood sampling is completed.

2.2.2 Circulatory Function Test

Depending on the purpose, ICG is injected into various vascular sites from the intracardiac cavity to the peripheral veins, usually through the forearm vein. The dose per adult is 5–10 mg of ICG, or about 1–2 mL, and the dose for children is reduced according to their body weight.

2.2.3 Evaluation of Blood Flow in Blood Vessels and Tissues

Indocyanine green is dissolved in 5–10 mL of water for injection at a dose of 25 mg and administered intravenously at a dose of 0.04–0.3 mg/kg, depending on the intended use. In the case of cerebral angiography during neurosurgery, 25 mg of ICG is dissolved in 5 mL of water for injection, and the usual dose is 0.1–0.3 mg/kg administered intravenously.

2.2.4 Identification of Sentinel Lymph Node

In the identification of sentinel lymph nodes in breast cancer, 25 mg of ICG is dissolved in 5 mL of water for injection, and 5 mL or less is usually administered in the vicinity of the malignant tumor or subcutaneously in the nipple area in divided doses as appropriate. In the identification of sentinel lymph nodes in malignant melanoma, 25 mg of ICG is dissolved in 5 mL of water for injection, and 1 mL is usually administered in several intradermal locations near the malignant tumor in divided doses as appropriate.

Point

- Fluorescence imaging using indocyanine green for identification of liver cancer

In our center, the ICG retention rate at 15 minutes (ICG R15) is routinely measured by injecting ICG at a dose of 0.5 mg/kg for the estimation of liver function [1, 2]. For intraoperative identification of liver cancer, the ICG that was administered for measurement of ICGR15 can also be used as a fluorescent source, if patients underwent the liver function test within 14 days before surgery.

3 5-Aminolevulinic Acid

3.1 Indications (Insurance Coverage)

5-aminolevulinic acid, a precursor of porphyrin, is a naturally occurring amino acid in the body and a precursor of heme. In cancer cells, an increase in porphobilinogen deaminase activity and a decrease in ferrochelatase activity cause intracellular accumulation of protoporphyrin IX (PpIX), which emits red fluorescence with a peak at 635 nm upon excitation with 405 nm blue-violet light (Table 1.1). 5-ALA has been clinically applied as a photosensitizer in photodynamic diagnosis using this mechanism, and is covered by insurance in Japan for "visualization of tumor tissue during tumor resection of malignant glioma" in neurosurgery and "visualization of non-muscle layer invasive bladder cancer during transurethral resection of bladder tumor" in urological surgery.

3.2 Dosage and Administration

The usual adult dose of 20 mg/kg of aminolevulinic acid hydrochloride dissolved in water is orally administered 3 hours (range: 2–4 hours) before induction of anesthesia during surgery.

Point
- Fluorescence imaging using 5-aminolevulinic acid for identification of liver cancer

For this purpose, we orally administered 5-ALA hydrochloride (1 g) 3 hours before surgery. Care must be taken in hepatic dysfunction, which has been reported as an adverse event associated with 5-ALA [3].

4 Expected Future Applications

The current applications of fluorescence imaging include (1) angiography and evaluation of blood flow, (2) lymph node mapping and lymphography, (3) visualization of anatomical structures, and (4) cancer localization. The applications will be expanded to various fields of surgery by adjusting fluorescent reagents and administration methods according to surgical situations.

For example, Mitsuhashi et al. reported that intraoperative ICG fluorescence imaging in hepatobiliary surgery is useful for understanding the anatomy of the hepatic artery, portal vein, and bile ducts [4]. We also reported that fluorescence cholangiography using fluorescence signals emitted from ICG bounded with bile proteins [5] was effective for detecting the origins of bile leaks, which could not be identified by conventional methods [6]. Furthermore, Uchiyama et al. have applied ICG fluorescence imaging to intraoperative navigation during hepatectomy for liver cancer [7].

We have also reported that the combined use of ICG and 5-ALA fluorescence imaging is useful for intraoperative identification of small hepatic lesions (Fig. 1.1) and disseminated nodules (Fig. 1.2), which can provide essential information for surgical planning in patients with liver cancer [8].

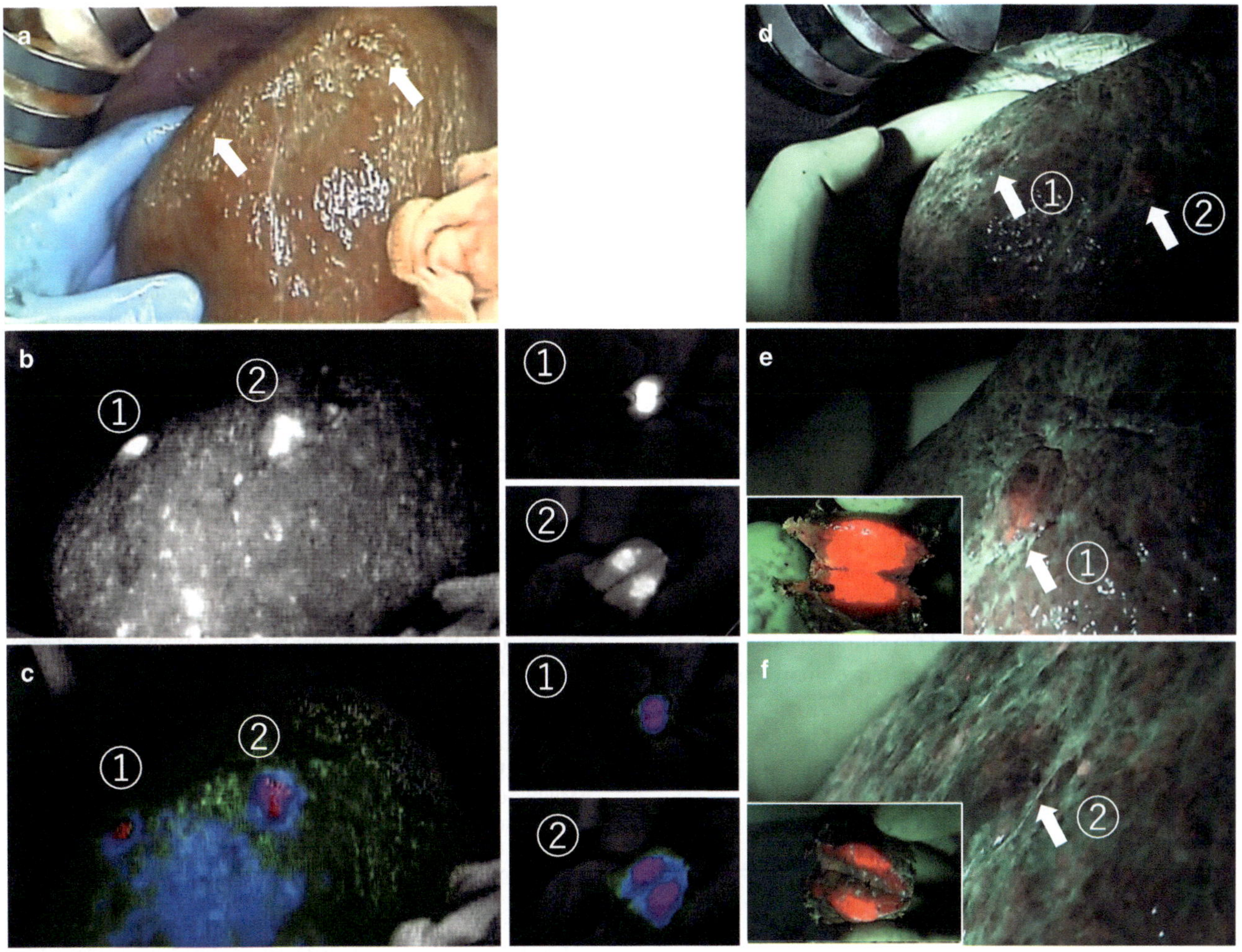

Fig. 1.1 Hepatocellular carcinoma identified by fluorescence imaging. (**a**) Two superficial malignant liver tumors with serosa (*arrows*) under conventional white light illumination. (**b**) Indocyanine green fluorescence imaging of the same liver tumors; insets show the incised lesions, which were diagnosed as hepatocellular carcinoma. (**c**) Indocyanine green fluorescence imaging of the same liver tumors using color mode; insets show the incised lesions. (**d**) The same liver tumors show 5-ALA fluorescence under blue light through an optical filter. (**e**) 5-aminolevulinic acid fluorescence imaging of one of the same liver tumors; inset shows the incised lesion. (**f**) 5-aminolevulinic acid fluorescence imaging of the other liver tumor; inset shows the incised lesion

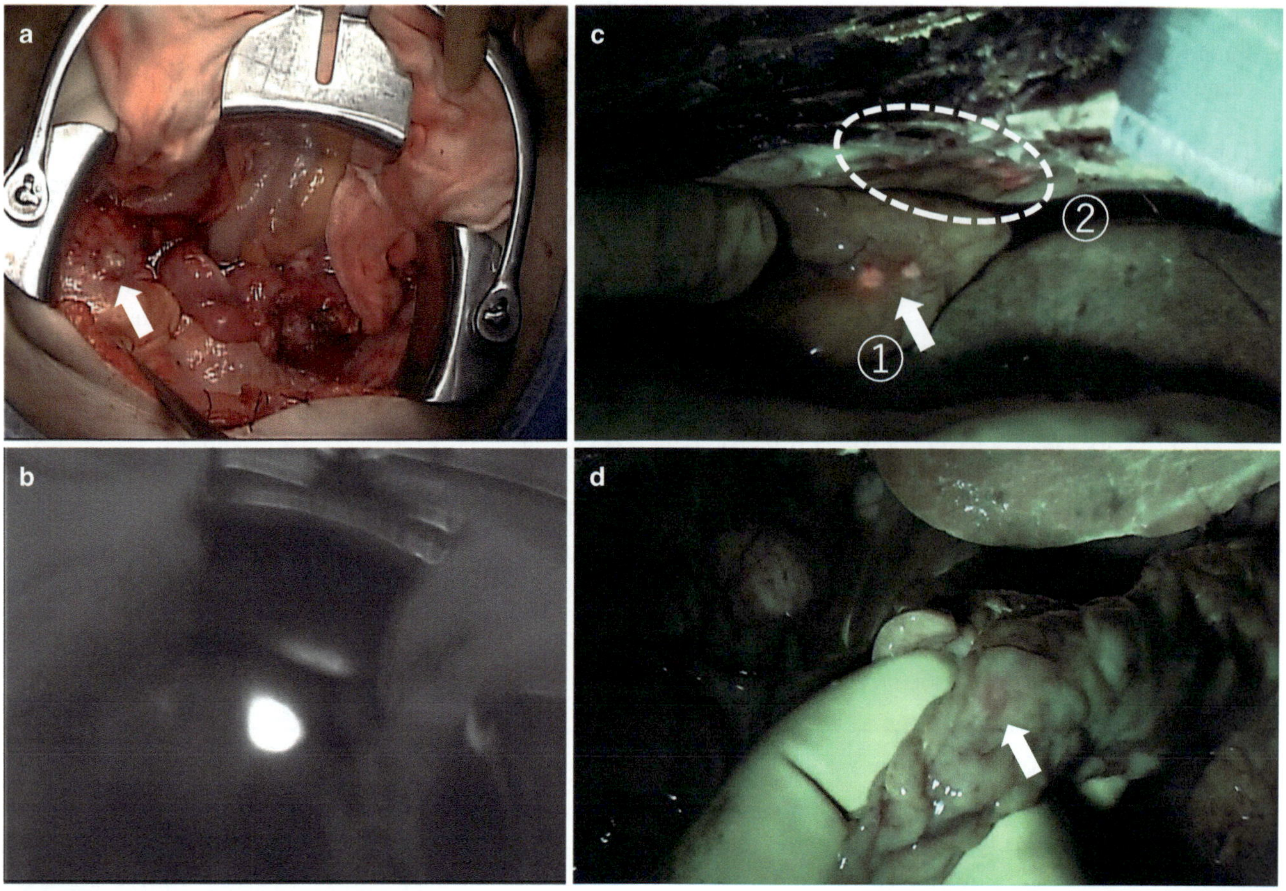

Fig. 1.2 Disseminated lesions identified by fluorescence imaging. (**a**) Ovarian and peritoneal metastasis from hepatocellular carcinoma, under conventional white light illumination. (**b**) Indocyanine green fluorescence imaging of peritoneal metastasis of hepatocellular carcinoma. (**c**) 5-aminolevulinic acid fluorescence imaging of colon ① and peritoneal ② metastases of intrahepatic cholangiocellular carcinoma. (**d**) 5-aminolevulinic acid fluorescence imaging of omental metastasis of hepatocellular carcinoma

Based on the experiences in liver surgery, we believe that fluorescence imaging can be applied to accurately determine the extent of tumor invasion and dissemination in various cancer surgeries.

Point

- It is important to understand the characteristics of fluorescent reagents for using intraoperative fluorescence imaging effectively.

The fluorescence imaging technique can be applied to intraoperative navigation.

5 Conclusions

The insurance coverage and dosage/administration of ICG and 5-ALA are summarized in this chapter. Intraoperative fluorescence imaging will be applied more widely with the development of fluorescence reagents and imaging techniques.

References

1. Makuuchi M, Kosuge T, Takayama T, et al. Surgery for small liver cancers. Semin Surg Oncol. 1993;9:298–304.
2. Ishizawa T, Hasegawa K, Aoki T, et al. Neither multiple tumors nor portal hypertension are surgical contraindications for hepatocellular carcinoma. Gastroenterology. 2008;134:1908–16.
3. Stummer W, Stocker S, Wagner S, et al. Intraoperative detection of malignant gliomas by 5-aminolevulinic acid-induced porphyrin fluorescence. Neurosurgery. 1998;42:518–26.
4. Mitsuhashi N, Kimura F, Shimizu H, et al. Usefulness of intraoperative fluorescence imaging to evaluate local anatomy in hepatobiliary surgery. Hepatobiliary Pancreat Surg. 2008;15:508–14.
5. Mulllock BM, Shaw LJ, Fitzharris B, et al. Sources of proteins in the human bile. Gut. 1985;26:500–9.
6. Kaibori M, Ishizaki M, Matsui K, et al. Intraoperative indocyanine green fluorescent imaging for prevention of bile leakage after hepatic resection. Surgery. 2011;150:91–8.
7. Uchiyama K, Ueno M, Ozawa S, et al. Combined intraoperative use of contrast-enhanced ultrasonography imaging using a sonazoid and fluorescence navigation system with indocyanine green during anatomical hepatectomy. Langenbeck's Arch Surg. 2011;396:1101–7.
8. Kaibori M, Matsui K, Ishizaki M, et al. Intraoperative detection of superficial liver tumors by Fluorescence Imaging Using Indocyanine Green and 5-aminolevulinic Acid. Anticancer Res. 2016;36:1841–9.

Indocyanine Green Fluorescence Imaging System for Open Surgery

2

Nobuyuki Takemura and Norihiro Kokudo

Summary
Currently in Japan, the following ICG fluorescence imaging systems are commercially available for laparotomy and body surface surgery.

- pde-neo® (Hamamatsu Photonics K.K.)
- SPY-PHI (Stryker)
- LIGHTVISION (Shimadzu Corporation)
- HyperEyeMedicalSystem Plus+ (Mizuho Corporation)
- VISIONSENSE system (Medtronic)
- LUOBEAM® (Fluoptics)
- MIPS (Mitaka Kohki Co., Ltd.)

1 Introduction

The development of indocyanine green (ICG) fluorescence imaging systems began with the identification of sentinel lymph nodes during breast cancer surgery by Dr. Kitai, with the use of a prototype of the Photo Dynamic Eye (PDE, Hamamatsu Photonics K.K.) [1]. Herein, we describe the features of the ICG fluorescence imaging systems for open surgery (laparotomy and body surface surgery) that are currently available for clinical use in Japan (Table 2.1).

N. Takemura (✉)
Hepato-Biliary Pancreatic Surgery Division, Department of Surgery, National Center for Global Health and Medicine, Shinjuku, Tokyo, Japan
e-mail: takemuranobu-tky@umin.ac.jp

N. Kokudo
National Center for Global Health and Medicine, Shinjuku, Tokyo, Japan

T. Ishizawa (ed.), *Fluorescence-Guided Surgery*, https://doi.org/10.1007/978-981-19-7372-7_2

Table 2.1 Characteristics of ICG fluorescence imaging system for open surgery commercially available in Japan (as of August 2020)

Product	Characteristics					Manufacturer/ distributor
	Image quality	Superimposed fluorescence	Excitation light and trend signal intensity adjustment	Recording device	Fluorescence intensity analysis	
pde-neo® Sales started: Aug. 2010~ Country of manufacture: Japan Price: 4800 Yen(net)	Non-HD	N/A	Available	External	Stored image analysis software available	Manufactured by Hamamatsu Photonics K.K., and distributed by IMI Co., Ltd.
SPY-PHI Sales started: Jan. 2015~ Country of manufacture: Canada Price: 7000–9000 Yen (actual price including display, etc.)	Full-HD	Available	N/A	Yes	N/A	Stryker Japan KK
LIGHTVISION Sales started: Aug. 2016~ Country of manufacture: Japan Price: 24,000 Yen	Full-HD	Available	Available	Yes	N/A	Shimadzu Corporation
HyperEyeMedicalSystem Plus+ Sales started: Aug. 2019~ Country of manufacture: Japan Price: 9000 Yen	Full-HD	Available	Available	Yes	N/A	Mizuho Corporation
EleVision™ Sales started: Dec. 2015~ Country of manufacture: Israel, USA Price: 10,000–12,000 Yen	Full-HD	Available	Available	Built-in	Available (numerical value relative to maximum fluorescence intensity)	Manufactured by Medtronic and distributed by Heiwa Medical Instruments Co., Ltd.
FLUOBEAM Sales started: Apr. 2019~ Country of manufacture: France Price: Open (10,000–12,000 Yen)	Full-HD	N/A	N/A	Yes	Available (relative evaluation of fluorescence intensity in the captured area)	Manufactured by Fluoptics and distributed by Vital Corporation
MIPS Sales started: Feb. 2020~ Country of manufacture: Japan Price: 24,000 Yen	Full-HD	Available (projection mapping directly on the affected area and display on the monitor)	Available	Yes	Available (multi-valued display according to fluorescence intensity)	Mitaka Kohki Co., Ltd.

Note: Prices are for the minimum configuration (e.g., rigid mirror, camera head, light guide, imaging system, and display)

2 pde-neo® (Hamamatsu Photonics K.K.)

The pde-neo® (Fig. 2.1) was developed based on the concept of measuring fluorescence with high sensitivity and is equipped with a fluorescence mapping function that displays the fluorescence intensity in color for clearer observation of fluorescence images.

pde-neo

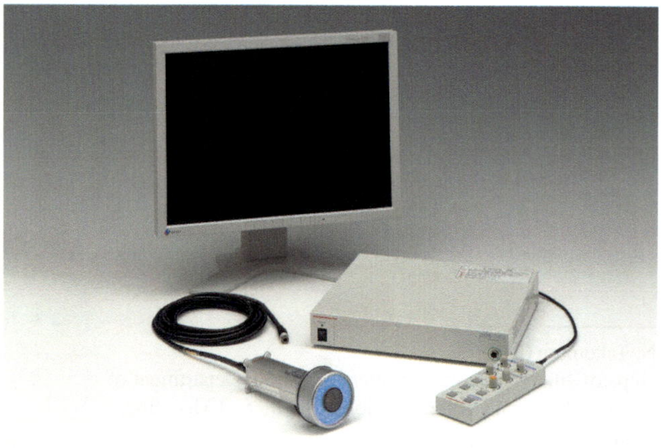

(Hamamatsu Photonics K.K.)

Fig. 2.1 pde-neo® (Hamamatsu Photonics K.K.)

3 SPY-PHI (Stryker)

The main body of this device (Fig. 2.2) is the same as that for laparoscopic surgery manufactured by Stryker. The features of SPY-PHI include "ICG black and white mode (SPY mode)" for clearer confirmation of the fluorescent area, "overlay mode (PINPOINT mode)" for displaying the fluorescent image obtained by fluorescence imaging superimposed in green on the white-light high-definition image, and "colorized mode (CSF mode)" for demonstrating trends of fluorescence intensity in a blue-red gradation. The surgeon can switch between the three modes at hand.

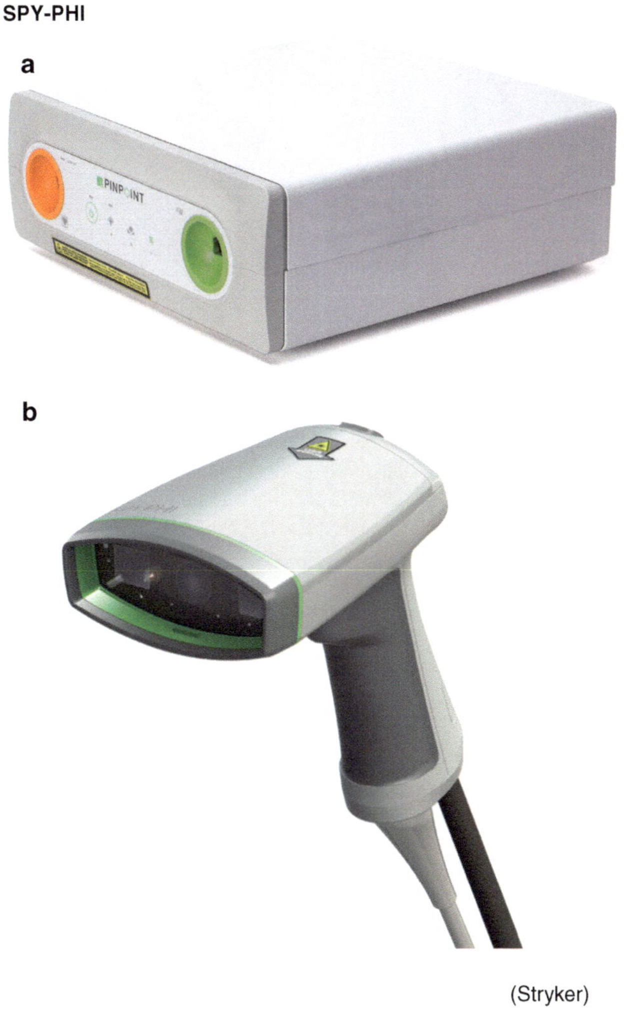

Fig. 2.2 SPY-PHI (Stryker): main body of the device (**a**) and handheld imager (**b**)

4 LIGHTVISION (Shimadzu Corporation)

LIGHTVISION (Fig. 2.3) is capable of high-definition imaging. Because it is an arm-type device, there is no need to hold the camera by hand during imaging, enabling surgeons to continue surgical procedures freely during fluorescence imaging. It is also equipped with a 10× zoom lens, which enables image magnification.

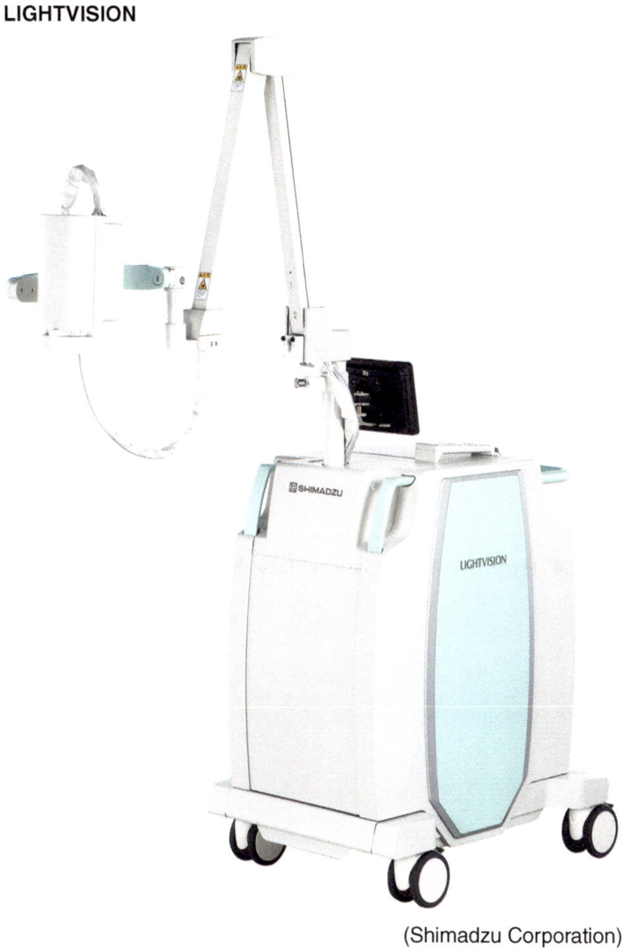

Fig. 2.3 LIGHTVISION (Shimadzu Corporation)

5 HyperEyeMedicalSystem Plus+ (Mizuho Corporation)

The HyperEyeMedicalSystem Plus+ (Fig. 2.4) is capable of high-definition imaging, and the fluorescence color can be selected between green and white. LEDs for auxiliary illumination are placed in the irradiation unit of the camera unit so that the near-infrared light can be captured with a bright field of view even in an environment where the illumination is turned off. Furthermore, as new functions, excitation LEDs for fluorescein fluorescence and also for 5-albuminate fluorescence are equipped.

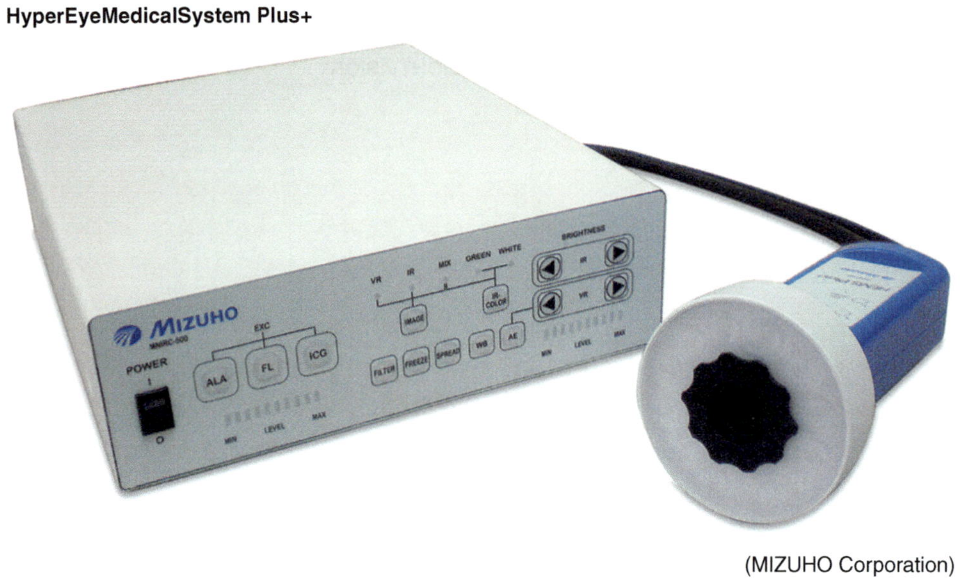

Fig. 2.4 HyperEyeMedicalSystem Plus+ (Mizuho Corporation)

6 EleVision™ IR Platform (Medtronic) (in Japan, Distributed as VISIONSENSE by Heiwa Medical Instruments Co., Ltd.)

The EleVision™ IR Platform (Fig. 2.5) is also capable of full high-definition imaging, and except for the lens part, everything including the camera can also be used for endoscopic surgery. The greatest feature of this device is its high sensitivity to detect fluorescence at a depth of up to 7 mm with a dual sensor and its ability to display fluorescence intensity numerically.

EleVision™ IR Platform

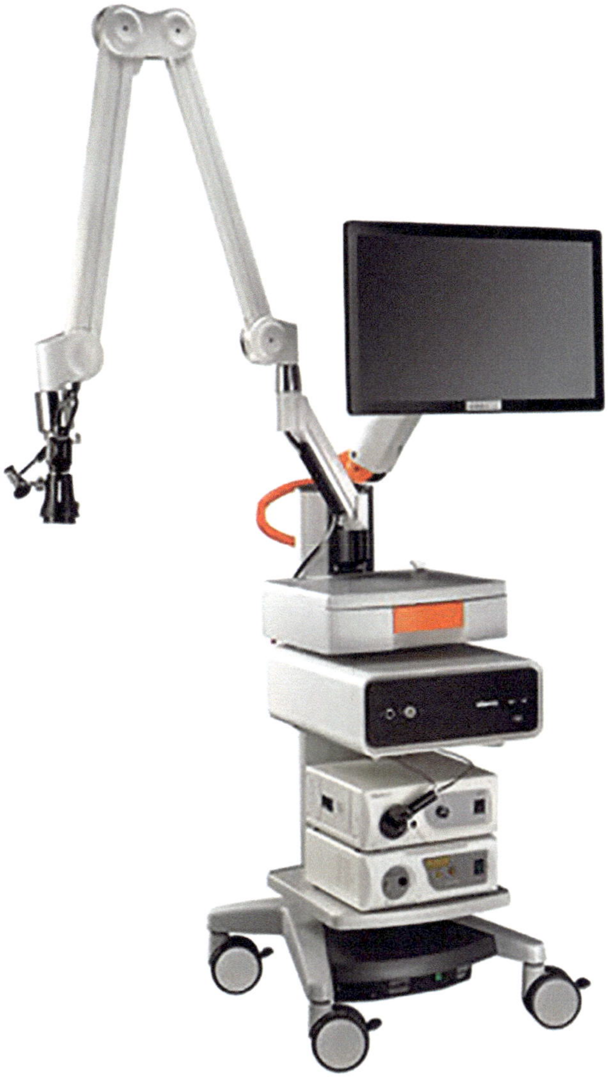

(Medtronic)

Fig. 2.5 EleVision™ IR platform (Medtronic)

7 FLUOBEAM® (Fluoptics)

In addition to the conventional functions for fluorescence imaging using ICG, FLUOBEAM® (Fig. 2.6) has a mode optimized for highly sensitive visualization of tissue autofluorescence under infrared irradiation, which can be used for the identification of parathyroid glands.

FLUOBEAM®

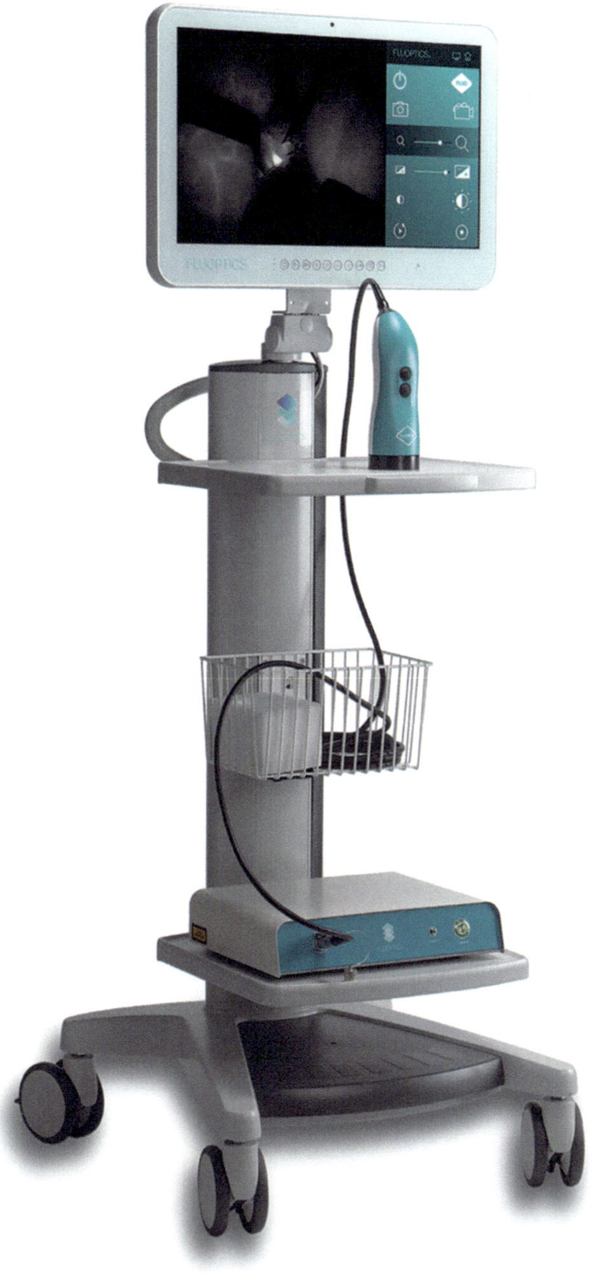

(Fluoptics)

Fig. 2.6 FLUOBEAM® (Fluoptics)

8 MIPS (Mitaka Kohki Co., Ltd.)

MIPS (Fig. 2.7) applies the projection mapping technology that has been widely used in the entertainment industry and projects the ICG fluorescence image directly onto the patient's organ in real time, eliminating the need for eye movement between the surgical field and the monitor. Another important feature of this system is that it uses an algorithm that converts non-fluorescent areas to white, which can provide a bright surgical field even when the surgical lights are turned off during fluorescence imaging.

MIPS

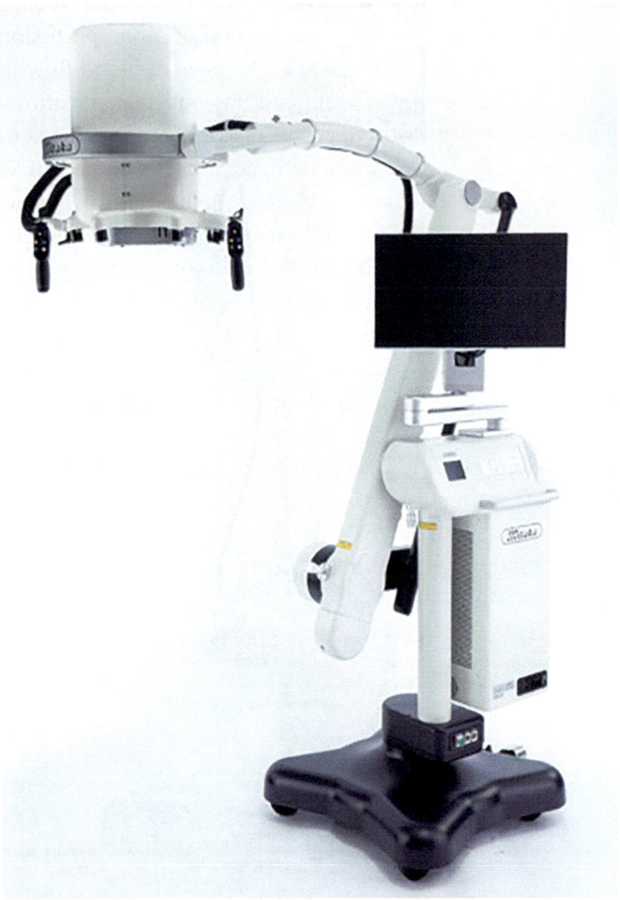

(Mitaka Kohki Co., Ltd.)

Fig. 2.7 MIPS (Mitaka Kohki Co., Ltd.)

9 Conclusions

Now that various fluorescence imaging systems for open surgery have become commercially available, it is essential to select the best device considering the characteristics of each system and the purpose of fluorescence imaging. With the advance in technology, image quality and functions of fluorescence imaging systems for open surgery will be further improved and expanded.

Reference

1. Kitai T, Inomoto T, Miwa M, et al. Fluorescence navigation with indocyanine green for detecting sentinel lymph nodes in breast cancer. Breast Cancer. 2005;12:211–5.

3 Indocyanine Green Fluorescence Imaging System for Endoscopic and Robot-Assisted Surgeries

Shuichi Watanabe, Toshiro Ogura, Hironobu Baba, Yusuke Kinugasa, and Minoru Tanabe

Summary

- Fluorescence imaging has become widely used during endoscopic surgery as the introduction of a variety of near-infrared endoscopic systems into clinical settings.
- The current models of the robot-assisted surgery system are also equipped with a near-infrared imaging system, which makes it even easier to use in clinical practice.

1 Introduction

In the early days of the indocyanine green (ICG) fluorescence imaging techniques, the types of near-infrared imaging devices that could be used for surgery, especially for endoscopic surgery, were quite limited, but nowadays, dedicated imaging devices have become commercially available from various manufacturers. In this chapter, we will outline the functions of the endoscopic fluorescence imaging system and show the features and practical applications of intraoperative fluorescence imaging in our department.

2 Basis of Near-Infrared Camera Systems

A near-infrared imaging system is designed to detect light (fluorescence) generated by irradiating an object with excitation light through a specific imaging filter. Among the fluorescence wavelengths, near-infrared light, which is often used clinically, is characterized by its excellent tissue permeability and properties of the fluorescence reagent. For example, ICG has the property of emitting fluorescence in a similar band by absorbing near-infrared light with 805 nm as the maximum absorption wavelength, and by observing it with an infrared imaging system, it is possible to detect the area where ICG exists (such as bile ducts where bile exists or tumors where ICG accumulates) with a penetration of about 1 cm.

The near-infrared laparoscopic systems designed for detecting weak fluorescence signals can have drawbacks in the brightness and resolutions of white-light color images, although these limitations are being improved in the latest models of laparoscopic imaging systems. When selecting a near-infrared laparoscopic imaging system, it is necessary to be familiar with the characteristics of fluorescence imaging as described above and to consider the situations in which the system will be used instead of normal observation based on white-light color imaging.

3 Characteristics of the Near-Infrared Laparoscopic Imaging System

Table 3.1 summarizes the infrared imaging systems for laparoscopic/thoracoscopic surgery and robot-assisted surgery that are currently available for clinical use in Japan. All of these systems allow observation in full HD, support scope diameters of 10 mm and 5 mm, and allow the use of oblique views in addition to direct views. At present, only rigid scopes are available for fluorescence observation. Therefore, in a situation where a flexible scope is often used (e.g., laparoscopic hepatectomy), surgeons may have to adjust port placement and scope manipulation. In addition, please note that only a few imaging systems support adjustment and measurement of fluorescence signal intensity in the field of laparoscopic surgery.

S. Watanabe (✉) · M. Tanabe
Department of Hepatobiliary and Pancreatic Surgery, Tokyo Medical and Dental University, Bunkyo-ku, Tokyo, Japan
e-mail: shumsrg@tmd.ac.jp

T. Ogura
Department of Gastrointestinal Surgery, Saitama Cancer Center, Ina, Kitaadachi District, Saitama, Japan

H. Baba · Y. Kinugasa
Department of Gastrointestinal Surgery, Tokyo Medical and Dental University, Bunkyo-ku, Tokyo, Japan

T. Ishizawa (ed.), *Fluorescence-Guided Surgery*, https://doi.org/10.1007/978-981-19-7372-7_3

Table 3.1 Characteristics of near-infrared fluorescence imaging systems for laparoscopic and robotic surgery, available in Japan

Product (manufacturer)	IMAGE1 S™ NIR/ ICG (KARL STORZ)	PINPOINT (Stryker)	1688 AIM4 K (Stryker)	VISERA ELITE II (Olympus)	VISIONSENSE Iridium (Medtronic)	da Vinci Xi/X (Intuitive Surgical)	CNOS-SK-1057 (Shinko Optical Co., Ltd.)
Image quality	Hull HD	Hull HD	4 K UHD (3840 × 2160)	Hull HD	Hull HD	3D HD (SXGA)	Full HD
Scope diameter	5 mm, 10 mm	5.5 mm, 10 mm	10 mm, 5.5 mm, 5.4 mm	10 mm, 5.4 mm	5 mm, 10 mm	8 mm	10 mm, 5.4 mm
Rigid/soft mirror	Rigid	Rigid	Rigid	Rigid	Rigid	Rigid	Rigid
View direction	0° / 30° / 45°	0° / 30° / 45°	0° / 30° / 45°	0° / 30°	0° / 30°	0° / 30°	0° / 30°
Superimposition	n/a	◎	◎	○	◎	Black and white display (standard mode)	n/a
Fluorescence signal intensity adjustment	n/a	n/a	Available	n/a	Available	Available	Available
Recording device	Built-in	n/a	n/a	n/a	Built-in	n/a	n/a
Fluorescence intensity analysis	n/a	n/a	n/a	n/a	Available	n/a	n/a
Sales started	Dec 2013~	Jan 2018~	Jan 2020~	Mar 2017~	Oct 2018~	Mar 2016~	Apr 2014~
Country of manufacture	Germany	Canada	USA	Japan	Israel, USA	USA	Japan
Price	JPY 12,000 k	JPY 15,000 ~ 18,000 k	JPY 12,000 ~ 15,000 k	JPY 11,500 k	JPY 11,500 k	Standard in the system	JPY 9400 k

Note: Prices are for the minimum configuration (e.g., rigid mirror, camera head, light guide, imaging system, and display)

4 Clinical Experiences of Using Laparoscopic Imaging Systems in Our Center

In our department, we have tried four near-infrared imaging systems for laparoscopy and confirmed their characteristics (Figs. 3.1 and 3.2).

Olympus and KARL STORZ install imaging systems that enable observation with the same light source and camera as their own conventional white-light endoscope systems, and it is possible to change from the field of view during normal white-light imaging to fluorescent imaging with one-touch action. On the other hand, fluorescence signals are displayed on a dark background without full-color information, which makes it difficult for surgeons to understand the spatial relationships with surrounding organs and the position of forceps during fluorescence imaging.

The Stryker and EleVision™ IR Platform (sold as VISIONSENSE in Japan) are especially focused on fluorescence imaging, and the visualizability of near-infrared fluorescence signals seemed to be very high. One of the features of these two systems is that they enable superimposition of fluorescence images onto white-light full-color images, which makes it easy for surgeons to go on surgical procedures using information obtained by fluorescence imaging.

In conclusion from our clinical experiences, it would be better to select a laparoscopic imaging system prioritizing the quality of white-light color imaging in surgical procedures where the time and frequency of near-infrared imaging are limited. In contrast, laparoscopic imaging systems with advantages in near-infrared fluorescence imaging (e.g., superimposition on full-color imaging) can be a promising option when fluorescence imaging is essential for assuring the efficacy of surgery.

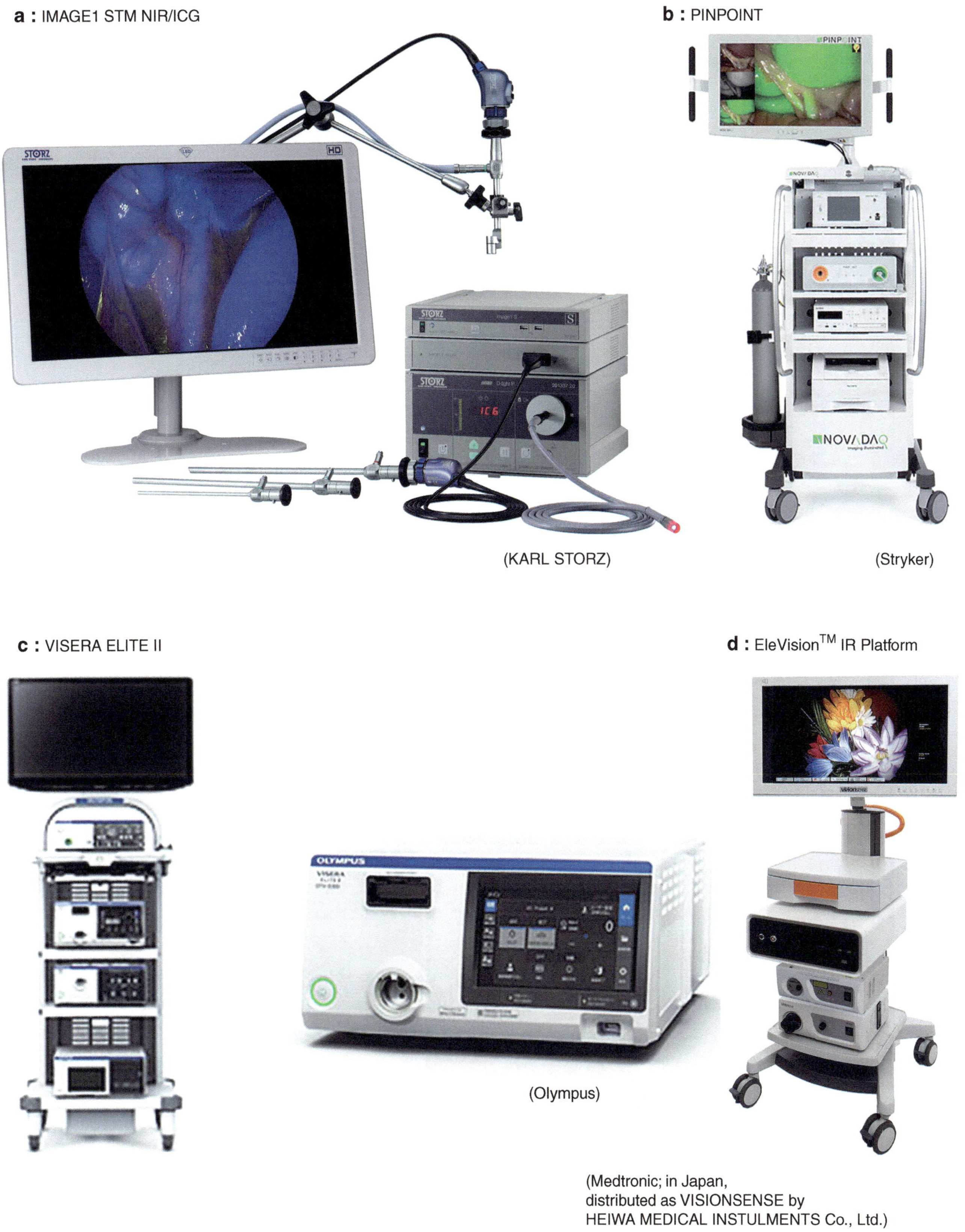

Fig. 3.1. Appearance of the near-infrared fluorescent imaging systems during laparoscopic surgery. (**a**) IMAGE1 S™ NIR/ICG (KARL STORZ). (**b**) PINPOINT (Stryker). (**c**) VISERA ELITE II (Olympus). (**d**) EleVision™ IR Platform (Medtronic; in Japan, distributed as VISIONSENSE by Heiwa Medical Instruments Co., Ltd.)

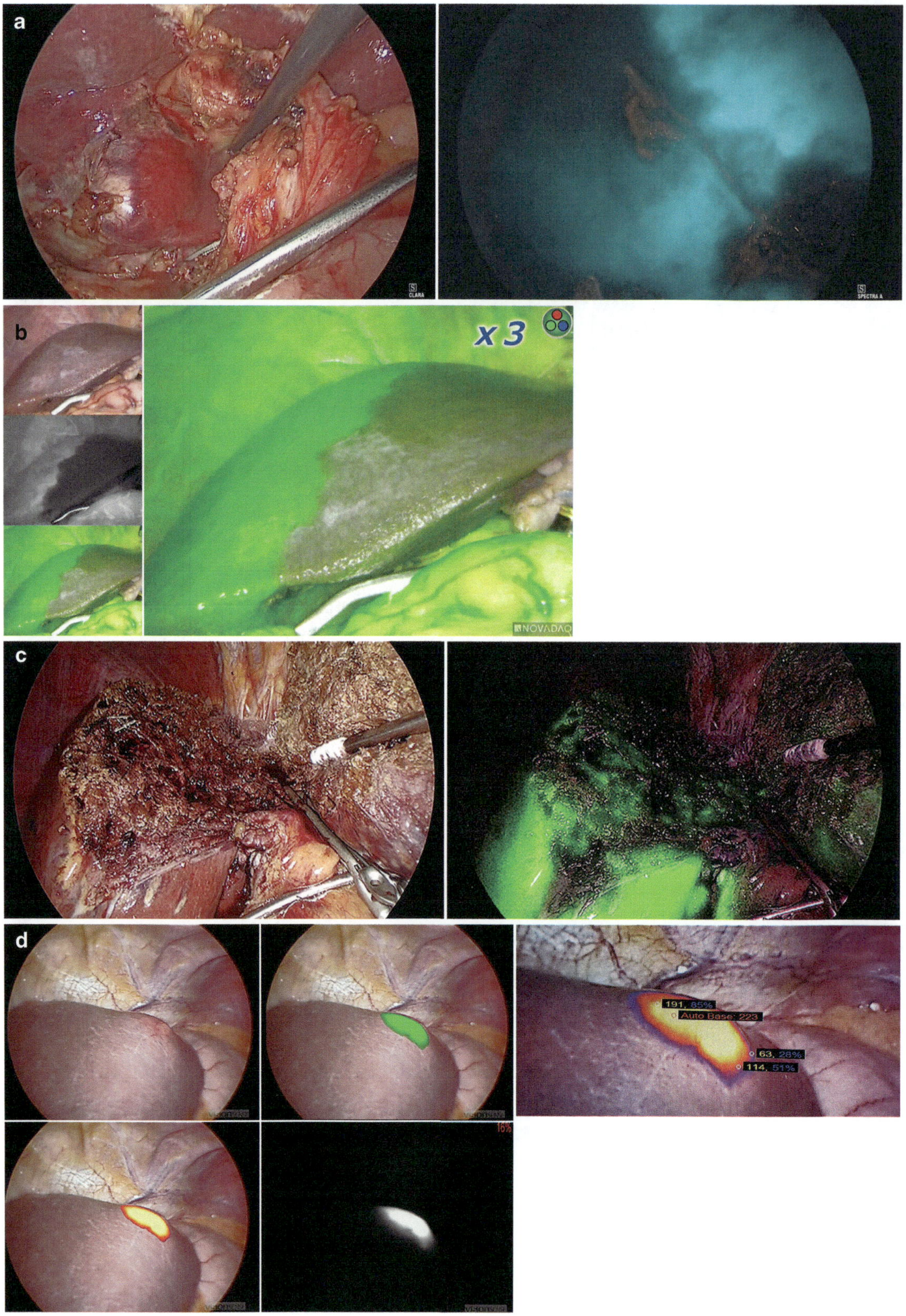

Fig. 3.2 Clinical applications of fluorescence imaging in laparoscopic surgery. (**a**) IMAGE1 S™ NIR/ICG. (**b**) PINPOINT. (**c**) VISERA ELITE II. (**d**) EleVision™ IR Platform (distributed as VISIONSENSE in Japan)

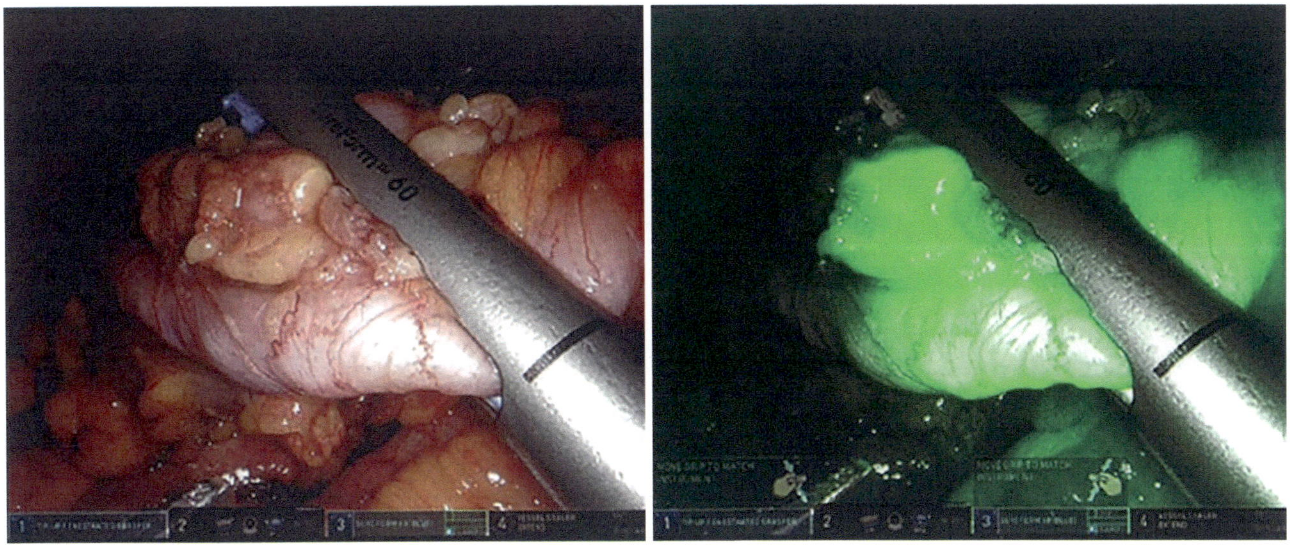

Fig. 3.3 Observation images by da Vinci® Xi/X Firefly mode

5 Characteristics of the Near-Infrared Imaging System in Robot-Assisted Surgery

The da Vinci® Xi/X (Intuitive Surgical, Inc.), which is widely used as a robot-assisted surgery system, is equipped with a near-infrared imaging system as standard and can be used in actual clinical practice. The near-infrared observation mode can be easily switched by on-screen operation. In the fluorescence observation mode, surgeons can manipulate a camera and forceps under sufficiently high fluorescence signals and background information (Fig. 3.3). Applications of fluorescence imaging will be further expanded, as robot-assisted surgery systems are indicated more widely to various surgical procedures.

Point

- We should select a near-infrared endoscopic imaging system based on the characteristics of imaging devices (quality and function of fluorescence imaging as well as white-light color imaging) and the significance of fluorescence imaging in each surgical procedure.
- Near-infrared fluorescence imaging can also be used as a standard feature in robot-assisted surgery with the da Vinci® Xi/X systems.

6 Conclusions

Characteristics of near-infrared fluorescence imaging systems for endoscopic surgery and robot-assisted surgery are summarized. With further improvement of signal detectability and feasibility, fluorescence imaging will be applied in minimally invasive surgery more widely and commonly.

5-Aminolevulinic Acid Fluorescence Imaging System

4

Tsutomu Namikawa and Kazuhiro Hanazaki

Summary

- The current 5-ALA endoscopic imaging system enables visualization of fluorescence images and white-light color images in high-definition, with adjustment of fluorescence signals.
- The 5-ALA microscope imaging system can be used to discriminate malignant tumor tissue from non-cancerous tissues in the eyepiece and on the display monitor.

1 Introduction

Photodynamic diagnosis (PDD) using 5-aminolevulinic acid (5-ALA) is a technique based on the biological feature of 5-ALA that can accumulate in cancerous tissues, which can be used to localize tumors in real time during surgery [1–3]. In this chapter, we demonstrate the characteristics and clinical applications of the 5-ALA fluorescence imaging system in the setting of open, endoscopic, and microscopic surgeries.

2 Characteristics of 5-Aminolevulinic Acid

5-aminolevulinic acid is a naturally occurring amino acid with a molecular weight of 131 that is commonly found in plants and animals. It is synthesized in vivo from succinyl CoA and glycine by 5-ALA synthase in mitochondria. 5-ALA is metabolized in the cytoplasm and biosynthesized in the mitochondria to protoporphyrin IX (PpIX), a photosensitive substance [1, 2]. In normal cells, intracellular iron is inserted and rapidly metabolized to heme and bilirubin. In cancer cells, however, PpIX specifically accumulates due to abnormalities in transporter activity and enzymes in the cell and mitochondrial membranes. When the cells are irradiated with blue visible light from 375 to 445 nm, PpIX is excited and emits red fluorescence from 600 to 740 nm. PDD is a diagnostic technique that applies this photochemical reaction [1, 3].

3 History of Fluorescence Imaging Method Using 5-Aminolevulinic Acid

In 1999, 18 hospitals in Germany formed "The 5-ALA-GLIOMA Study Group" and conducted a phase III clinical trial using a system with a built-in excitation light source in a conventional operating microscope. As a result, aminolevulinic acid hydrochloride was approved for marketing in Europe in 2007 as an intravitreal diagnostic agent for malignant glioma surgery.

In Japan, the use of 5-ALA for PDD of bladder cancer was approved as a "highly advanced medical treatment" in 2010. In 2012, an investigator-initiated phase II/III clinical trial of lyophilized 5-ALA solution in patients with non-muscle layer invasive bladder cancer was conducted, followed by a corporate phase III trial. In 2013, manufacturing and marketing approval was obtained in Japan for the indication of "visualization of tumor tissue during tumor resection of malignant glioma," and the lyophilized formulation (Araglio® 1.5 mg/kg) was launched as a PDD agent. In 2017, the indication of Araglio® was extended to the visualization of non-invasive bladder cancer during transurethral resection.

T. Namikawa (✉) · K. Hanazaki
Department of Surgery, Kochi Medical School, Kochi University, Nankoku, Kochi, Japan

Center for Photodynamic Medicine, Kochi Medical School, Nankoku, Kochi, Japan
e-mail: tsutomun@kochi-u.ac.jp

T. Ishizawa (ed.), *Fluorescence-Guided Surgery*, https://doi.org/10.1007/978-981-19-7372-7_4

4 Medical Photographing Equipment in Laparoscopic Surgery

A variety of 5-ALA fluorescence imaging systems have been developed, each of which has its own characteristics according to its clinical use (Table 4.1). In 2011, KARL STORZ developed the TRICAM SL II, a PDD-capable fluorescence imaging system, which evolved into the IMAGE1 S™ in 2017 to enable high-definition imaging (Fig. 4.1). The camera head (H3-Z FI) sensitively picks up fluorescence signals in the wavelengths at 635 nm, and PDD fluorescence observation is possible by using a telescope with a built-in filter

Table 4.1 Characteristics of 5-ALA fluorescence imaging systems available in Japan. (as of August 2020)

Product (manufacturer)	IMAGE1 S™ PDD (KARL STORZ)	Aladuck® (SBI Pharmaceuticals Co., Ltd.)	KINEVO 900® (Carl Zeiss)
Resolution	1920 × 1080p	–	2160 × 3840p
Superimposed fluorescence	n/a	n/a	n/a
Excitation light intensity adjustment	n/a	Available	Available
Fluorescence signal intensity adjustment	Available	n/a	Available
Recording device	Built-in	n/a	Built-in
Fluorescence intensity analysis	n/a	n/a	n/a
Sales started	Oct. 2017	Apr. 2014	Aug. 2018
Country of manufacture	Germany	Japan	Germany
Price	JPY 10,000 k	JPY 2400 k (incl. Light source unit, light guide, and cut filter)	JPY 55,000 k ~ (microscope body and 5-ALA module)

Note: Prices are for the minimum configuration (e.g., rigid mirror, camera head, light guide, imaging system, and display)

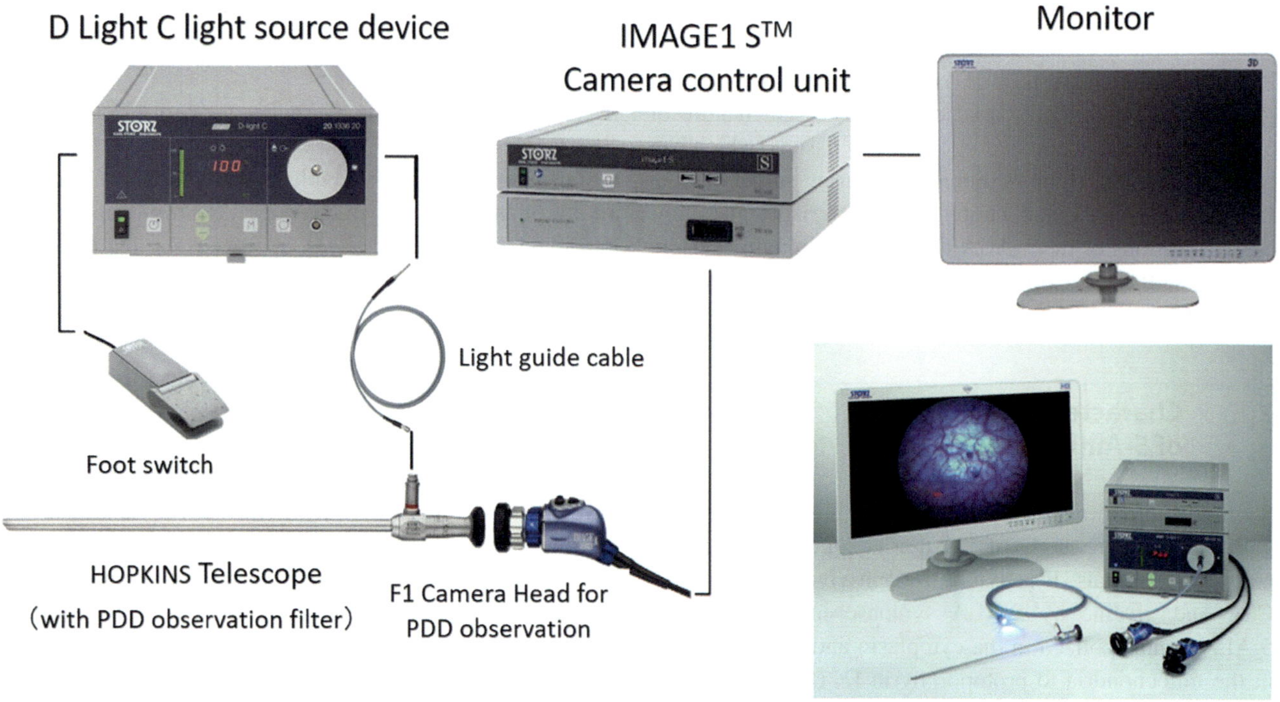

Fig. 4.1 5-aminolevulinic acid fluorescence imaging system for endoscopic surgery. The system consists of a dedicated light source device that irradiates blue light from 380 to 420 nm, a dedicated camera that captures red fluorescence from 600 to 740 nm, a control unit, and a monitor

that cuts wavelengths below 440 nm and a camera head without a filter. It is also possible to immediately switch between conventional white light and fluorescence observation, facilitating the observation of the same lesion.

The two-color LED light source Aladuck LS-DLED, manufactured by KD-CLOUT Co., Ltd. and launched by SBI Pharmaceuticals Co., Ltd. in 2014, is a light source system that enables the use of 5-ALA fluorescence imaging for endoscopic, laparotomy, and craniotomy procedures (Fig. 4.2). It can be used together with commercially available endoscopic camera systems and is equipped with white light and blue light LEDs with a peak wavelength of 400–410 nm. The optical tube is equipped with a cut filter to block blue light, and red fluorescence can be visualized with high sensitivity. This system is also applicable to craniotomy, and by attaching a collimator lens (focusing lens) to the tip of this device set at a 15–20 cm distance from operation fields, it is possible to visualize fluorescence images assuring sufficient working space for surgeons.

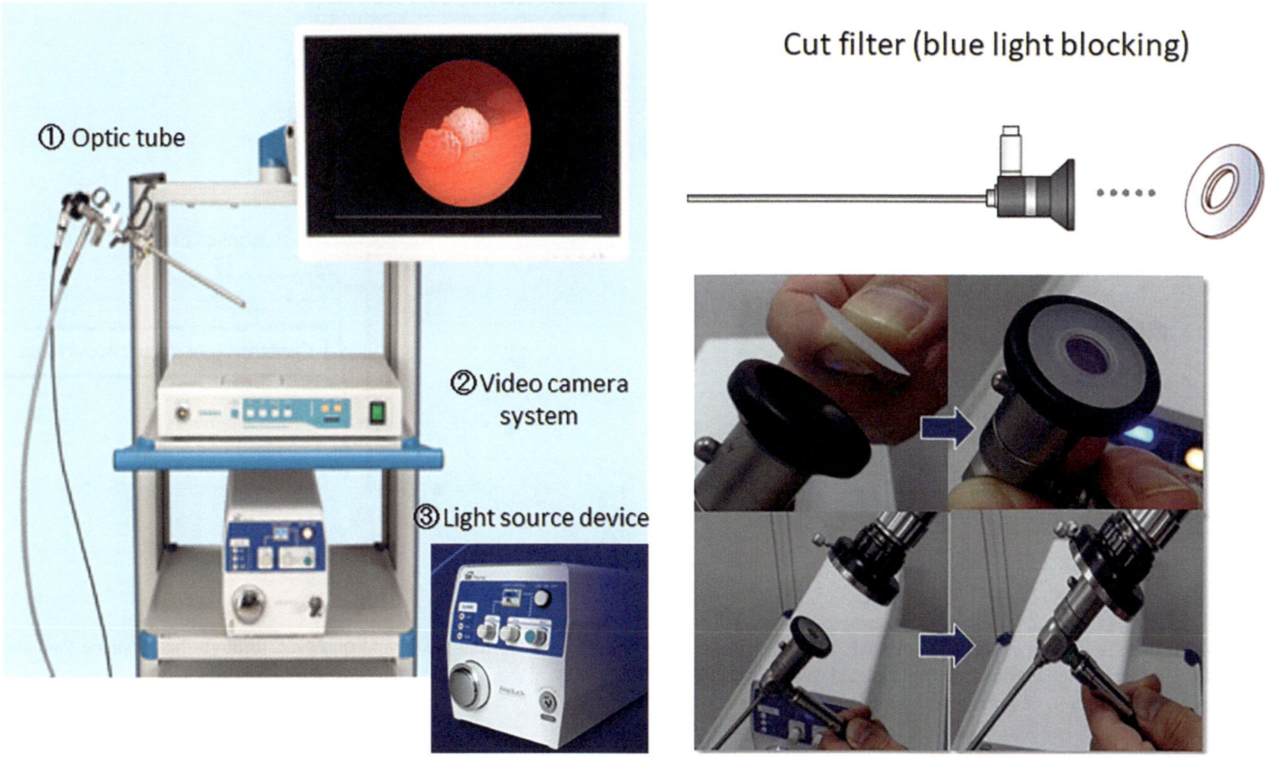

Fig. 4.2 Two-color LED light source system. This light source device can be used in conjunction with existing endoscope systems, by attaching a cut filter that blocks blue light to the optical viewing tube

5 Medical Photographing Equipment Using a Surgical Microscope

Intraoperative observation of malignant brain tumor tissue using 5-ALA has been recognized as a method to distinguish malignant brain tumor tissue from surrounding normal tissue and to easily observe the extent of tumor growth. In 2005, Carl Zeiss launched Pentero, the first surgical microscope in the world with a fully integrated intraoperative tumor observation module. It has evolved from the Pentero to the Pentero® 900, and then to the KINEVO® 900, which can distinguish clear tumor boundaries, and can be used for 4 K imaging and 3D external viewing. These systems are equipped with a light source and a camera. All light sources and filters are built into these systems, and normal and excitation light can be adjusted at the touch of a button, allowing direct observation of normal and malignant tumor tissues in the eyepiece and on the monitor (Fig. 4.3).

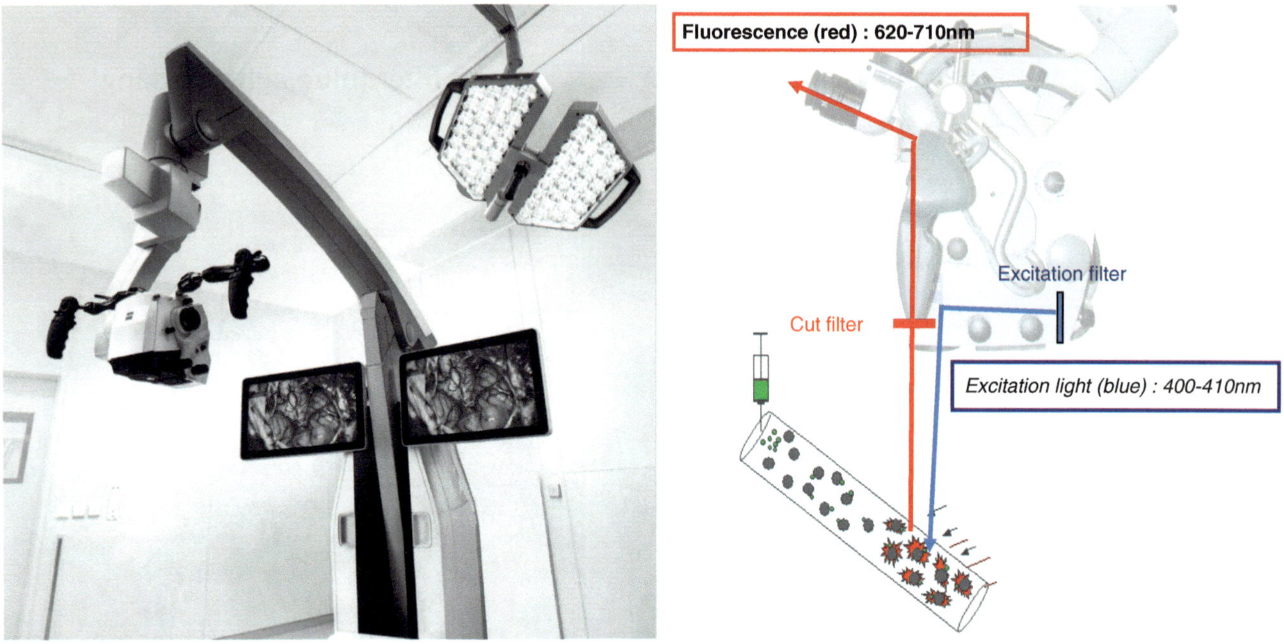

Fig. 4.3 Surgical microscope with built-in 5-ALA fluorescence imaging module. When blue excitation light is irradiated, red fluorescence is generated from the tumor, and malignant tumor tissue can be discriminated from normal tissue and observed through the eyepiece lens and monitor

6 Clinical Applications of 5-Aminolevulinic Acid Fluorescence Imaging Using 5-Aminolevulinic Acid

In the field of neurosurgery, 5-ALA-PDD has been approved by the pharmaceutical affairs bodies in Japan for visualization of tumor tissue during tumor resection of malignant glioma and has been put to practical use in daily clinical practice. In the field of urology, 5-ALA-PDD can be used to enhance the diagnosis rate of bladder intraepithelial carcinoma and reduce the recurrence rate by decreasing the residual tumor during resection [1, 3–5].

On the other hand, in the field of gastrointestinal surgery, several studies have suggested the efficacy of 5-ALA-PDD in improving intraoperative identification of peritoneal dissemination and lymph node metastases [6–8]. In our center, we have used a PDD endoscope system (IMAGE1 S™, KARL STORZ) with a 300 W Xenon lamp as the light source, blue light of 380–440 nm as the excitation light, and 50 mW as the tip output [7, 8]. Currently, a multicenter investigator-initiated clinical trial is underway for evaluating the efficacy of 5-ALA-PDD in improving the diagnostic accuracy of peritoneal dissemination from advanced gastric cancer (Fig. 4.4).

Points

- 5-aminolevulinic acid fluorescence imaging systems for open, endoscopic, and microscopic surgery have become commercially available.

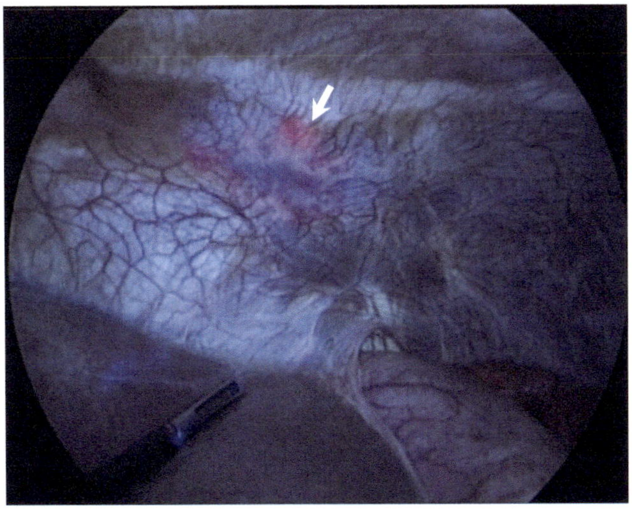

Fig. 4.4 Diagnostic laparoscopy with 5-ALA-PDD. Peritoneal metastasis of gastric cancer emitting red fluorescence in the left diaphragm (*arrow*), which is expected to improve the accuracy of diagnostic laparoscopy

- It is essential to consider the characteristics of the imaging device before purchasing it for clinical use.
- Appropriate use of fluorescence imaging based on a good understanding of the characteristics of 5-ALA and imaging devices improves diagnostic accuracy.

7 Conclusion

The image quality and functions of 5-ALA fluorescence imaging have been evolving with the development of information-processing technology in recent years, and they are being developed day by day in response to requests from surgeons in each field. For better identification of cancer tissues, it is important to understand the characteristics of the imaging device and 5-ALA, select the right device, and use it appropriately according to the purpose of the surgical procedure.

References

1. Inoue K, Fukuhara H, Shimamoto T, et al. Comparison between intravesical and oral administration of 5-aminolevulinic acid in the clinical benefit of. Cancer. 2012;118:1062–74.
2. Hagiya Y, Endo Y, Yonemura Y, et al. Pivotal roles of peptide transporter PEPT1 and ATP-binding cassette (ABC) transporter ABCG2 in 5-aminolevulinic acid (ALA)-based photocytotoxicity of gastric cancer cells in vitro. Photodiagn Photodyn Ther. 2012;9:204–14.
3. Inoue K, Karashima T, Kamada M, et al. Regulation of 5-aminolevulinic acid-mediated protoporphyrin IX accumulation in human urothelial carcinomas. Pathobiology. 2009;76:303–14.
4. Fukuhara H, Inoue K, Satake H, et al. Photodynamic diagnosis of positive margin during radical prostatectomy: preliminary experience with 5-aminolevulinic acid. Int J Urol. 2011;18:585–91.
5. Inoue K, Fukuhara H, Kurabayashi A, et al. Photodynamic therapy involves antiangiogenic mechanism and is enhanced by ferrochelatase inhibitor in in urothelial carcinoma. Cancer Sci. 2013;104:765–72.
6. Namikawa T, Yatabe T, Inoue K, et al. Clinical applications of 5-aminolevulinic acid- mediated fluorescence for gastric cancer. World J Gastroenterol. 2015;21:8769–75.
7. Namikawa T, Inoue K, Uemura S, et al. Photodynamic diagnosis using 5-aminolevulinic acid during gastrectomy for gastric cancer. J Surg Oncol. 2014;109:213–7.
8. Namikawa T, Fujisawa K, Munekage E, et al. Clinical application of photodynamic medicine technology using light-emitting fluorescence imaging based on a specialized luminous source. Med Mol Morphol. 2018;51:187–93.

How to Introduce Fluorescence Imaging to the Operating Room

5

Sunao Uemura, Tsutomu Namikawa, and Kazuhiro Hanazaki

Summary

- Fluorescence imaging is widely used in a variety of surgical fields and is becoming an indispensable support device.
- The fluorescence imaging system can be managed smoothly in collaboration with co-medical staff, especially clinical engineers.
- Methods and outcomes of fluorescence imaging should be recorded for claiming costs and also for further evaluation in medical research.

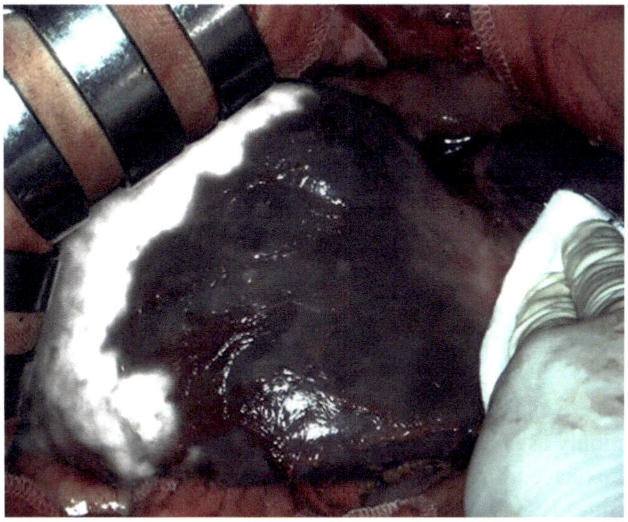

Fig. 5.1 Example of ICG fluorescence imaging during surgery. The demarcation line between the left and right liver is clearly visualized by ICG fluorescence imaging

1 Introduction

In recent years, fluorescence imaging has been widely used in a variety of surgical fields. At our hospital, indocyanine green (ICG) fluorescence imaging systems were introduced in 2010 in the Department of Breast Surgery and the Department of Cardiovascular Surgery, and as of March 2020, they are being used in many departments including the Department of Gastrointestinal Surgery (Fig. 5.1). In anatomical drawings in textbooks of surgery, blood vessels and organs are color-coded and clearly depicted. The ICG fluorescence imaging is an epoch-making invention that vividly depicts blood vessels in a bright field, and it has become an indispensable surgical support device in modern surgical treatment not only for dissection but also for organ blood flow evaluation and tumor localization. While fluorescence imaging has become increasingly common, there will still be many facilities that are considering introducing fluorescence imaging systems. We herein demonstrate how clinical engineers, nurses, and surgeons should share information on the installation and storage of the fluorescence imaging system, management of reagents in the operating room, and operation methods for imaging and recording, as well as cost management.

2 Fluorescence Imaging Systems Used at Kochi University

At Kochi University, Prof. Takayuki Sato of the Department of Cardiovascular Physiology succeeded in identifying the fluorescence of ICG in bright fields with high sensitivity and color. In 2010, the HyperEye Medical System (HEMS, Mizuho Corporation) was introduced in the Department of Breast and Cardiovascular Surgery at our hospital [1]. After that, we introduced the LIGHTVISION (Shimadzu Corporation), and as of March 2020, we are performing ICG fluorescence imaging mainly with the LIGHTVISION in many fields.

S. Uemura (✉) · T. Namikawa · K. Hanazaki
Department of Surgery, Kochi Medical School, Kochi University, Nankoku, Kochi, Japan
e-mail: suemura@kochi-u.ac.jp

T. Ishizawa (ed.), *Fluorescence-Guided Surgery*, https://doi.org/10.1007/978-981-19-7372-7_5

3 Installation and Storage of Imaging Equipment

Both HEMS and LIGHTVISION are managed by clinical engineers in the operating department as well as other medical devices. However, they are not always available in each operating room, so it is important to share information on which surgery the ICG fluorescence method is used in the operating department. For example, our hospital has a "HEMS" tag in the "Special Equipment" section when entering operations in the electronic medical record. By selecting this tag, the clinical engineer can know which surgery will use the ICG fluorescence method by at least the week before the surgery, and nurses and anesthesiologists can also obtain this information. Furthermore, by clearly stating that the ICG fluorescence method will be used at the time of surgical timeout, the team will be able to recognize it again at the start of surgery. This will speed up the setup of the imaging system. And when the system is actually in use, we contact the clinical engineer, and within a few minutes, the system is connected and ready to use. In other words, close cooperation with clinical engineers is essential for smooth operation.

4 Management of Reagents

Indocyanine green, a typical fluorescence agent, is commonly used in liver function tests in Japan, and many hospitals have a fixed number of ICG in their drug departments. Since ICG is used in a wide variety of surgeries at our hospital, we also keep a fixed number of 5 in the operating room. We have also reported the usefulness of the ICG fluorescence method in confirming intestinal blood flow during emergency surgery [2]. Unlike aminolevulinic acid hydrochloride (5-ALA), ICG does not require special management, so it is desirable to keep it in the operating room so that it can be promptly administered during emergency surgery. The ICG is dissolved in 5–10 mL of water for injection and administered intravenously at a dose of 0.04–0.3 mg/kg. However, since the dose of ICG varies depending on the intended use, it is important to have sufficient communication with the anesthesiologist. The use of ICG is contraindicated in patients with hypersensitivity or iodine hypersensitivity. It is necessary to confirm that the patient is not allergic to contrast media before actual use.

5 Operation and Recording Methods of Fluorescence Imaging

Both HEMS and LIGHTVISION are operated by clinical engineers. In our hospital, the operating room was completely newly built in 2015. In the past, it was necessary to record the images on the internal hard disk and retrieve them when necessary, but now, by connecting to the recording monitors installed in all operating rooms, the images can be viewed remotely by many electronic medical records even outside the operating rooms, and automatic recording is possible. Figure 5.2 shows an image taken during esophageal surgery at our hospital. A clinical engineer is operating the system, the anesthesiologist is administering medication, and the surgeon is watching the monitor.

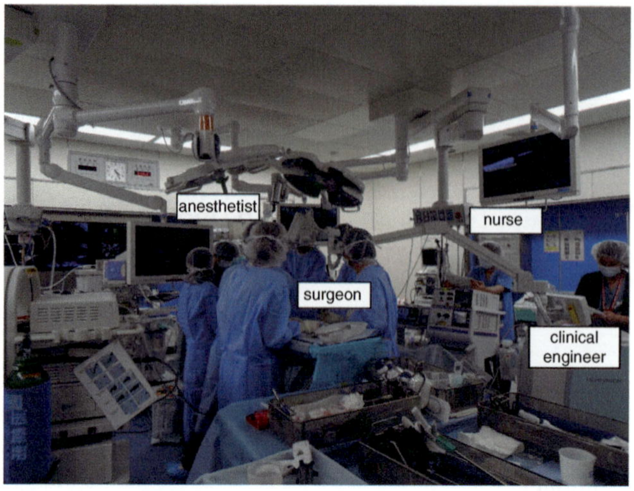

Fig. 5.2 Setup of an operating room during ICG fluorescence imaging

6 Cost Control

Currently in Japan, ICG is marketed as "a drug for liver and circulatory function tests, a fluorescent angiographic agent, and a drug for sentinel lymph node identification" at a drug price of 575 yen. In addition, a procedure fee can be calculated for "confirmation of blood vessels or tumors by fluorescence measurement, or confirmation of blood flow in the gastrointestinal tract" using ICG or 5-ALA in neurosurgery, coronary revascularization, or bladder malignancy surgery. For claiming medical costs to the Japanese insurance system, it is necessary to specify the details in the surgical record. In our hospital, we collaborate with co-medical staff such as clinical engineers to manage fluorescence imaging equipment and costs smoothly.

Points

- Cooperation with co-medical staff, especially clinical engineers, promotes the introduction, operation, and management of fluorescence imaging.
- Surgeons should notice the potential use of fluorescence imaging at the beginning of surgery for smooth operation.
- ICG should be kept in the OR for emergency use.

7 Conclusions

The fluorescence imaging system is becoming an indispensable surgical navigation system for the evaluation of organ blood flow and tumor imaging. It is important to collaborate with clinical engineers and other co-medical staff in the introduction, operation, and management of fluorescence imaging.

References

1. Handa T, Katare RG, Nishimori H, et al. New device for intraoperative graft assessment: HyperEye charge-coupled device camera system. Gen Thorac Cardiovasc Surg. 2010;58:68–77.
2. Namikawa T, Uemura S, Kondo N, et al. Successful preservation of the mesenteric and bowel circulation with treatment for a ruptured superior Mesenteric artery aneurysm using the HyperEye Medical System. Am Surg. 2014;80:E359–61.

Recording of Intraoperative Fluorescence Imaging

6

Koshi Kumagai

Summary

- Storage of operative images and videos is beneficial for keeping accurate information on fluorescence imaging used in each procedure.
- The linkage of surgical videos with other electric medical records is already in practical use.

1 Introduction

Recording and storage of surgical images and videos are important issues that need to be addressed as fluorescence-guided surgery becomes more widespread. For example, when fluorescence cholangiography is used during cholecystectomy, records of operative images and videos in addition to conventional surgical documents will enhance the objectivity of operative information. In this chapter, the current medical system for recording and storing visual information during surgery linked to other electronic medical records was introduced.

2 Legal Regulations for the Recording and Storage of Surgical Videos

It goes without saying that surgical images and videos are important medical information for the treatment of patients and the education of surgeons/students. In the event of an accident caused by an intraoperative medical act, the surgical video can be objective information, and by reviewing it, both the medical staff and the patient can reconcile their perceptions. However, the current Japanese laws and regulations do not have direct provisions on the filming, recording, and storage of surgical images. In addition, even in the guidelines published by related academic societies, there are no specific operational standards for the filming, recording, and storage of surgical images. In other words, there are no standardized operational procedures for filming and storing surgical images, such as when and how they should be filmed, what media they should be recorded on, and where they should be stored, and so they are largely left to the judgment of each medical institution.

3 Selection of Images to Be Kept as Surgical Records

With the widening indication of endoscopic surgery, the question arises as to where and how to store surgical videos. From an educational standpoint, it would be ideal to store all surgical videos so that trainees can easily view them at any time. However, the storage method has become a difficult problem in many institutions because the data volume is increasing with the improvement of image quality. If only the intraoperative fluorescence imaging images are to be kept as part of the surgical record, the data volume will not be so large, so it is of course important to appropriately select scenes that are appropriate for keeping as a record.

4 Linking Surgical Images/Videos and Electronic Medical Records

Surgeons often reflect on the details of surgery when treating postoperative patients. Currently, the surgeons refer to the written surgical records and hand-drawn pictures attached to them to review the details of the surgery. What would be ideal for surgeons to be able to refer to fluorescence imaging images when reviewing surgical details? Many surgeons have been doing this for a long time, noting the patient's medical record number, returning to the doctor's office, and finding and viewing the hard disk of the relevant period from the date of surgery. Wouldn't it be ideal for surgeons to be

K. Kumagai (✉)
Division of Gastric Surgery, Department of Gastroenterological Surgery, The Cancer Institute Hospital of JFCR, Koto, Tokyo, Japan
e-mail: koshi.kumagai@jfcr.or.jp

T. Ishizawa (ed.), *Fluorescence-Guided Surgery*, https://doi.org/10.1007/978-981-19-7372-7_6

able to click on a certain part of the operation record stored in the electronic medical record in the outpatient consultation room and view the operation video on the spot? Technically, this is quite possible, but if the video is stored as part of the medical record, it will occupy a considerable amount of memory size. If the video is attached to the electronic medical record, it is necessary to install video playback software in the electronic medical record system, which raises the hurdle in cooperation with the electronic medical record vendor. For these reasons, attaching videos to electronic medical records is not the best method. The ideal method is to access videos stored on an external server via a network.

5 In the Case of the Cancer Institute Hospital

In our hospital, we have introduced Contents Management System (CMS, SONY Corporation), which allows us to import images of surgical field cameras, endoscopes, and surgical microscopes from recorders and manage them centrally on a server (Fig. 6.1). The CMS can be used to record and organize surgical videos and related documents. When recording images, patient information can be registered as well, enabling efficient management and utilization. The videos and related documents recorded on the server can be searched based on the date of surgery and patient information. In addition, you can easily select the scene you want to watch from the video file of a certain patient's operation so that you can easily access an important scene of a long operation. Furthermore, the recorded video can be easily edited by splitting and merging on the viewing terminal PC. The video files recorded and managed on the server can be accessed and viewed from a PC connected via the network, and by pasting a link to the electronic medical record, the video and still images can be called up from the electronic medical record.

Point

- In order to establish an in-hospital system enabling easy access to surgical videos in high quality, it seems to be better to store intraoperative imaging data on an external server linked to the electronic medical record.

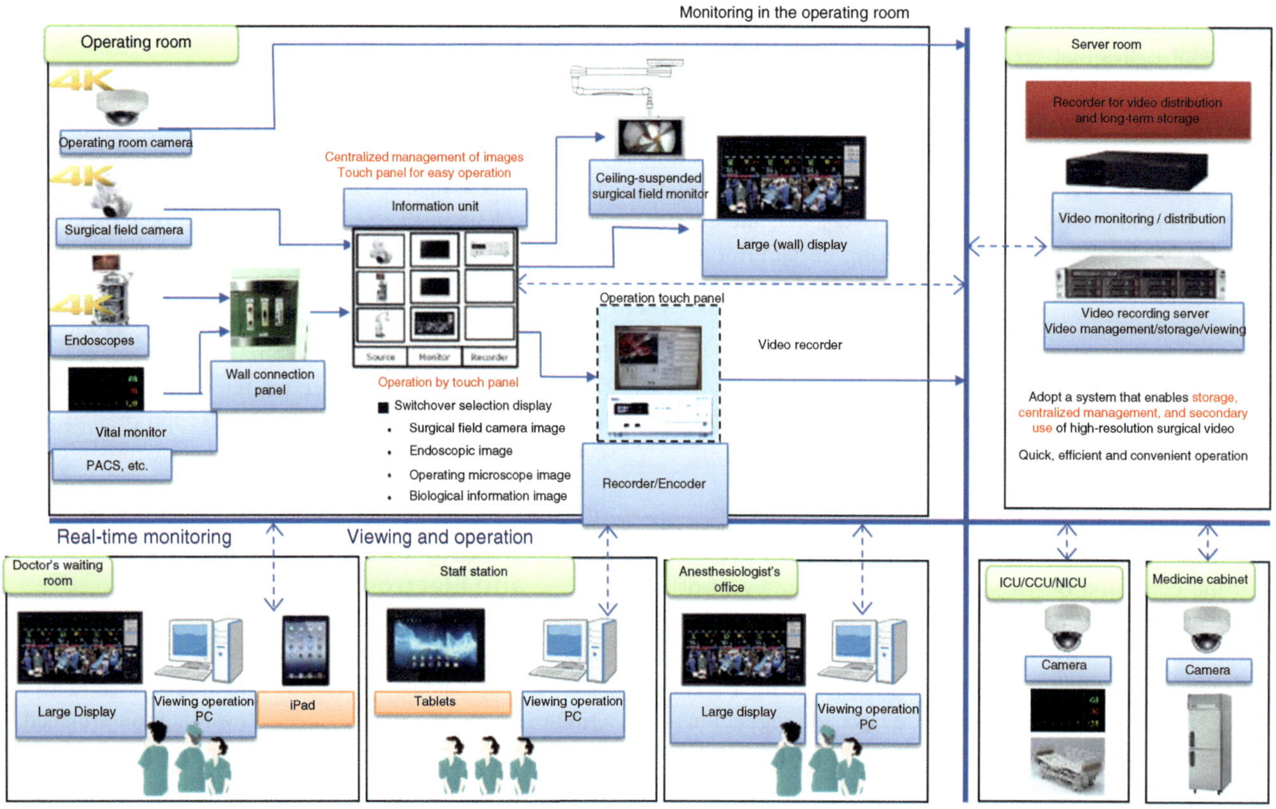

Fig. 6.1 System overview of CMS

6 Conclusions

Intraoperative fluorescence imaging is one of the most important image findings during surgery, and it is important to keep it as still images and videos, if possible, together with the operative record.

Column 1: Establishment and Activities of JSFGS (Japanese Society for Fluorescence Guided Surgery)

Takeaki Ishizawa

The pioneering research group on intraoperative fluorescence imaging was the ICG Fluorescence Navigation Surgery Study Group (later to become the Fluorescence Navigation Surgery Study Group), founded by Dr. Mitsuo Kusano. At the first seminar held in 2008 (Fig. 1). I, myself, while working at the Cancer Institute Hospital, had the opportunity to organize a workshop on fluorescence imaging-guided surgery as part of the activities of the Medical Device Development Center (director, Dr. Naoki Hiki; chairman. Dr. Yoshiaharu Yamaguchi). Since it was a unique opportunity to integrate disciplines, we invited lecturers from various fields, mainly those who had been involved in ISFGS activities (Part II). The medical device manufacturers kindly agreed to our unusual request to exhibit their equipment for imaging "common fluorescent phantoms" and to have their development and sales representatives give presentations on their products. As a result, more than 50 participants attended the event and hot information exchange took place despite it being a single facility event in the middle of the severe winter season (January 2017) (Fig. 2).

Shortly thereafter, Dr. Masashi Yoshida suggested that we establish a research organization for the further development of intraoperative fluorescence imaging by utilizing the network established this time. We continued our discussions under the guidance of Dr. Masaki Kitajima (deceased) and Dr. Norihiro Kunido (National Center for Global Health and Medicine), who were our mentors. At the same time, we were informed that the aforementioned "Fluorescence Navigation Surgery Study Group" would be terminated after the tenth meeting. We proposed to Dr. Masakazu Toi (Kyoto University) and Dr. Mitsuo Kusano, the representative sponsors, to establish a successor research group, and they agreed to continue their guidance in the new organization. I also created a logo of this society as a common icon (Fig. 3). After this process, the Japanese Society for Fluorescence Guided Surgery was finally established in 2018, and we have continued to hold annual meetings since the first meeting (April of the same year, see Preface).

The objectives of the society are "to share information on fluorescence imaging across specialties, to promote technological development and clinical implementation" and "to contribute to improving the safety and effectiveness of diagnosis and treatment of surgical patients through the spread of image-guided surgery." In addition to the 60 doctors who have agreed with the purpose of the association and are participating in its management as board members, a major feature of the association is that 20 companies are actively participating as supporting members based on the policy that "the development of research in the field is of common interest." The formation of such a consortium is a major force in expanding insurance coverage and regulatory approval for reagents, medical devices, and surgical procedures. In order to intro-

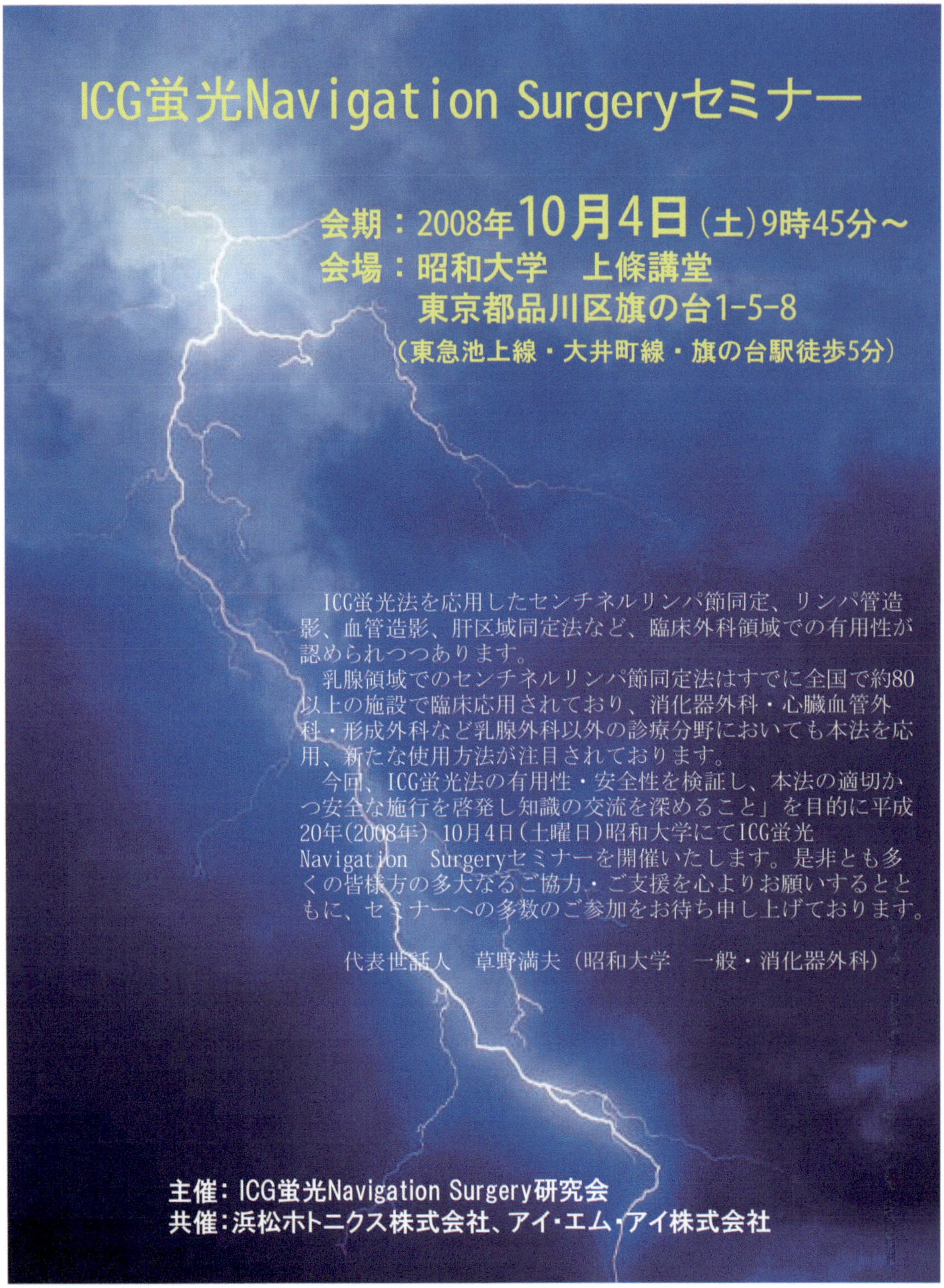

Fig. 1 ICG Fluorescence Navigation Surgery Seminar Brochure

duce fluorescence imaging to patients and the general public, the study group's website (http://plaza.unim.ac.jp/jsfgs/index.html) lists specific examples of surgeries and facilities (supporting members) that are active in this technology. The society will continue to promote the exchange of information between physicians, researchers, and engineers at and contribute to the development of intraoperative fluorescence imaging.

Fig. 2 Cancer Institute Fluorescence Image-Guided Surgery Workshop Poster

Fig. 3 Logo of the Japan Society for Fluorescence Guided Surgery (JSFGS)

It represents how fluorescence imaging can be used to identify targets and approach treatment

It is anticipated that similar research organizations will be established in other countries and regions in the future, and the fact that it is a "Japanese" research group is also sympathetic

Part II

Intraoperative Fluorescence Imaging [Practice] – Perfusion Assessment

7 Introduction

Masashi Yoshida

1 History of Perfusion Assessment by Indocyanine Green Fluorescence Imaging

The first paper on intraoperative perfusion assessment using indocyanine green (ICG) fluorescence imaging traces its origin to a study in Miami (University of Miami), where Kogure reported a method for evaluating ocular fundus blood flow by ICG infrared absorption in 1970 [1]. Citing this paper, Flower reported choroidal angiography by ICG infrared absorption spectroscopy [2]. In 1973, he reported the first human application of the ICG fluorescence method [3]. A PubMed search using the keywords "ICG fluorescence" and "blood flow" revealed 4 articles in the 1970s, 4 in the 1980s, 19 in the 1990s, 62 in the 2000s, and 251 in the 2010s. We can say that the research developed in the 2010s, probably due to the development of the fluorescence camera systems. Clinical application of ICG fluorescence imaging to perfusion assessment of skin flap was first reported by Still et al. [4] in the United States in 1999, and the first clinical report in coronary artery bypass grafting and in neurosurgery was published in Germany by Detter et al. [5] in 2002 and by Raabe et al. [6] in 2003, respectively. In the field of liver surgery, Sekijima et al. [7] reported the evaluation of blood flow after vascular reconstructions in liver and kidney transplantations in 2004, and Aoki et al. [8] reported the perfusion mapping of liver segments by ICG portal injection in 2008. One of the initial reports on the lower gastrointestinal tract would be that of Kudszus et al. [9] in 2010, which developed into the multicenter prospective study (PILLAR II) reported by Jafari et al. [10] in 2015. The PILLAR II Study is a key paper that contributed to the generalization of perfusion assessment by fluorescence imaging in gastrointestinal surgery. The first report in gastric surgery was a case of gastric cancer developed in the remnant stomach after esophageal surgery, in which ICG fluorescence imaging was effective for preserving the oral side of the stomach (Saito et al. [11] in 2012). The first evaluation of blood flow in esophageal surgery was performed by Shimada et al. [12] in 2011, which means that it has taken about 40 years since the first report in ophthalmology in 1973. The development history indicates that information sharing between scientists, surgeons, and engineers is essential for further development of fluorescence-guided surgery.

2 Effective Use of Indocyanine Green Fluorescence Imaging

Indocyanine green fluorescence imaging is based on the fact that ICG binds to alpha-1 lipoprotein [13] when it enters the body and visualizes the biodistribution of alpha-1 lipoprotein by its near-infrared fluorescence, which characterizes the advantages and limitations of the use of this technique for perfusion assessment. First, ICG fluorescence imaging enables "objective" evaluation of blood perfusion in a sense that surgeons can share images of blood flow in real time during surgery, although imaging systems allowing quantitative measurements of organ fluorescence intensity are still limited. Another advantage of the ICG fluorescence technique is the tissue penetration of near-infrared fluorescence. Second, this technique can theoretically detect fluorescence signals by covering tissues up to 10 mm in depth, although imaging conditions (direction between the camera head and target organs, elimination of light diffusion, etc.) should be optimized to have the best performance. Lastly, ICG fluorescence imaging can be used repeatedly for perfusion assessment (before and after the anastomosis, e.g.), because more than 90% of the ICG is excreted from blood into bile within 15 minutes in patients with normal liver function. But please note that alpha-1-lipoprotein-bound ICG can seep into tissues with

M. Yoshida (✉)
Department of Surgery, International University of Health and Welfare Hospital, Nasushiobara, Tochigi, Japan
e-mail: masashi@iuhw.ac.jp

T. Ishizawa (ed.), *Fluorescence-Guided Surgery*, https://doi.org/10.1007/978-981-19-7372-7_7

hypoperfusion or ischemia-reperfusion [14, 15], which may lead to unclear fluorescence images and inaccurate assessment of organ blood perfusion. In the future, these limitations may be solved with the use of fluorescent reagents that do not bind to lipoproteins or albumin.

References

1. Kogure K, David NJ, Yamanouchi U, et al. Infrared absorption angiography of the fundus circulation. Arch Ophthalmol. 1970;83:209–14.
2. Flower RW. Infrared absorption angiography of the choroid and some observations on the effects of high intraocular pressures. Am J Ophthalmol. 1972;74:600.
3. Flower RW, Hochheimer BF. A clinical technique and apparatus for simultaneous angiography of the separate retinal and choroidal circulations. Investig Ophthalmol. 1973;12:248–61.
4. Still J, Law E, Dawson J, et al. Evaluation of the circulation of reconstructive flaps using laser-induced fluorescence of indocyanine green. Clinical trial. Ann Plast Surg. 1999;42:266–74.
5. Detter C, Russ D, Iffland A, et al. Near-infrared fluorescence coronary angiography: a new noninvasive technology for intraoperative graft patency. Heart Surg Forum. 2002;5:364–9.
6. Raabe A, Beck J, Gerlach R, et al. Near-infrared indocyanine green video angiography: a new method for intraoperative assessment of vascular flow. Neurosurgery. 2003;52:132–9.
7. Sekijima M, Tojimbara T, Sato S, et al. An intraoperative fluorescent imaging system in organ transplantation. Transplant Proc. 2004;36:2188–90.
8. Aoki T, Yasuda D, Shimizu Y, et al. Image-guided liver mapping using fluorescence navigation system with indocyanine green for anatomical hepatic. World J Surg. 2008;32:1763–7.
9. Kudszus S, Roesel C, Schachtrupp A, et al. Intraoperative laser fluorescence angiography in colorectal surgery: a noninvasive analysis to reduce the rate of anastomotic leakage. Langenbeck's Arch Surg. 2010;395:1025–30.
10. Jafari MD, Wexner SD, Martz JE, et al. Perfusion assessment in laparoscopic left- sided/anterior resection (PILLAR II): a multi-institutional study. J Am Coll Surg. 2015;220:82–92.
11. Saito T, Yano M, Motoori M, et al. Subtotal gastrectomy for gastric tube cancer after esophagectomy: a safe procedure preserving the proximal part of the. J Surg Oncol. 2012;106:107–10.
12. Shimada Y, Okumura T, Nagata T, et al. Usefulness of blood supply visualization by indocyanine green fluorescence for reconstruction during esophagectomy. Esophagus. 2011;8:259–66.
13. Baker KJ. Binding of sulfobromophthalein (BSP) sodium and indocyanine green (ICG) by plasma alpha-1 lipoproteins. Proc Soc Exp Biol Med. 1966;122:957–63.
14. Yoshida M, Wakabayashi G, Ishikawa H, et al. A protease inhibitor attenuates gastric erosions and microcirculatory disturbance in the early period. J Gastroenterol Hepatol. 1998;13:104–8.
15. Yoshida M, Kurose I, Wakabayashi G, et al. Suppressed production of nitric oxide as a cause of irregular constriction of gastric venules induced by thermal injury in rats. J Clin Gastroenterol. 1997;25:S56–60.

Coronary Angiography

8

Tohru Asai

Summary

- Indocyanine green fluorescence angiography has been applied as an ideal method for intraoperative graft evaluation in coronary artery bypass grafting.
- A small amount of ICG is administered intravenously, and a clear angiogram can be obtained instantaneously by a CCD video camera.
- Graft failure can be detected with higher sensitivity than the conventional Doppler-based test (TTFM).
- Indocyanine green fluorescence angiography is attracting attention as a method enabling functional evaluation of bypass grafting as well as an assessment of graft patency.

1 Introduction

Coronary artery bypass grafting (CABG) not only relieves anginal pain in severe ischemic heart disease but also prevents possible future myocardial infarction and ischemic heart failure and prolongs life expectancy. There are two methods of coronary artery bypass surgery: "on-pump CABG," in which a bypass anastomosis is constructed under cardiac arrest with a cardioplegic solution under the heart-lung machine, and "off-pump CABG," in which the position of the heart is controlled and the anastomosis is performed with a local stabilizer while the heart is beating without circulatory support. In Japan, 50–60% of CABG are currently performed as off-pump CABG. The graft vessels used for bypass include arterial grafts (the internal thoracic artery, ITA; right gastroepiploic artery, GEA; and radial artery, RA) and the great saphenous vein. Types of grafts and surgical procedures are determined by surgeons and surgical teams according to the conditions of each patient.

In CABG, the diameter of the anastomotic site is 1.0–2.0 mm. If a subtle intraoperative problem causes stenosis and/or occlusion at the anastomosis, the value of bypass surgery itself can disappear. In fact, previous reports have indicated that 4% of bypass grafts (8% of cases) are occluded intraoperatively [1], and 5–20% of grafts are occluded before hospital discharge [2]. These results suggest that intraoperative detection of invisible problems enables re-anastomosis, leading to the improvement of graft patency after surgery, as performed in current coronary artery bypass surgery. In Japan, we can claim medical expenses for intraoperative graft evaluation, which has played an important role in the quality control of coronary artery bypass surgery.

2 Conventional Graft Evaluation Methods and Problems

The most common method for intraoperative graft evaluation is transit time flow measurement (TTFM), which uses the Doppler principle to measure graft blood flow, because of its simplicity and repeatability. Limitations of TTFM lie in the fact that it does not visualize the actual graft vessel and some of the measurements require understanding and interpretation by surgeons. Recently, the morphology of the graft anastomosis can be confirmed by using TTFM with high-frequency ultrasound images of the cardiac surface. However, this technique requires a surgeon's skills, and it is still difficult to understand the gross and spatial status of blood perfusion like radiographic coronary angiography. A conventional coronary angiography following injection of radiographic contrast materials provides information on the coronary artery and graft patency with the highest resolution. Radiographic angiography can be performed intraoperatively in some medical centers with a hybrid operating room. In daily clinical practice, however, it is rarely performed because of the lack of equipment and the increasing number of patients with renal dysfunction.

T. Asai (✉)
Department of Cardiovascular Surgery, Juntendo University, Bunkyo-ku, Tokyo, Japan
e-mail: t.asai.jj@juntendo.ac.jp

T. Ishizawa (ed.), *Fluorescence-Guided Surgery*, https://doi.org/10.1007/978-981-19-7372-7_8

In contrast, intraoperative fluorescence imaging allows the visualization of the bypass graft only with a small amount of intravenous indocyanine green (ICG), enabling real-time and on-site evaluation of the blood supply. It also has the advantage of preventing renal dysfunction due to angiography. In addition, unlike radiographic coronary angiography following bolus injection of contrast materials, ICG fluorescence imaging can reflect the balance of blood flows between the coronary artery and the bypass with information on collateral vessels under the natural cardiac circulation. In this chapter, we will describe methods and outcomes of ICG fluorescence imaging in CABG mainly with a SPY system.

3 Development History of Indocyanine Green Fluorescence Imaging in Cardiac Surgery

The application of intraoperative contrast-enhanced fluorescence imaging for CABG was developed by Novadaq Technology in Toronto, Canada, in 2000 as the "SPY Intraoperative Imaging System." Detter et al. [3] reported that ICG fluorescence angiography was useful for clear angiographic images in a porcine coronary artery bypass model. The first clinical use in humans was reported in 2002 by Rubens et al. [4] in 20 bypass operations, resulting in graft re-anastomosis in one patient. In 2005, Balacumaraswami et al. [5] compared TTFM with the SPY system and found that TTFM alone may lead to unnecessary graft re-anastomosis. In 2005, the US FDA approved the use of the SPY system for perfusion assessment in coronary artery surgery. In addition, in 2006, Desai et al. [6] conducted a randomized trial on the usefulness of the TTFM and SPY systems and reported that the SPY system detected intraoperative graft failure with greater accuracy than the TTFM. In Japan, the SPY system was introduced in 2002, and the number of facilities using the system has been gradually increasing since then; as of 2018, approximately 50 SPY systems nationwide have become widely available for intraoperative evaluation of coronary artery bypass surgery.

Point

- Indocyanine green fluorescence angiography has been used as an ideal method for the detection of graft failure in CABG.

4 Methods of Fluorescence Imaging

Recently in Japan, the majority of patients undergo CABG as off-pump surgery. Off-pump CABG has advantages over on-pump CABG in terms of operation time and duration of cardiac arrest required for revisions of bypass when insufficient blood flow is detected by cardiac angiography. Intraoperative fluorescence angiography using the SPY system is particularly useful in off-pump CABG.

Following the bypass grafting, the camera arm of the SPY system is introduced from the opposite side of the operating surgeon. The camera head is covered with sterilized plastic and set just above the operation field. The surgical lights should be turned off. Then, ICG is administered through a central vein to assess the graft patency and tissue perfusion. The advantage of ICG fluorescence imaging lies in the fact that surgeons can understand the outcomes of fluorescence angiography clearly and easily within 2–3 minutes. Although the graft anastomosis can be located on the side, back, or bottom of the heart, we can obtain fluorescence images just after the anastomosis because the heart has already been mobilized sufficiently. Even if there are five or six coronary anastomoses, all evaluations can be performed in two or three imaging sessions. When fluorescence imaging is used repeatedly, it is necessary to wait at least 3 minutes per acquisition for the ICG to disappear from the bloodstream. The images (movies) of fluorescence angiography can be played back immediately, enabling surgeons to review graft perfusion whenever necessary. In addition, ICG fluorescence angiography has potential advantages over conventional radiographic angiography in that it enables assessment of graft perfusion under normal circulation (without bolus infusion of contrast materials) with fewer people, without catheter insertion or a risk of radiation exposure and renal disorder.

The SPY system can be used easily at any point during surgery. In our off-pump CABG, we have used ICG fluorescence angiography when all anastomoses are completed, because it enables us to redo the problematic bypass anastomosis under a beating heart immediately, although the incidence is very low.

In Figs. 8.1, 8.2, and 8.3, specific examples of intraoperative fluorescence angiography are demonstrated. The RITA was anastomosed to the left anterior descending artery (LAD) of the left coronary artery. The GEA was sequentially anastomosed through the posterior descending artery (4PD) of the right coronary artery to the posterior lateral branch (PL). As soon as the last anastomosis was constructed, the camera arm of the SPY system was covered with a sterile cover and fluorescence angiography was performed. The examination was recorded as a clear movie in three shots. This movie could be played back repeatedly after imaging. In both the internal thoracic artery graft and the gastric major artery graft, the skeletonized arterial graft allows us to clearly follow the movement of blood flow through the vessel wall as in the venous graft. Examples of problematic cases are shown in Fig. 8.4. In this case, the RITA graft to the LAD

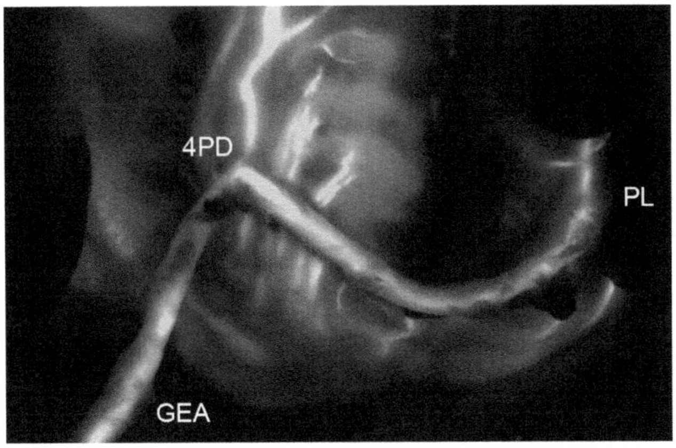

Fig. 8.1 Sequential anastomosis with the right gastroduodenal artery (GEA-PDA-PL). Indocyanine green fluorescence angiography of a five-vessel beating coronary artery bypass using only an in situ arterial graft; this imaging was performed with intact cardiac deployment and stabilization after the last anastomosis

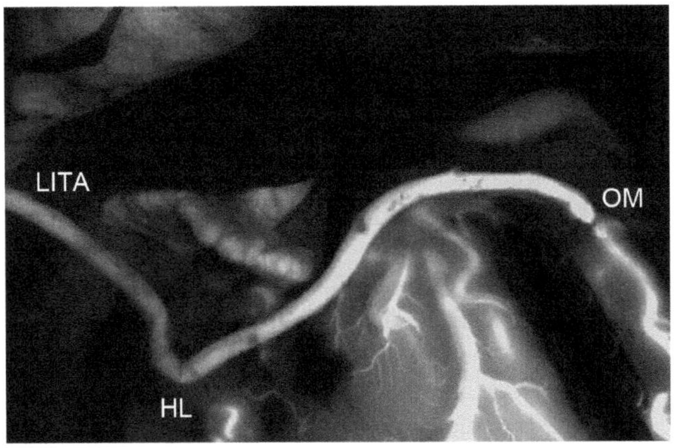

Fig. 8.2 Sequential anastomosis with the left internal thoracic artery (LITA-HL-OM). Indocyanine green fluorescence angiography of five-vessel beating coronary artery bypass with in situ arterial graft alone

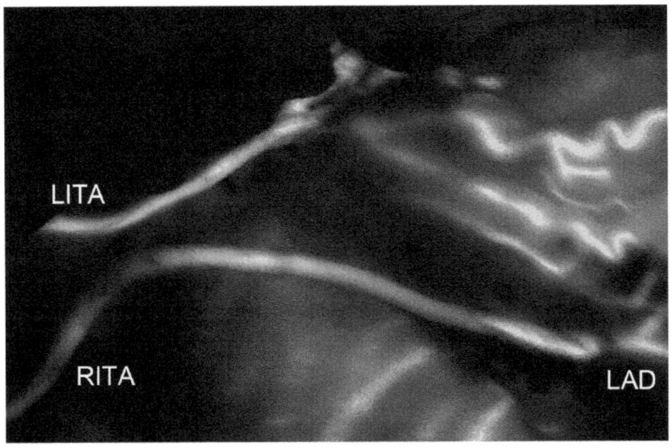

Fig. 8.3 Single anastomosis of the right internal thoracic artery to the left anterior descending branch (RITA-LAD). Indocyanine green fluorescence angiography of five-vessel beating coronary artery bypass with in situ arterial graft alone

was not visualized at all, so the anastomosis was revised. As shown in Fig. 8.4b, the graft blood flow after re-anastomosis was well depicted.

Although the incidence is low, graft occlusion can develop during CABG because of technical errors in the anastomosis, vessel dissection, lumen occlusion due to hematoma, wandering of surrounding tissue into the anastomotic vessel, and thrombus formation. The SPY system allows visualization of the actual blood flow inside an apparently normal bypass anastomosis and is a valuable tool for resolving problems before the end of surgery.

Our initial experiences with intraoperative fluorescence angiography were as follows [7, 8]: From April 2009 to November 2011, we performed intraoperative fluorescence angiography using the SPY system in 159 patients. In this series, fluorescence imaging was used after the completion of all anastomoses based on conventional techniques (TTFM and redo, if necessary). Although the TTFM detected abnormal values in 12/142 RITA, 13/155 LITA, 20/88 GEA, and 10/50 SVG, ICG fluorescence angiography visualized sufficient blood flow in all anastomoses; but all intraoperative fluorescence images (IFI) showed contrast enhancement. Confirmatory angiography (CT angiography in 128 patients and direct angiography in 31 patients) at about 1 week after surgery showed that all arterial grafts were open and only two venous grafts were occluded.

Some authors have also reported the detection rate of intraoperative graft failure by the SPY system: Taggart et al. [9] reported 4 of 213 grafts (1.9%), Reuthebuch et al. [10] reported 4 of 107 grafts (3.7%), Desai et al. [11] reported 5 of 348 grafts (1.4%), Balacumaraswami et al. [5] reported 8 of 533 grafts (1.5%), Takahashi et al. [12, 13] reported 4 of 290 grafts (1.9%), and Kishimoto et al. [14] reported 4 of 533 grafts (1.9%). Since some of these abnormal findings might have been undetected by conventional techniques, ICG fluorescence angiography is expected to contribute greatly to the improvement of surgical outcomes of CABG.

Point

- Intraoperative fluorescence angiography using the SPY system provides clear images of the coronary artery and anastomosis only with a small amount of ICG through a central vein.
- ICG fluorescence angiography detects abnormal blood flows sensitively at any time during surgery, enabling surgeons to revise the anastomosis immediately.

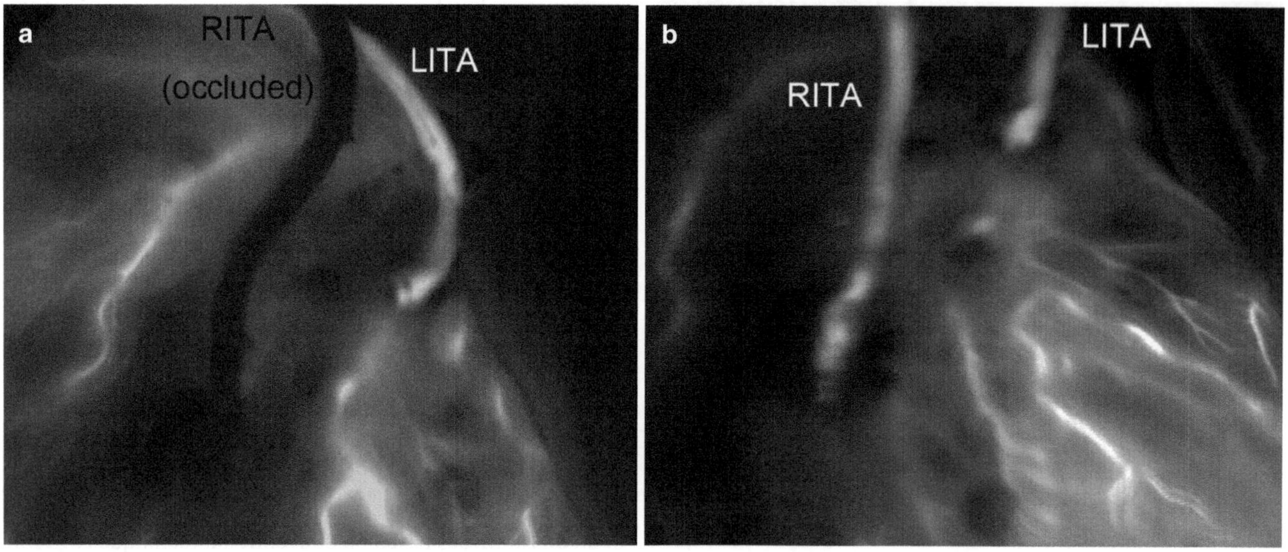

Fig. 8.4 A graft problem detected by intraoperative fluorescence angiography. (**a**) The left internal thoracic artery was well delineated, but the blood flow from the right internal thoracic artery was not contrasted. (**b**) The anastomosis was immediately revised, the right internal thoracic artery was slightly shortened, and the anastomosis was reconstructed. Fluorescence angiography identified sufficient blood flows through the anastomosis

5 Expected Roles of Indocyanine Green Fluorescence Angiography in Coronary Artery Bypass Grafting

5.1 Evaluation of Blood Flow in Bypass Grafts

The real-time images on the monitor clearly demonstrate the blood flow from the bypass vessel into the coronary artery, which can easily be interpreted and shared by surgeons. If there is no contrast effect at all, this may indicate a problem due to blood clots, dissection, kinking, or bending of the anastomosis or graft. In such a case, the coronary anastomosis can be redone immediately, followed by repeated fluorescence angiography. A possible limitation of the SPY system is that when bypass vessels are harvested with surrounding connective tissues, blood flow can be invisible because of the limited tissue permeability of near-infrared light. In addition, blood flow from the anastomosis to the peripheral coronary artery itself cannot be visualized by fluorescence imaging because the fluorescence signals are blocked by fatty tissues on the cardiac surface. These points are considered to be the limitations of the current imaging system.

5.2 Assessment of the Competing Blood Flows Between the Graft and Host Coronary Artery

When the arterial graft is used for mild stenosis of the coronary artery, the blood flow through the graft may compete with the native blood flow, causing the bypass vessel to become thin and lose blood supply in some cases. Unlike conventional catheter angiography, in which contrast materials are forcefully injected from the target vessel, ICG fluorescence angiography can delineate the real status of blood flows under natural circulation, including the competing blood flows. Therefore, intraoperative fluorescence angiography may become an important method for future research on the proper use of arterial grafts and the long-term postoperative effects of the competing blood flows.

6 Technical Notes

When ICG fluorescence angiography is used for visualization of the internal thoracic artery and the right gastroduodenal artery, fluorescence signals can be identified slowly because they are far from the origin of the coronary artery. This should not be misunderstood as insufficient blood supply. In addition, there can be a time lapse in the visualization of the host coronary artery, the vein graft from the ascending aorta, and the arterial grafts, which may affect the incidence of the competing blood flows. It is unclear what the clinical significance of these phenomena is, how they affect subsequent bypass function, or whether they are at all problematic, but assessments of blood flow status by ICG fluorescence imaging may provide clues to the clinical significance of these phenomena.

The most important issue in this technique would be the quantitative assessment of blood flows. A semi-quantitative assessment of graft vessels and corresponding myocardial regions before and after bypass grafting may be useful to propose new criteria for the intraoperative evaluation of the graft patency, as suggested by Ferguson et al. [14].

One of the limitations of intraoperative fluorescence angiography lies in its tissue permeability especially due to connective tissues attached to the graft and cardiac muscles. Further improvement of signal detectability by fluorescence imaging and combination with other diagnostic modalities such as TTFM and fractional flow reserve (FFR) [15] would clarify unknown functions and long-term changes of the graft vessels.

7 Conclusions

Although the mainstream of intraoperative graft evaluation in Japan remains TTFM, applications of ICG fluorescence angiography in CABG will be expanded as a safe, easy, and reliable diagnostic tool, with the improvement of imaging systems.

References

1. D'Ancona G, Karamanoukian HL, Ricci M, et al. Graft revision after transit time flow measurement in off-pump coronary artery bypass grafting. Eur J Cardiothorac Surg. 2000;17:287–93.
2. Balacumaraswami L, Taggart DP. Intraoperative imaging techniques to assess coronary artery bypass graft patency. Ann Thorac Surg. 2007;83:2251–7.
3. Detter C, Russ D, Iffland A, et al. Near-infrared fluorescence coronary angiography: a new noninvasive technology for intraoperative graft patency. Heart Surg Forum. 2002;5:364–9.
4. Rubens FD, Ruel M, Fremes SE. A new and simplified method for coronary and graft imaging during CABG. Heart Surg Forum. 2002;5:141–4.
5. Balacumaraswami L, Abu-Omar Y, Choudhary B, et al. A comparison of transit-time flowmetry and intraoperative fluorescence imaging for assessing. J Thorac Cardiovasc Surg. 2005;130:315–20.
6. Desai ND, Miwa S, Kodama D, et al. A randomized comparison of intraoperative indocyanine green angiography and transit-time flow measurement to detect technical errors in coronary bypass grafts. J Thorac Cardiovasc Surg. 2006;132:585–94.
7. Kuroyanagi S, Asai T, Suzuki T, et al. Advantages of intraoperative fluorescence imaging during coronary artery bypass grafting. J Jpn Coron Assoc. 2013;19:223–7.
8. Kuroyanagi S, Asai T, Suzuki T. Intraoperative fluorescence imaging after transit-time flow measurement during coronary artery bypass grafting. Innovations (Phila). 2012;7:435–40.
9. Taggart DP, Choudhary B, Anastasiadis K, et al. Preliminary experience with a novel intraoperative fluorescence imaging technique to evaluate the patency of bypass grafts in total arterial revascularization. Ann Thorac Surg. 2003;75:870–3.
10. Reuthebuch O, Häussler A, Genoni M, et al. Novadaq SPY: intraoperative quality assessment in off-pump coronary artery bypass grafting. Chest. 2004;125:418–24.
11. Desai ND, Miwa S, Kodama D, et al. Improving the quality of coronary bypass surgery with intraoperative angiography: validation of a new technique. J Am Coll Cardiol. 2005;46:1521–5.
12. Takahashi M, Ishikawa T, Higashidani K, et al. SPYTM: an innovative intra-operative imaging system to evaluate graft patency during off-pump coronary artery bypass grafting. Interact Cardiovasc Thorac Surg. 2004;3:479–83.
13. Waseda K, Ako J, Hasegawa T, et al. Intraoperative fluorescence imaging system for on-site assessment of off-pump coronary artery bypass graft. JACC Cardiovasc Imaging. 2009;2:604–12.
14. Ferguson TB Jr. Physiology of in-situ arterial revascularization in coronary artery bypass grafting: preoperative, intraoperative and postoperative factors and influences. World J Cardiol. 2016;8:623–37.
15. Hatada A, Okamura Y, Kaneko M, et al. Comparison of the waveforms of transit-time flowmetry and intraoperative fluorescence imaging for assessing. Gen Thorac Cardiovasc Surg. 2011;59:14–8.

Cerebral Angiography (Cerebral Aneurysm)

9

Yasuo Murai, Fumihiro Matano, and Akio Morita

Summary

- Intraoperative ICG videoangiography (ICGVAG) is useful to confirm the occlusion of cerebral aneurysms and patency of bypass vessels.
- Semi-quantitative analysis based on the brightness of the region of interest (ROI) and its time trend remains a problem to solve.
- Surgeons should understand the characteristics of each microscope instrument and the limitations of ICGVAG.

1 Introduction

In this chapter, we first describe how indocyanine green videoangiography (ICGVAG) came to be used in neurosurgery. Intraoperative indocyanine green (ICG) imaging for cerebrovascular surgery, which was covered by insurance in 2016, has become an essential intraoperative examination for neurosurgeons throughout Japan, and we have accumulated more than 700 cases. We introduce here the improvement of the accuracy of the anatomical understanding and the quantitative evaluation of blood flow in the observation by various methods, based on the previous reports.

Supplementary Information The online version contains supplementary material available at https://doi.org/10.1007/978-981-19-7372-7_9.

Y. Murai (✉) · F. Matano · A. Morita
Department of Neurological Surgery, Nippon Medical School, Bunkyo, Tokyo, Japan
e-mail: ymurai@nms.ac.jp

2 Development History of Indocyanine Green Videoangiography

Indocyanine green was a fluorophore approved for clinical use by the U.S. Food and Drug Administration (FDA) in 1956. Among a variety of applications, ICG was first used for the assessment of blood flow in ophthalmology (fundus and retinal angiography) at the outpatient level since the 1970s and in vascular surgery since around 2002. Since the initial experience of ICG videoangiography in the field of neurosurgery in 2003 [1], this technique has widely been used with the term "ICG videoangiography," because fluorescence images are usually recorded and assessed quantitatively with a microscopic imaging system [2–4]. The use of ICG videoangiography in cerebrovascular surgery is similar to fundus/retinal angiography in that blood flow in the target vessels can be visualized in real time within 1–2 minutes after intravenous injection of ICG.

In the first report of ICGVAG in the field of neurosurgery, a quite primitive imaging system in which an infrared light filter was attached to the lens of a consumer video camera was used [1]. However, the simplicity of their imaging techniques as well as the principle of on-site angiography greatly appealed to neurosurgeons so that the ICGVAG system was quickly installed in various surgical microscopes around the world. I myself witnessed the world's first clinical study of a microscope in which this system was installed while studying in the USA in 2005 and remember being shocked by the simplicity and minimally invasive nature of ICGVAG. I also had the opportunity to use the Carl Zeiss PENTERO® 900 microscope equipped with ICGVAG for the first time in Japan and to make an initial report. Since then, almost all surgical microscopes, endoscopes, and exoscopes used in the field of neurosurgery have been equipped with the ICGVAG system [5].

T. Ishizawa (ed.), *Fluorescence-Guided Surgery*, https://doi.org/10.1007/978-981-19-7372-7_9

3 Advantages of Indocyanine Green Videoangiography in Cerebral Aneurysm Surgery

Conventional radiographic cerebral angiography (digital subtraction angiography, DSA), which is usually performed by the Seldinger method from the femoral or the radial artery, is considered to be the golden standard for confirming anastomotic patency and complete closure of cerebral aneurysm. However, DSA has not been used routinely for intraoperative assessment because it is invasive to patients and technically demanding, especially in neurosurgery, where the head is fixed with metallic fixation devices and the patient may be placed in the lateral or prone position. ICGVAG, on the other hand, is administered by peripheral vein and is the same method as intravenous digital subtraction angiography (IVDSA), making it extremely simple. In addition, since the cerebral blood vessels to be observed exist in the subarachnoid space and are not buried in the brain parenchyma, direct and clear observation is possible. This would be the main cause of the spread of ICGVAG in the field of neurosurgery. In addition, the high number of cerebral aneurysm surgery and severity of postoperative complications, the affinity between microscopic surgery and fluorescence imaging, and assurance for the safety of ICG would promote the use of ICGVAG.

4 Case Presentations

4.1 Basic Technique of Indocyanine Green Videoangiography Imaging (Fig. 9.1)

Once the target cerebral aneurysm, its parent vessel, and peripheral vessels are exposed and captured in the field of view, the microscope used is changed to "ICG mode" or "near-infrared light mode." ICG (2.5 mg/mL, usually at a dose of 0.10–0.25 mg/kg) is injected intravenously via a peripheral vein with a bolus. Surgical lighting in the OR should be turned off. The time from intravenous injection of ICG to visualization of blood flows depends on the heart rate, blood pressure, and other factors, but it usually takes a few seconds to 30 seconds for ICG to reach the intracranial area. Fluorescence images can be visualized clearly for the first 30 minutes, although this technique can be used repeatedly after waiting for the washout of ICG from background structures. ICGVAG images are automatically recorded and can be reviewed in the OR when needed. In the imaging mode for ICGVAG, fluorescent blood vessels are contrasted in white, yellow, blue, etc., according to the microscopic system to be used.

4.2 Confirmation of Complete Occlusion Following Aneurysm Clipping

In the clipping technique, which is the golden standard for the treatment of cerebral aneurysms, the aneurysm neck is occluded with a clip to block blood flow into the aneurysm and prevent rupture. Before the development of the ICGVAG, intraoperative radiographic angiography (DAA) and/or the Doppler method were used for this purpose. In the use of ICGVAG for confirmation of aneurysm occlusion, it is necessary to observe the presence or absence of reentry of blood flow for about 1 minute and the retention of the contrast medium injected before clip occlusion (Fig. 9.1d). It is not clear whether additional clip placement is necessary when minute blood flow in the aneurysm neck is detected by ICGVAG, and whether the aneurysm will thrombose if left untreated, but we do perform additional clips.

4.3 Confirmation of Anastomotic Patency in the Treatment of Cerebral Aneurysms (Movies 9.1, 9.2, 9.3, and 9.4)

Although clipping is the standard surgical treatment for cerebral aneurysms, sometimes aneurysms are treated by closing the aneurysm together with the parent vessel and reconstructing the peripheral vessel (anastomosis). ICGVAG can also be used to confirm the patency of the reconstructed vessel in aneurysm surgery as well as in the treatment for cerebral ischemia. Although most of the recipients are 1–2 mm in diameter, they are depicted with blood flow from the donor because the peripheral side flows only from the donor. ICGVAG may visualize retrograde blood flow in recipient arteries through intracranial peripheral blood vessels even if the forward blood flow from the donor is poor. In such a case, ICGVAG should be performed again after the recipient is temporarily closed. Sometimes, the direction of blood flow is unclear in ICGVAG (Movie 9.5) due to a variety of technical factors. Since the direction of inflow is important information for detecting central stenosis of the donor, some microscopic imaging systems are equipped with a function to delineate the direction of blood flow by indicating the increase of fluorescence intensities with color codes and by measuring the timing of fluorescent increase in each region of interest (ROI) set on the vessels.

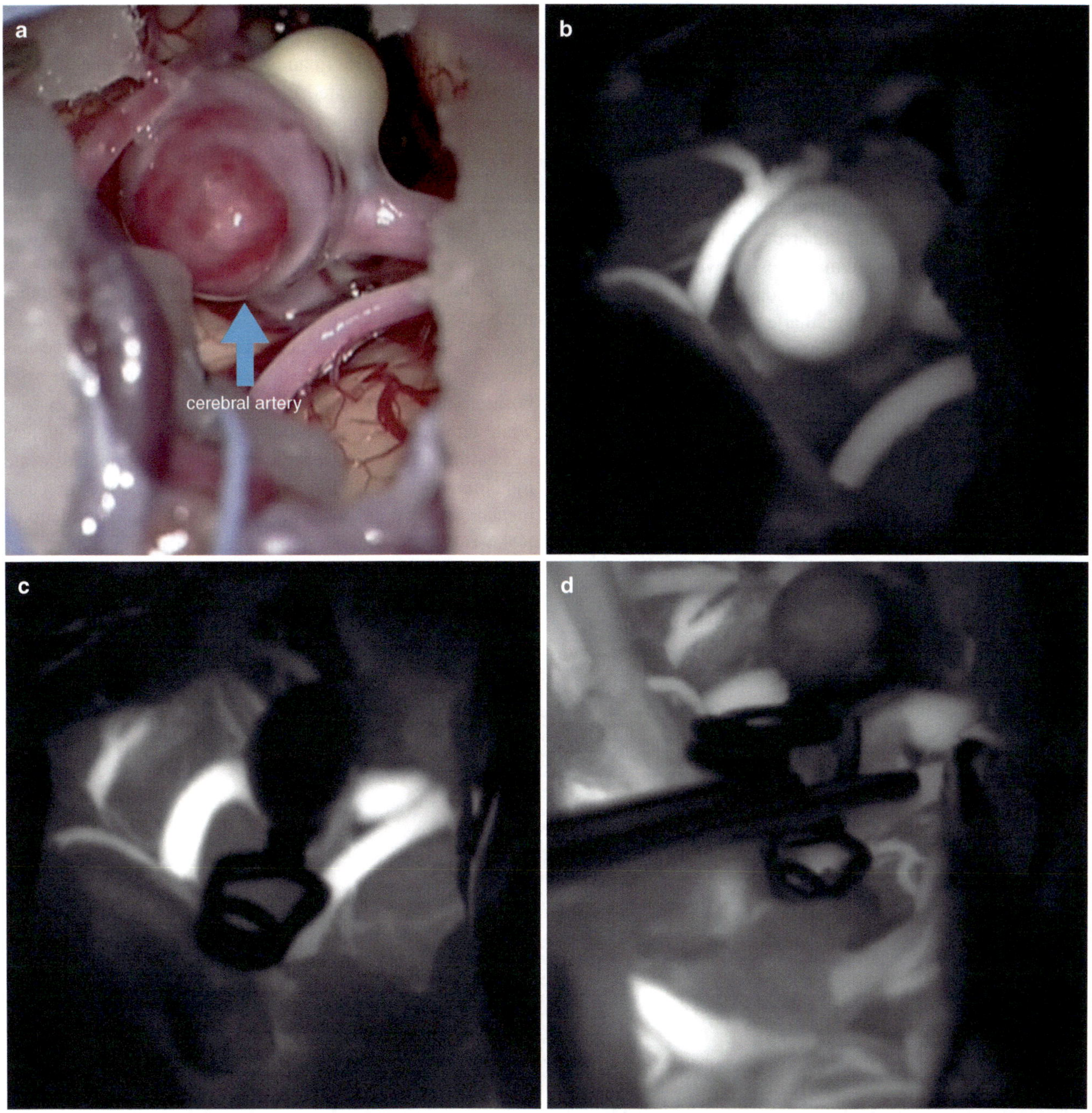

Fig. 9.1 Indocyanine green videoangiography imaging of clipping for middle cerebral artery. (**a**) After opening the Sylvian fissure, the middle cerebral artery is exposed just before clipping. There is a thickened area in the aneurysm wall. (**b**) Pre-clipping ICGVAG findings. The thick wall of the aneurysm is less stained by the contrast medium. (**c**) ICGVAG findings immediately after clipping. There is no contrast within the aneurysm. (**d**) ICGVAG findings after the addition of the second clip. Contrast material remains in the aneurysm, suggesting complete occlusion

Point

- Indocyanine green videoangiography has clear advantages over conventional radiographic angiography (DSA) in terms of simplicity and safety.
- Fluorescence images should be observed for 1 minute after intravenous injection of ICG and visualization of the target vessel/aneurysm.
- Efficacy and limitations of ICGVAG for confirmation of aneurysm occlusion and graft patency have been reported.
- Understanding the characteristics of ICG and imaging systems is essential for obtaining clear and accurate information.

5 Clarification of Indocyanine Green Fluorescence Images by Advanced Image Processing

Since the beginning of the application of ICG fluorescence imaging for intracranial diseases, limitations in image definition due to anatomical situations have been pointed out. In other words, ICGVAG is limited in its ability to delineate small vessels (0.1–0.3 mm) in narrow and deep areas. In the intracranial space, there are many vessels with a diameter of around 0.2 mm that can cause serious complications when occluded (the anterior choroidal artery, the lateral lenticulostriate artery, etc.). These vessels are in contact with the preferred site of cerebral aneurysms and may be occluded as a result of aneurysm closure. Another difficult situation is the treatment for the bifurcations of the anterior and posterior communicating arteries. They are often located on the back of the aneurysm, narrowing at greater depths, and the clips are often placed in front of the observation field, which makes it difficult to illuminate target regions from an appropriate orientation [1]. Aneurysms developing in these regions should be clipped completely because of the higher risk of postoperative rupture. In addition, ICG fluorescence images are usually demonstrated in a separate field of view rather than in the microscopic field of view in the eyepiece (Movie 9.2), which makes it difficult for surgeons to differentiate the target vessel and evaluate its patency.

Advances in image processing technology may overcome these problems in ICGVAG. A possible solution is to superimpose translucent fluorescence images on full-color images within the field of view of the eyepiece (Fig. 9.2). Another method is to brighten the entire surgical field to enhance the identifiability of anatomical structures (Movie 9.2). The former approach still has limitations in a discrepancy in the pixel position and framerate between fluorescence images and white-light color images. Brightening fluorescence images makes simultaneous observation of the surgical field by white light imaging difficult. Further development of image processing technology is supposed to solve the remaining problems of ICGVAG for the use of complicated surgical procedures.

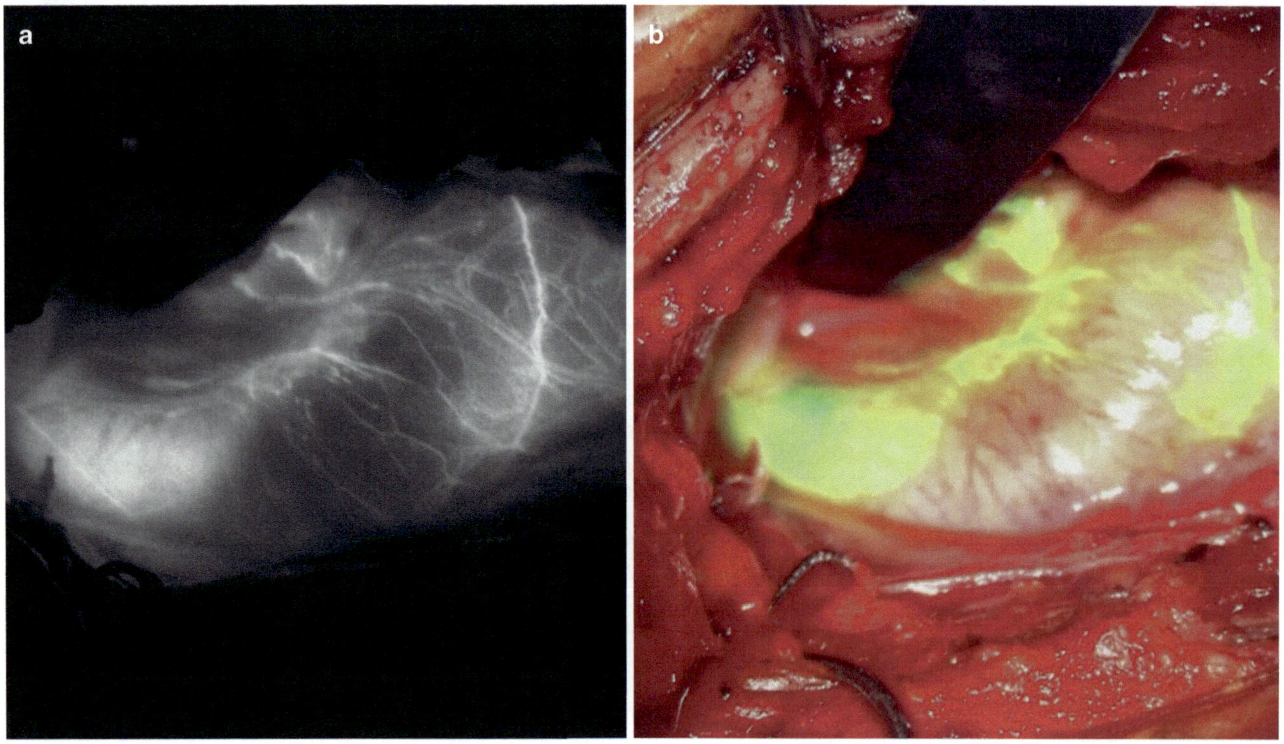

Fig. 9.2 Indocyanine green videoangiography superimposed image during carotid endarterectomy. (**a**) Grayscale fluorescence images. (**b**) Superimposed image of ICGVAG using the KINEVO 900 (Carl Zeiss) during left carotid endarterectomy. In this model, the ICGVAG image is drawn in yellow tone and superimposed in the mirror field of the eyepiece to prevent the surgeon from shifting his field of view

6 Difference Between Indocyanine Green and Fluorescein as a Fluorophore

Fluorescein is another fluorescent contrast agent for cerebral blood flow analysis [4], although it is not covered by insurance in Japan except for its use in ocular angiography. In the field of cerebrovascular surgery, a shorter half-life of ICG is suitable for repeated use during surgery, but on the other hand, ICG is contraindicated in patients with a history of iodine hypersensitivity. Characteristics of ICG and fluorescein are summarized in Table 9.1. As shown in Fig. 9.3 and Movie 9.3, our conclusion from the comparative study is that ICG and fluorescein are useful for the visualization of thick vessels and thin vessels, respectively.

Table 9.1 Comparison of indocyanine green and fluorescein (FC)

	Half-life	Visualization of microscopic blood vessels	Repeated inspection	Iodine	Insurance coverage	Wavelength of maximum fluorescence	Wavelength of maximum absorption
ICG	Short: 3 ~ 4 min	Bad	15 min	Include	Available	835 nm	805 nm
FC	Long	Good	Over 30 min	Not include	N/A	525 nm	480 nm

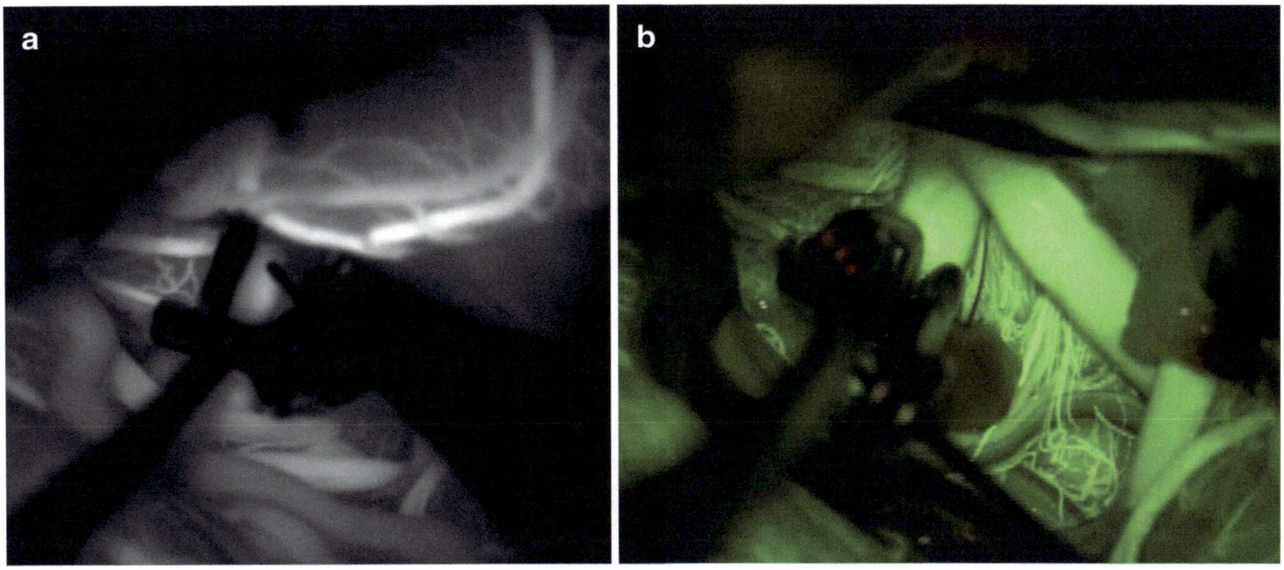

Fig. 9.3 Indocyanine green and fluorescein imaging findings in the same field of view (the Sylvian fissure). (**a**) ICG. (**b**) Fluorescein

7 Applications of Quantitative Evaluation of Fluorescence Signals

There are two methods of quantitative evaluation of cerebral blood flow: one is to measure the transit time in perfusion images, that is, the timing of drawing the target tissue, and the other is to measure the maximum fluorescence intensity of the target tissue or the fluorescent increase per unit time (Figs. 9.4, 9.5, 9.6, and Movie 9.6). Each of these measurement methods has its own problems to be solved [3–6], and at the present stage, it is often only possible to observe relative changes in the same operation.

Assessment of tissue perfusion by ICG fluorescence imaging involves several problems. The infusion rate of ICG, blood pressure, heart rate, and cardiac output may affect the inflow of ICG into intracranial space. The influence of light scattering, such as indirect illumination from surrounding tissues, may also affect the degree of intracranial flow. For example, even non-perfused regions can show fluorescence signals when surrounded with highly perfused tissues [6] (Fig. 9.6). In addition, since the evaluation of fluorescence intensity is based on the average values of the target areas, shifting the target areas will affect the calculations (Fig. 9.5). In the current software for quantitative assessment of blood flows during neuro-

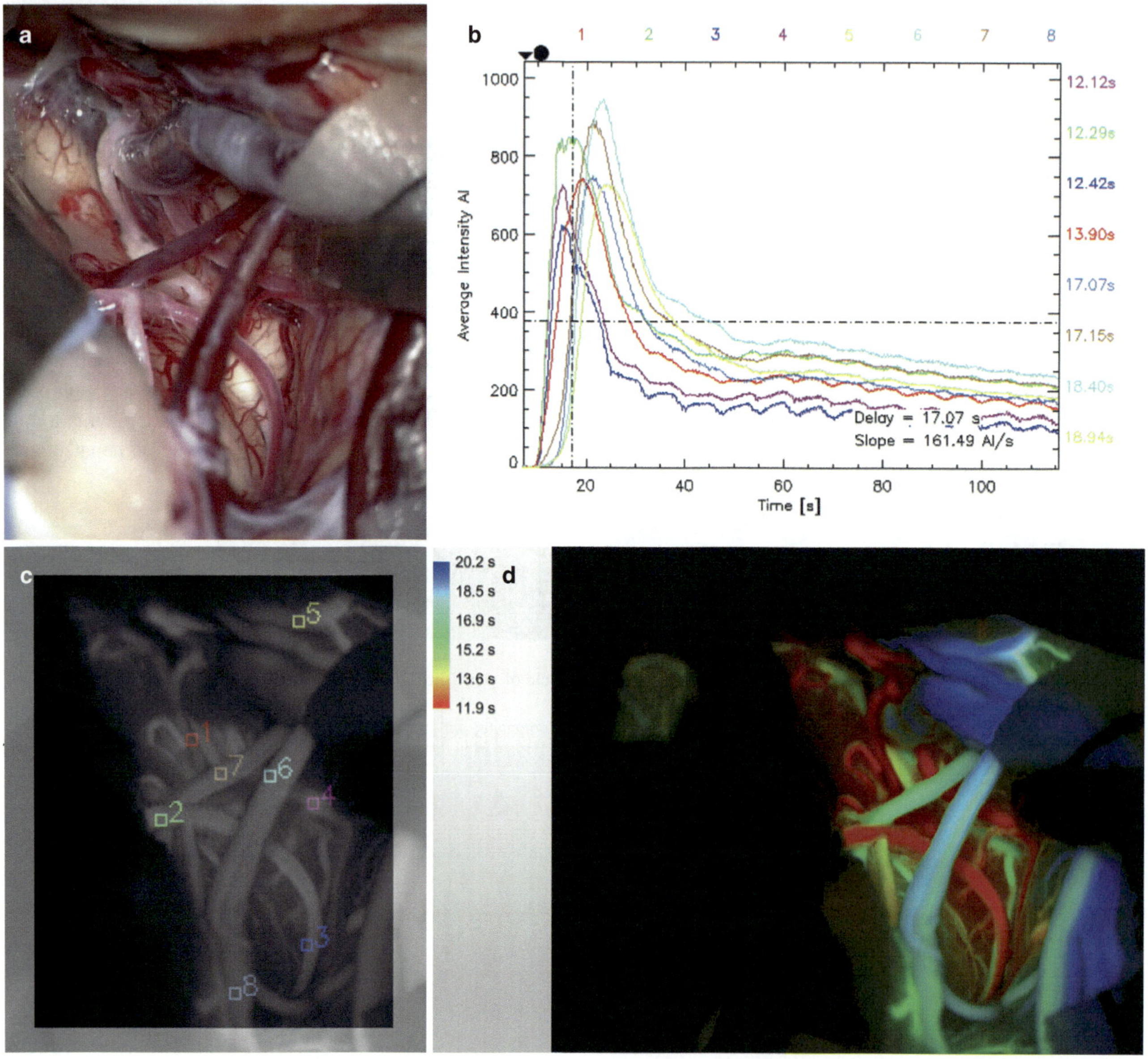

Fig. 9.4 Indocyanine green videoangiography and Color Code Map installed in PENTERO® 900 (Carl Zeiss). (**a**) After the opening of the Sylvian fissure. Middle cerebral artery, Sylvian vein, frontal lobe, temporal lobe. (**b**) Trends of fluorescence intensities in each ROI. Differences between arteries and veins are observed in the time to maximum intensity and the maximum fluorescence intensity. (**c**) ICGVAG image and ROI setting site. (**d**) Color Code Map image. Early-phase areas are depicted in red tone and late-phase areas are depicted in blue tone

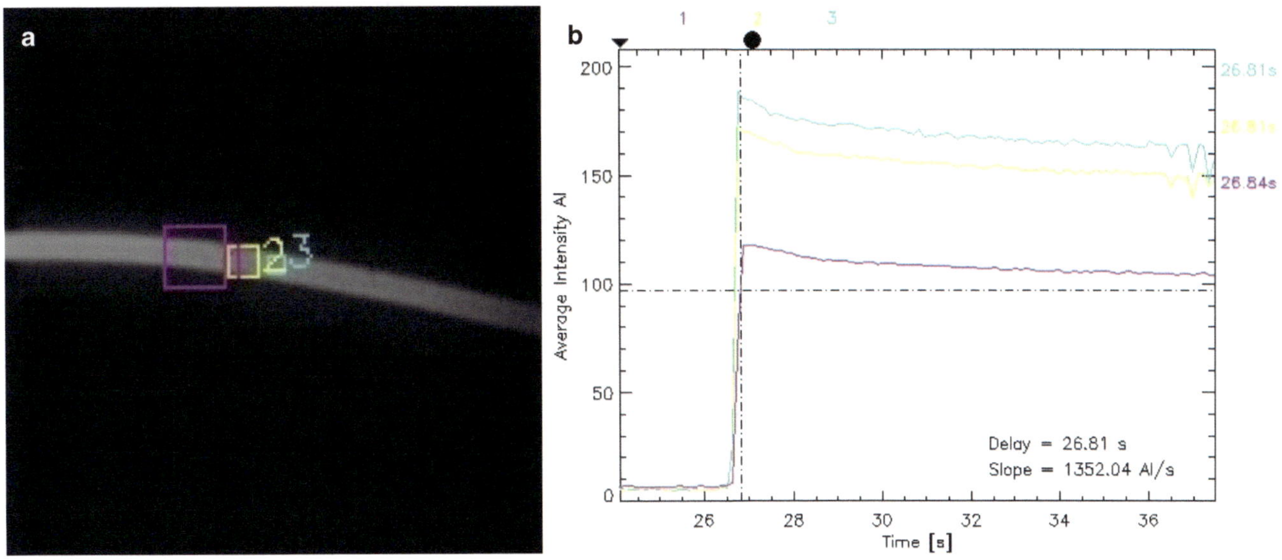

Fig. 9.5 Size and location of the ROI and the outcome of quantitative assessment. (**a**) After injecting ICG into the artificial blood vessel of the phantom, a square ROI was set with three different sizes. (**b**) The red number 1, in which the ROI extends outside the simulated artery, has the lowest luminance, and the light blue number 3, in which the ROI is limited within the simulated artery, has the highest luminance evaluation. This indicates that the fluorescence intensity is measured based on the average signal intensity of the ROI

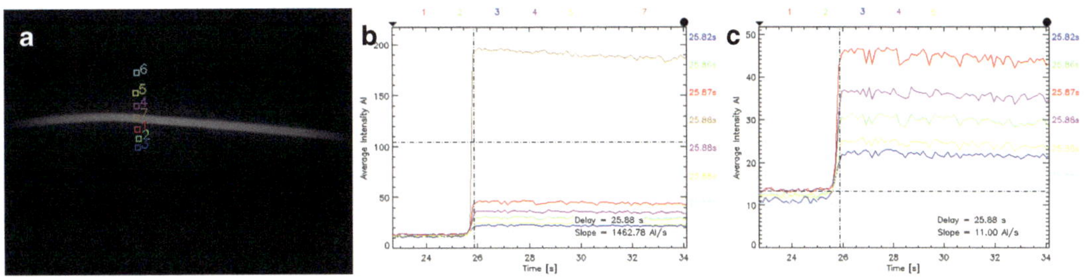

Fig. 9.6 Relationships between distance from the ROI and values of fluorescence intensity. (**a**) After injection of ICG into the simulated artery, a square ROI of the same size was set on the simulated artery and at a distance of 2 mm from it. (**b**) The ROI on the simulated artery in brown No. 7 has the highest intensity, but there is also an increase in intensity in other areas where ICG was not injected at all. (**c**) Expanded image of the middle graph, excluding brown number 7. The ROI close to the simulated artery has high evaluated luminance, and as the distance increases, the fluorescence intensity decreases to 4, 5, and 6 due to diffusion of fluorescence signal in the areas free from ICG

surgery, fluorescence signals are calculated based on the mean fluorescence intensities of the ROI set on the target region [2–4, 6]. For this reason, when ICG fluorescence imaging is used for quantitative evaluation, the position, focus, shooting range, and visual axis (orientation) of the microscope cannot be moved at all. Even if the same surgical field is captured, quantitative measurement of fluorescence signals can differ according to the size of the ROI. Therefore, in the quantitative assessment of cerebral blood flow by fluorescence imaging, surgeons should consider possible bias associated with the dose and speed of ICG injection, cardiac output, serum albumin levels, and the size of the ROI [7].

8 Evaluation of the Patency by Temporary Occlusion of the Supplying Vessel

Indocyanine green fluorescence angiography can be used to confirm vascular patency by temporary occlusion or the target vessel prior to intravenous injection of ICG and reopening during the observation [2, 3]. For example, ICG fluorescence imaging during temporary occlusion of the proximal side of a vessel buried in the sulcus can be used to evaluate the patency of blood flow to the distal side and the development status of collateral arteries. Alternatively, a delayed reopening of the vessel after intravenous ICG infusion can also be used to confirm the presence or absence of blood flow in the target regions.

9 Effect of Pharmacological Properties of Indocyanine Green on Fluorescence Imaging

Indocyanine green videoangiography imaging can be repeated in about 15 minutes after a single intravenous injection because of its half-life [1, 2, 7]. On the other hand, when the aneurysm neck is closed immediately after the first imaging, ICG is retained in the aneurysm and shows fluorescence signals after about 20 minutes, which can be used for confirmation of complete closure of the aneurysm neck. ICG is a water-soluble tricarbocyanine dye that binds to plasma proteins (mainly β-lipoprotein) after intravenous administration and fluoresces. Therefore, in a patient with low LDL-cholesterol or blood dilution by intraoperative fluid infusion, the sensitivity of ICGVAG can decrease.

10 Conclusions

Indocyanine green videoangiography is useful to confirm the occlusion of cerebral aneurysms and patency of bypass vessels. The clinical impact of intraoperative fluorescence imaging on surgical outcomes has also been reported recently, although we still need large prospective studies for the standardization of imaging techniques and patient selection. Further development of image processing technology will enable clearer visualization of minute vessels in 3D.

References

1. Raabe A, Beck J, Gerlach R, et al. Near-infrared indocyanine green videoangiography: a new method for intraoperative assessment of vascular flow. Neurosurgery. 2003;52:132–9.
2. Nakagawa S, Murai Y, Matano F, et al. Evaluation of patency after vascular anastomosis using quantitative evaluation of visualization time in indocyanine green video angiography. World Neurosurg. 2018;110:e699–709.
3. Murai Y, Nakagawa S, Matano F, et al. The feasibility of detecting cerebral blood flow direction using indocyanine green video angiography. Neurosurg Rev. 2016;39:685–90.
4. Matano F, Mizunari T, Murai Y, et al. Quantitative comparison of the intraoperative utility of indocyanine green and fluorescein video angiographies in cerebrovascular curgery. Oper Neurosurg (Hagerstown). 2017;13:361–6.
5. Murai Y, Sato S, Yui K, et al. Preliminary clinical microneurosurgical experience with the 4K3-dimensional micro video scope (ORBEYE) system for microneurological surgery: observation study. Oper Neurosurg (Hagerstown). 2019;16:707–16.
6. Tsukiyama A, Murai Y, Matano F, et al. Optical effects on the surrounding structure during quantitative analysis using indocyanine green video angiography: a phantom vessel study. J Biophotonics. 2018;11:e201700254.
7. Guo Z, Ishii T, Hasegawa Y, et al. Usefulness and pitfalls of intraoperative Indocyanine Green fluorescence angiography, from engineering and clinical perspectives. Cerebral Craniofac Surg J. 2008;17:865–9.

Evaluation of Blood Perfusion in Skin Flaps

10

Keisuke Okabe and Kazuo Kishi

Summary

- Indocyanine green (ICG) fluorescence imaging is becoming an indispensable tool for safe and reliable skin flap surgery.
- Fluorescence imaging enables mapping and real-time evaluation of blood perfusion in the skin flap.
- Improvement of accuracy and standardization of evaluation methods are needed for further development of fluorescence imaging in this field.

1 Introduction

Reconstructive surgery using the skin flap is used in a variety of situations, including trauma and extended resection of malignant tumors. In order to assure surgical safety, it is important to confirm sufficient blood supply in the skin flap during surgery. In this chapter, we outline methods for the evaluation of graft perfusion by intraoperative fluorescence imaging using indocyanine green (ICG).

2 Basis and Limitations of Conventional Methods for the Evaluation of Skin Flap Perfusion

Skin flap grafting is a procedure in which the graft tissue is elevated from the donor site and moved to the intended recipient site. There is a trade-off between graft mobility and blood perfusion, as a thinner vascular pedicle is more advantageous for increasing flap mobility, but it also poses the risk of ischemia and congestion of the graft. Therefore, surgeons have to balance blood perfusion and mobility of the skin flaps. Because postoperative graft necrosis can lead to serious complications such as exposure of organs, their functional damage, and severe infection, it is necessary to confirm graft perfusion throughout the treatment using appropriate methods.

Preoperatively, contrast-enhanced computed tomography (CT), ultrasonography, and Doppler stethoscope are used to evaluate the vessel anatomy and the number and location of perforating branches that flow into the skin graft. During the harvesting of the skin flap, the exact location of the blood vessels can be checked with a Doppler stethoscope if necessary. For the perfusion assessment after flap grafting, surgeons can rely on physical examinations such as color tone (pale/congestive), temperature, capillary refilling after compression, and tissue elasticity, as well as auscultation of blood flow at the graft pedicle and the presence and nature of bleeding from the graft edge. If abnormal blood perfusion is suspected, the presence or absence of bleeding from the graft can be confirmed by puncture with a needle or scalpel (pinprick technique) or by wiping the margins with gauze. Although these conventional methods of evaluation by visual inspection are simple and easy, they require a certain amount of experience based mainly on subjective findings.

Objective assessment of graft perfusion includes transcutaneous partial pressure of oxygen (tcPO2) [1], plethysmography [2], and laser Doppler flowmetry [3, 4]. However, these methods are not widely used in the current clinical practice because they do not provide stable results for intraoperative use. In addition, these conventional techniques have limitations in providing detailed information on graft perfusion, such as determining the demarcation between ischemic and non-ischemic regions, the exact location of the main feeder, and hemodynamics in the skin graft. With the advent of fluorescence imaging, it has become possible to delineate graft perfusion in real time and is now widely used as a reliable intraoperative modality complementing conventional methods.

K. Okabe (✉) · K. Kishi
Department of Plastic and Reconstructive Surgery, Keio University School of Medicine, Shinjuku, Tokyo, Japan
e-mail: dawndawn@keio.jp

T. Ishizawa (ed.), *Fluorescence-Guided Surgery*, https://doi.org/10.1007/978-981-19-7372-7_10

Point

- The dilemma in skin flap grafting is that the greater the mobility of the graft, the higher the risk of instable blood perfusion.
- Fluorescence imaging allows for mapping and real-time assessment of blood perfusion in the skin graft.
- Complementary use of fluorescence imaging with conventional techniques based on visual inspection and palpation is highly recommended.

3 Development History of Fluorescence Imaging for Perfusion Assessment of Skin Graft

In 1962, Myers applied fluorescence imaging using fluorescein to surgery for the estimation of the extent of skin necrosis [5]. Silverman et al. reported in 1980 that the contrast effect detected by the dermofluorometer was correlated with postoperative graft survival [6]. Subsequently, several studies have shown a correlation between the contrast-enhancing effects of fluorescein-fluorescence imaging and graft survival [7, 8]. In this technique, about 50% of the fluorescein adheres to the surface of blood albumin and erythrocyte membranes, and the rest is dissolved in plasma and leaks from capillaries to the stroma. Therefore, fluorescence imaging using fluorescein also enhances extravascular areas, and the drug remains there for more than 7–8 hours, which makes it difficult to use fluorescein imaging repeatedly during surgery. Furthermore, since the absorption wavelength is shorter than that of ICG, fluorescence imaging using fluorescein has limitations in delineating deeply located vessels, leading to the decline of this technique in the field of skin flap surgery. In contrast, the molecular weight of ICG is relatively large (775), and most of it binds to plasma proteins and remains in blood vessels, so it is discharged from the bloodstream with a half-life of 3–4 minutes, enabling fluorescence imaging with higher contrast compared with fluorescein. In 1973, Flower and Hochheimer reported the application of ICG to choroidal angiography in the ophthalmologic field [9]. Later, in 1994–1995, Eren and Rübben et al. evaluated blood flow by ICG fluorescence imaging in rat skin flap models and lower limb skin of human ischemic limb patients [10, 11]. Still et al. reported that ICG fluorescence imaging correlated with graft survival [12], and Holm et al. showed that the extent of intraoperative lack of ICG contrast in free and pedicled flap corresponded well with the area of postoperative graft necrosis [13, 14]. Since then, the evaluation of blood flow by ICG fluorescence imaging has been applied to various types of skin flaps. The use of fluorescence imaging has been shown to reduce the overall complication rate of breast cancer surgery, including postmastectomy skin flap necrosis [15–17], and to improve the results of breast reconstruction surgery [18]. Other reported applications include the selection of perforating vessels in anterolateral thigh flap elevation [19], elucidation of skin flap hemodynamics [20, 21], and prediction of the extent of skin necrosis after trauma [22].

4 Clinical Practice

4.1 Indocyanine Green Administration

First of all, make sure that the patient is not allergic to iodine, which is a known contraindication to ICG. Intraoperatively, ICG is administered intravenously as a bolus for evaluation of blood perfusion in the skin flap. In the measurement of tissue circulation, ICG is administered at a dose of 0.1–0.3 mg/kg in daily clinical practice, while most of the previous literature suggest 0.1–0.2 mg/kg. The total daily dose is limited to less than 5 mg/kg/day for patients 11 years of age and older, less than 2.5 mg/kg/day for children 2–10 years of age, and less than 1.25 mg/kg/day for children 0–1 year of age, according to the UK and German approvals. When ICG is used, it is dissolved in the supplied water for injection (2.5 mg/mL) and administered as a bolus dose of 2–4 mL in an adult patient weighing 50 kg, resulting in a dose of 0.1–0.2 mg/kg. To reduce the total dose in cases where repeated administration is expected, a single dose of about 2 mL can be used for observation in most cases.

4.2 Evaluation of Blood Perfusion in the Skin Flap

Blood perfusion in the skin flap is evaluated with a near-infrared fluorescence camera system for open surgery, like PDE (Hamamatsu Photonics) or SPY fluorescence imaging system (Stryker). The signal sensitivity can be increased by dimming the room illumination.

In cases of good perfusion, a reticular pattern along the arteries within the skin flap is seen within 1–2 minutes after ICG administration, following the increase of fluorescence signals in the surrounding healthy skin. After that, the fluorescence pattern becomes uniform as ICG enters the capillaries. Krishnan et al. [23] described a “delay in uptake” when it takes more than 4 minutes for the skin to fluoresce, and a “delay in clearance” when the fluorescence intensity of the graft does not decay even after 8 minutes following the decrease of fluorescence intensity in the background healthy skin. In general, it is reasonable to assume that the area is ischemic if there is no fluorescence signal in the graft more than 4 minutes after ICG administration, and we evaluate blood flow at about 4 minutes in our department. If there is an area around the edge of the skin flap that is not enhanced at that time during fluorescence imaging, we can revise surgical procedures such as trimming to prevent postoperative complications. However, neither the sensitivity nor the specificity of ICG fluorescence imaging is 100%, so it is necessary to make a comprehensive judgment, including visual and palpation findings. It should also be noted that fluorescence intensity cannot be measured as an absolute value because it varies depending on the condition of the background tissues and the imaging environment in each surgery. For the quantitative assessment of blood flow, it is necessary to measure the fluorescence intensities of the target region as a relative value to the surrounding healthy tissues.

5 Expected Role of Fluorescence Imaging in Skin Flap Surgery

Indocyanine green fluorescence imaging provides a variety of information that cannot be obtained with conventional examinations and can be used for surgical decision-making.

5.1 Estimation of the Extent of Graft Necrosis

Non-fluorescing areas in the skin flap detected by ICG fluorescence imaging are associated with a high risk of necrosis after surgery. ICG fluorescence imaging enables surgeons to mark and remove the non-fluorescing areas to prevent postoperative complications.

Figure 10.1 shows a patient with an esophagobronchial fistula who was scheduled to undergo closure of the fistula with a pedicle latissimus dorsi flap, but a part of this muscle had been dissected during a previous surgery. When ICG fluorescence imaging was performed after the elevation of the flap, little fluorescence signal was seen in the distal side of the graft. Therefore, the non-fluorescing regions were resected, and the fistula was successfully closed using the proximal part of the latissimus dorsi flap combined with the intercostal muscle flap. In this particular case, it might be impossible to estimate the risk of postoperative necrosis based only on conventional techniques such as inspection of

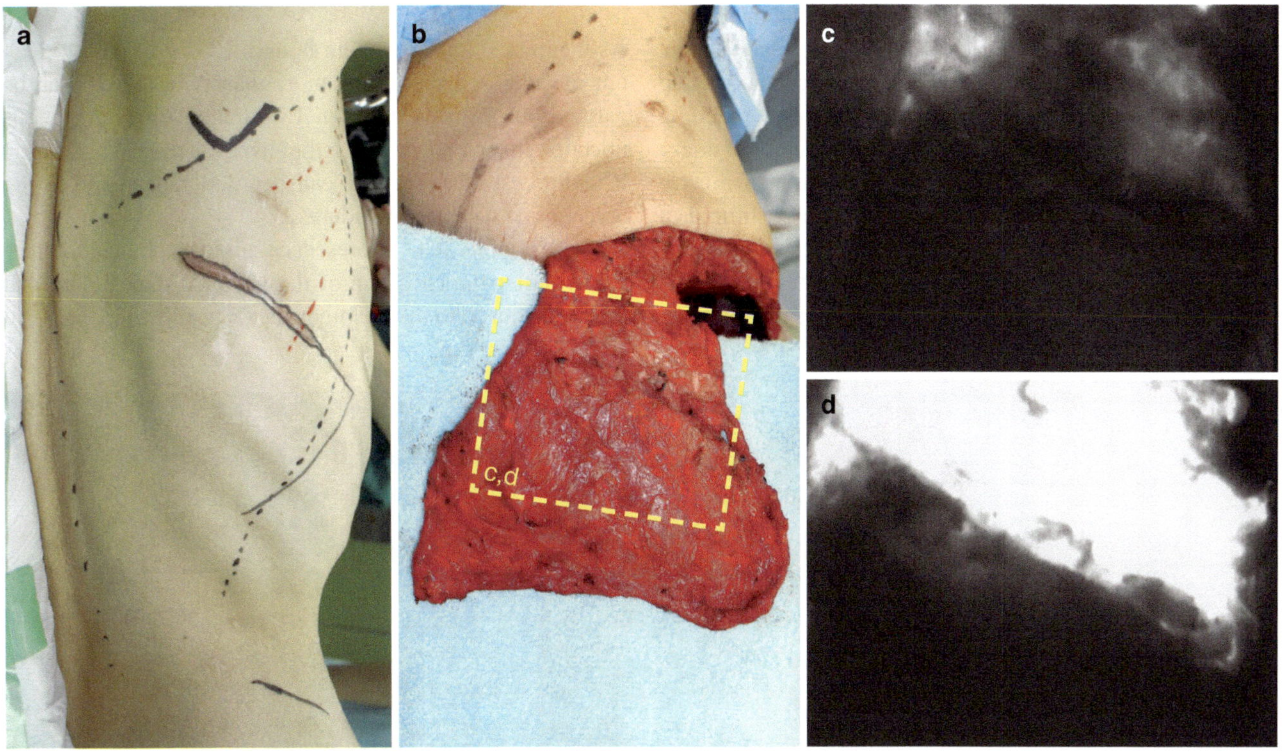

Fig. 10.1 Postoperative esophagobronchial fistula after esophageal cancer surgery. (**a**) In this patient, a part of the latissimus dorsi muscle was divided by a lateral thoracotomy in the previous surgery. (**b**) After elevation of the pedicle latissimus dorsi flap. (**c**) The latissimus dorsi muscle is being enhanced proximally approximately 1 minute after ICG administration. (**d**) Approximately 1.5 minutes after ICG administration. There is little contrast distal to the dissected area

color changes and the presence or absence of bleeding. Similarly, when the graft should be extended beyond the anatomical areas of blood supply (extended skin flap), ICG fluorescence imaging is recommended to evaluate the exact extent of well-perfused areas, which can differ among the patients.

5.2 Understanding of Microcirculation and Hemodynamics in the Skin Flap

Regarding hemodynamics inside the skin flap, it is estimated that blood flow enters the graft through the thin pedicle and spreads along the arteries. However, it is impossible to confirm the details of this process with the naked eye, and the direction of blood flow cannot be determined by contrast-enhanced CT images. On the other hand, ICG fluorescence imaging has the great advantage of allowing us to observe the blood flow into the flap in real time.

Figure 10.2 indicates a case of rectus abdominis musculocutaneous flap used for treating infection of the occipital bone after surgery for meningioma. The rectus abdominis muscle is fed by the superior epigastric artery on the cephalad side and the inferior epigastric artery on the caudal side, and when a ribbed rectus abdominis musculocutaneous flap is elevated using the inferior epigastric artery as the feeding artery, it is known that blood flow enters the intercostal artery from the superior epigastric artery through inter-muscle communication with the inferior artery. However, little data are available regarding the amount and velocity of blood flow in the intercostal artery and its supplying tissues. In this case, fluorescence angiography confirmed that the intercostal arteries were sufficiently identified within 2 minutes after administration of ICG, assuring the safety of the procedure.

5.3 Evaluation of Anastomotic Patency and Identification of Thrombus

In free flap transplantation, the graft artery and vein are microscopically anastomosed with the recipient vessels. If either artery or vein is occluded, postoperative graft necrosis is almost inevitable. In order to confirm anastomotic patency during surgery, a "patency test" is often used by compressing the anastomosed vessels with micro forceps to confirm blood passage through the anastomosis. In some cases, anastomotic problems undetected by the conventional method like a patency test can be detected by ICG fluorescence angiography [24]. It is also suggested that fluorescence imaging can be used to identify the exact location of the thrombus around the anastomotic site.

5.4 Repeated Evaluation of Graft Perfusion

As mentioned above, protein-bound ICG is difficult to leak out of the blood vessel and more than 90% of ICG can be washed out from the body in 15 minutes after intravenous injection in a patient with normal liver function. That is why ICG fluorescence imaging can be used repeatedly during revisions of surgical procedures.

For example, ICG fluorescence angiography can be performed after the elevation of the pedicled flap, before and after it is moved to the recipient site. When the flap is moved

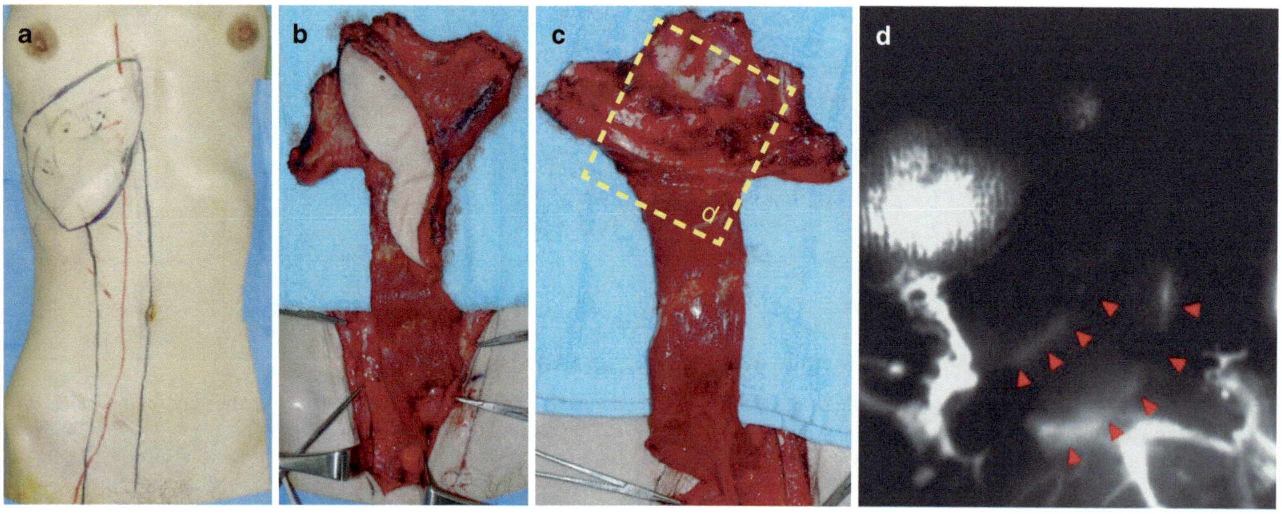

Fig. 10.2 Postoperative infection of occipital bone after surgery for meningioma. (**a**) Design of a ribbed rectus abdominis musculocutaneous flap with the inferior abdominal wall artery. (**b**) Ventral aspect of the elevated flap. (**c**) Dorsal aspect. (**d**) After ICG administration, sufficient blood flow to the intercostal artery was confirmed (*red arrowhead*) in about 2 minutes

and sutured in place, hemodynamic changes may occur due to torsion of the vessel pedicle and/or tension on the graft. If fluorescence angiography identifies a change in blood flow after the grafting, the flap can be repositioned to correct the abnormal blood perfusion and prevent postoperative complications. In a skin flap with multiple vascular pedicles, ICG fluorescence angiography can be used to determine the dominant region of each vessel, by obtaining fluorescence images repeatedly with one of the vessels clamped. Furthermore, when separating the flap pedicle several days after the initial grafting such as forehead flap and cross finger flap, ICG fluorescence imaging after clamping the pedicle can provide useful information on the evaluation of sufficient blood perfusion from the recipient tissues surrounding the graft [21].

5.5 Determination of the Extent of Debridement

Necrosis of the skin may develop in the first few days after trauma with complex soft tissue injuries such as open fractures of the foot. In order to prevent severe infection, the soft tissues with a high risk of necrosis should be removed at an early phase of the treatment. On the other hand, excessive removal of the soft tissues may make the subsequent reconstruction difficult. Therefore, in trauma management, it is effective to identify the appropriate areas for debridement based on discrimination of the well- and non-perfused regions by ICG fluorescence imaging [22].

Figure 10.3 shows a case of intractable skin ulceration on the left elbow due to leakage of an intravenous anticancer

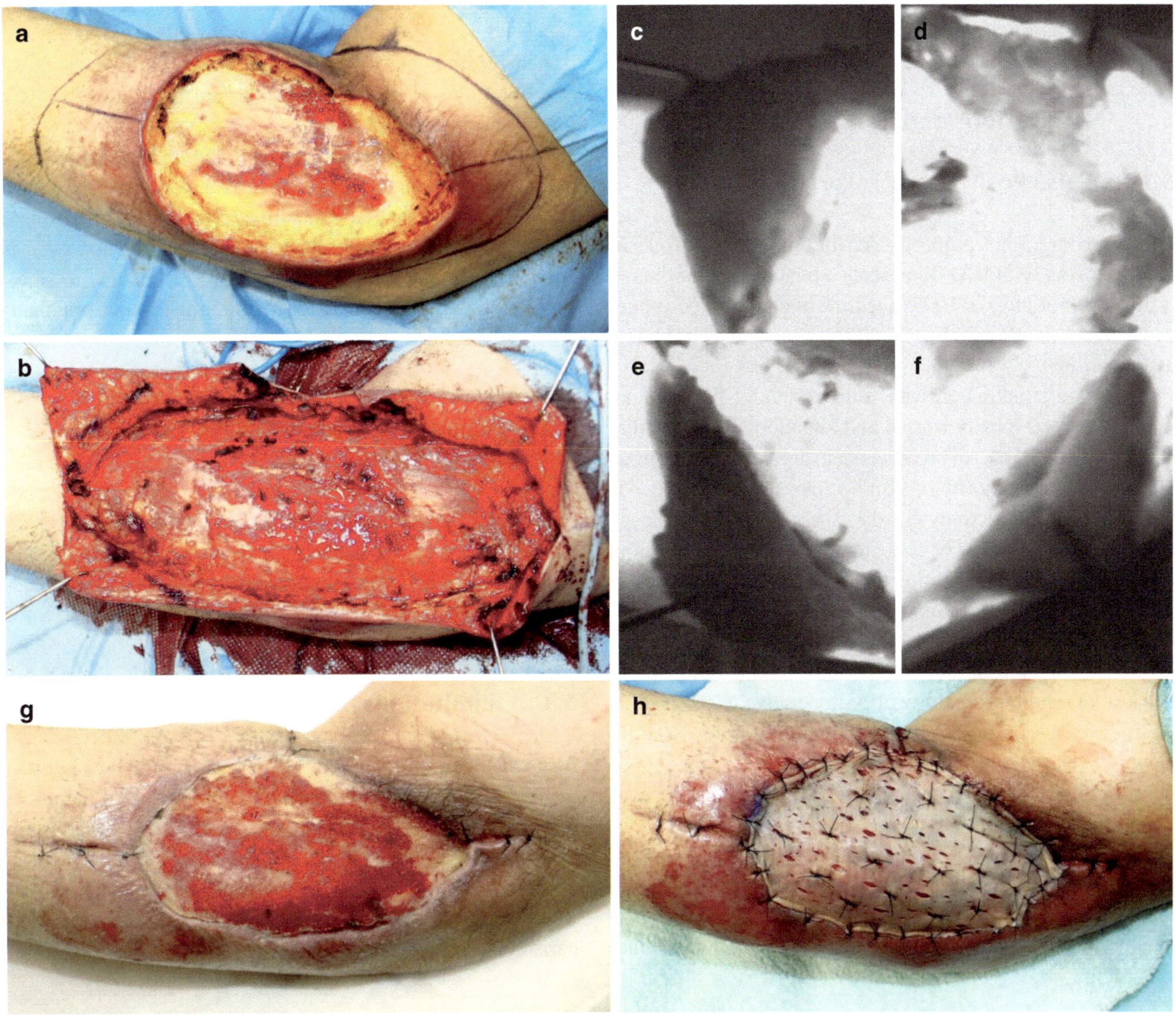

Fig. 10.3 Intractable skin ulcer on the left upper extremity after anticancer drug leakage. (**a**) A refractory ulcer was found on the left elbow. While granulation was observed in the center, necrosis at the ulcer margin was advanced. (**b**) Debridement surgery. (**c–f**) ICG fluorescence angiography showed that none of the marginal skin was contrasted, indicating ischemia. (**g**) After debridement and NPWT, granulation was formed. (**h**) Split-thickness skin grafting was performed successfully

drug. Because the area of necrosis was enlarged, surgical debridement was performed. According to the intraoperative ICG angiography, the skin at the edge of the pocket incision was excised because it was thought to progress to necrosis due to poor blood flow. The remaining necrotic tissue at the base of the ulcer was also excised using fluorescence imaging followed by local negative pressure wound therapy (NPWT) and successful skin grafting 3 weeks after the debridement.

Point

The expected roles of ICG fluorescence angiography in skin grafting are:

- Estimation of the extent of graft necrosis.
- Understanding the hemodynamics in the flap.
- Evaluation of anastomotic vessel patency.
- Repeated evaluation of changes in tissue perfusion.
- Determination of the area for debridement.

6 Limitations and Future Challenges

6.1 False Positive and False Negative

The most important point to keep in mind when assessing flap perfusion with ICG fluorescence imaging is that the test results are not always 100% accurate and should be judged by referring to other findings, including visual and palpatory examinations.

There are situations in which the results of fluorescence imaging can be mostly trusted and situations in which they cannot. For example, in breast cancer surgery, when estimating whether the majority of skin necrosis occurs due to skin thinning after mastectomy [15–17], the sensitivity and specificity are considered to be close to 100%, and complications can be prevented by trimming the non-fluorescing tissues. In addition, if fluorescence imaging delineates a clear demarcation between well- and non-enhanced regions in the same skin flap, the non-fluorescing regions should be removed because of a high probability of postoperative necrosis.

In the case of a skin flap in which arterial inflow is sufficient but venous return is weak, it is difficult to predict postoperative congestion by fluorescence imaging because even the graft with a risk of postoperative congestion can be enhanced very well through arterial inflow at the timing of elevation for grafting. It may be possible to solve this problem by continuing imaging until the fluorescence intensity declines, or by confirming the disappearance of fluorescence signals at a 15- to 20-minute interval, but there is no reliable standard at present. How to determine and interpret the factor of congestion is an issue for the future.

Conversely, it should be noted that even if the entire valve is hardly enhanced by fluorescence imaging, the skin flap may not always be ischemic. Figure 10.4 shows a case in which a forehead flap was used for a defect after resection of a basal cell carcinoma in the right nasal ala. It is known that the pedicle can be extended beyond the left supratrochlear artery to the angular artery to increase the mobility of the skin flap [25]. When ICG fluorescence angiography was performed in this case, little enhancement was observed throughout the graft, but the flap was successfully grafted without any problems. Similarly, when ICG fluorescence angiography was performed on a flap that was fed by a small artery that accompanies a vein or nerve [26], we sometimes experienced a discrepancy between low fluorescence signals and good graft survival. The reason for this is not clear, but in general, the skin flap with slow arterial inflows as those described above may not be enhanced accurately by ICG fluorescence imaging.

In the use of ICG fluorescence imaging in clinical practice, it is important to evaluate perfusion of the skin flap, keeping in mind that false-positive results due to inadequate venous return (leading to necrosis from congestion even if contrast is obtained) and false-negative results due to slow-flow arteries (no ischemia even if contrast is not obtained) are both possible.

6.2 About Allergy

As mentioned earlier, patients with iodine allergy may experience allergic reactions, including anaphylactic shock with ICG [27]. It is important to interview them about their history of allergy and to be prepared to deal with allergic reactions if they occur.

6.3 Limitations in Quantitative Evaluation

At present, the evaluation of blood perfusion in the skin flap is often performed by comparing the brightness of the target region with that of the surrounding healthy skin, which is not perfect for accurate quantification. The development of versatile software to quantify and analyze fluorescence signals would enable standardized evaluation and inter-patient comparison of tissue perfusion.

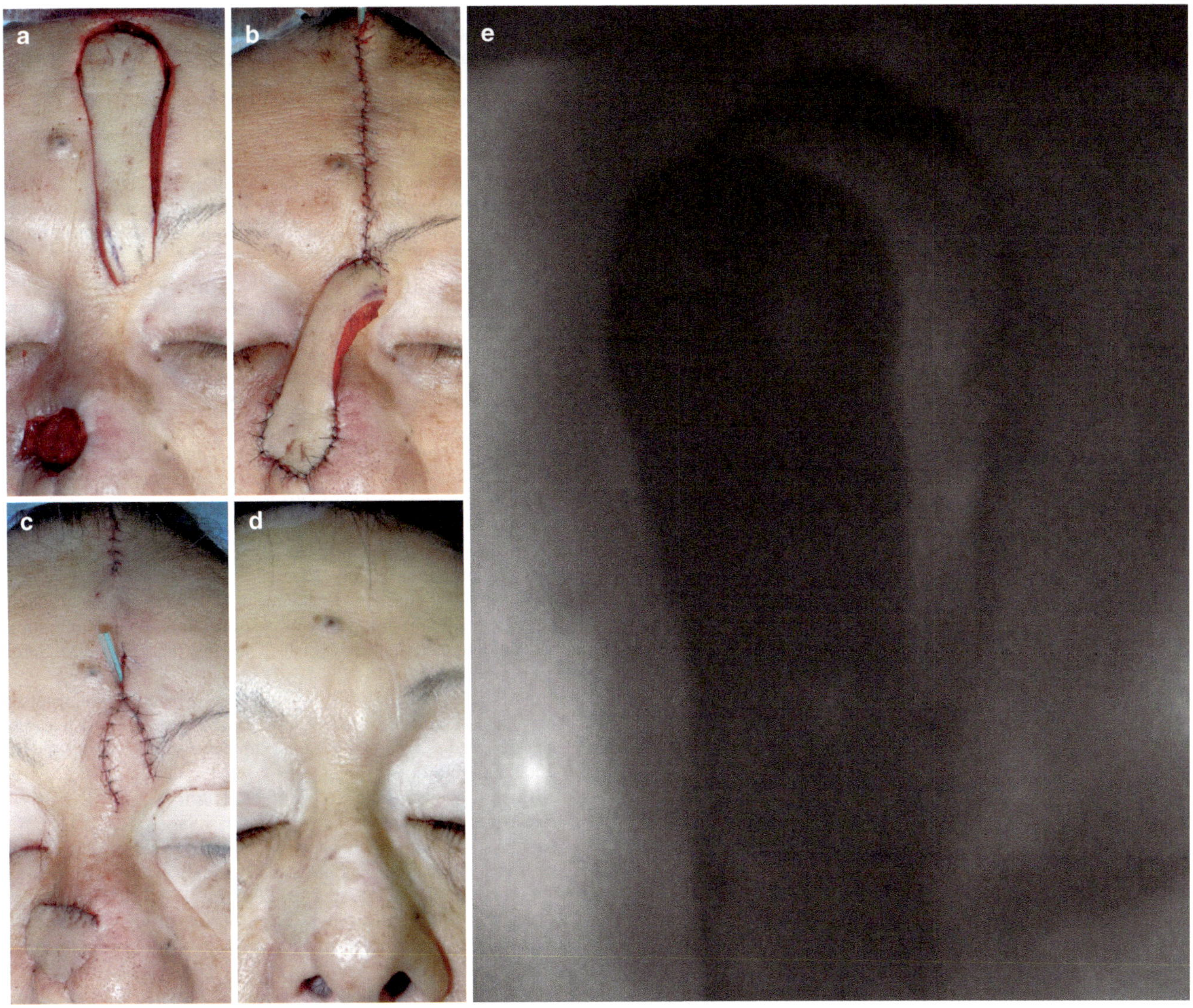

Fig. 10.4 A case of basal cell carcinoma of the nasal ala. (**a**) For the treatment of a defect in the right nasal ala, a forehead flap with a pedicle of the left angular artery beyond the supratrochlear artery was elevated. (**b**) After skin flap grafting. (**c**) The pedicle was detached 2 weeks after skin flap grafting. (**d**) One-year postoperative status. The flap has engrafted without problems. (**e**) ICG fluorescence angiography after the elevation of the skin flap. Even after 5 minutes of ICG administration, the graft is barely enhanced

7 Conclusions

As described above, the evaluation of the skin flap by fluorescence imaging using ICG has many indispensable advantages and will be more widely applied in the future. It will be necessary to standardize the interpretation of imaging outcomes through prospective, randomized, comparative studies using the software enabling quantitative evaluation so that it can be used more easily and widely in the field of reconstructive surgery.

Recently, it has been reported that ICG can be used for shortwave infrared (SWIR) or near-infrared (NIR)-II imaging in the wavelengths of 1000 nm or higher, and it can be used for the identification of deeply located (20–30 mm) biological structures [28, 29]. These novel technologies may also be installed in clinical systems to enhance the sensitivity of ICG fluorescence imaging in the near future.

References

1. Hirigoyen MB, Blackwell KE, Zhang WX, et al. Continuous tissue oxygen tension measurement as a monitor of free-flap viability. Plast Reconstr Surg. 1997;99:763–73.
2. Futran ND, Stack BC Jr, Hollenbeak C, et al. Green light photoplethysmography monitoring of free flaps. Arch Otolaryngol Head Neck Surg. 2000;126:659–62.
3. Jones BM, Mayou BJ. The laser Doppler flowmeter for microvascular monitoring: a preliminary report. Br J Plast Surg. 1982;35:147–9.
4. Svensson H, Pettersson H, Svedman P. Laser doppler flowmetry and laser photometry for monitoring free flaps. Scand J Plast Reconstr Surg. 1985;19:245–9.
5. Myers MB. Prediction of skin sloughs at the time of operation with the use of fluorescein dye. Surgery. 1962;51:158–62.
6. Silverman DG, LaRossa DD, Barlow CH, et al. Quantification of tissue fluorescein delivery and prediction of flap viability with the fiberoptic. Plast Reconstr Surg. 1980;66:545–53.
7. Graham BH, Walton RL, Elings VB, et al. Surface quantification of injected fluorescein as a predictor of flap viability. Plast Reconstr Surg. 1983;71:826–31.
8. Thompson JG, Kerrigan CL. Dermofluorometry: thresholds for predicting flap survival. Plast Reconstr Surg. 1989;83:859–64.
9. Flower RW, Hochheimer BF. Indocyanine green dye fluorescence and infrared absorption choroidal angiography performed simultaneously with fluorescein angiography. Johns Hopkins Med J. 1978;138:33–42.
10. Eren S, Rübben A, Krein R, et al. Assessment of microcirculation of an axial skin flap using indocyanine green fluorescence angiography. Plast Reconstr Surg. 1995;96:1636–49.
11. Rübben A, Eren S, Krain R, et al. Infrared videoangiofluorography of the skin with indocyanine green – rat random cutaneous flap model and results in man. Microvasc Res. 1994;47:240–51.
12. Still J, Law E, Dawson J, et al. Evaluation of the circulation of reconstructive flaps using laser-induced fluorescence of indocyanine green. Ann Plast Surg. 1999;42:266–74.
13. Holm C, Mayr M, Hoefter E, et al. Intraoperative evaluation of skin-flap viability using laser-induced fluorescence of indocyanine green. Br J Plast Surg. 2002;55:635–44.
14. Holm C, Tegeler J, Mayr M, et al. Monitoring free flaps using laser-induced fluorescence of indocyanine green: a preliminary experience. Microsurgery. 2002;22:278–87.
15. Phillips BT, Lanier ST, Conkling N, et al. Intraoperative perfusion techniques can accurately predict mastectomy skin flap necrosis in breast reconstruction: results of a prospective trial. Plast Reconstr Surg. 2012;129:778e–88e.
16. Sood M, Glat P. Potential of the SPY intraoperative perfusion assessment system to reduce ischemic complications in immediate postmastectomy Breast Reconstruction. Ann Surg Innov Res. 2013;7:9.
17. Duggal CS, Madni T, Losken A. An outcome analysis of intraoperative angiography for postmastectomy breast reconstruction. Aesthet Surg J. 2014;34:61–5.
18. Casey WJ, Connolly KA, Nanda A, et al. Indocyanine green laser angiography improves deep inferior epigastric perforator flap outcome following abdominal suction lipectomy. Plast Reconstr Surg. 2015;135:491–497e.
19. La Padula S, Hersant B, Meningaud JP, et al. Intraoperative use of indocyanine green angiography for selecting a more reliable perforator of the anterolateral thigh flap: a comparison study. Microsurgery. 2018;38:738–44.
20. Losken A, Zenn MR, Hammel JA, et al. Assessment of zonal perfusion using intraoperative angiography during abdominal flap breast reconstruction. Plast Reconstr Surg. 2012;129:618e–24e.
21. Woodard CR, Most SP. Intraoperative angiography using laser-assisted indocyanine green imaging to map perfusion of forehead flaps. Arch Facial Plast Surg. 2012;14:263–9.
22. Kamolz LP, Andel H, Auer T, et al. Evaluation of skin perfusion by use of indocyanine green video angiography: rational design and planning of trauma surgery. J Trauma. 2006;61:635–41.
23. Krishnan KG, Schackert G, Steinmeier R. The role of near-infrared angiography in the assessment of post-operative venous congestion in a random pattern, pedicled Island and free flaps. Br J Plast Surg. 2005;58:330–8.
24. Holm C, Mayr M, Höfter E, et al. Assessment of the patency of microvascular anastomoses using microscope-integrated near-infrared angiography: a preliminary study. Microsurgery. 2009;29:509–14.
25. Kishi K, Imanishi N, Shimizu Y, et al. Alternative 1-step nasal reconstruction technique. Arch Facial Plast Surg. 2012;14:116–21.
26. Kishi K, Nakajima H, Imanishi N. Distally based greater saphenous venoadipofascial- sartorius muscle combined flap with venous anastomosis. Plast Reconstr Surg. 2007;119:1808–12.
27. Obana A, Miki T, Hayashi K, et al. Survey of complications of indocyanine green angiography in Japan. Am J Ophthalmol. 1994;118:749–53.
28. Starosolski Z, Bhavane R, Ghaghada KB, et al. Indocyanine green fluorescence in the second near-infrared (NIR- II) window. PLoS One. 2017;12:e0187563.
29. Carr JA, Franke D, Caram JR, et al. Shortwave infrared fluorescence imaging with the clinically approved near-infrared dye indocyanine green. Proc Natl Acad Sci U S A. 2018;115:4465–70.

Evaluation of Blood Perfusion in the Upper Gastrointestinal Tract

11

Kazuo Koyanagi, Soji Ozawa, Yamato Ninomiya, Kentaro Yatabe, Itaru Higuchi, and Miho Yamamoto

Summary

- Indocyanine green (ICG) fluorescence imaging is rapidly becoming a popular method for evaluating blood perfusion in the upper gastrointestinal tract.
- Blood perfusion in the gastric tube can be evaluated quantitatively by ICG fluorescence imaging.
- Measuring the blood flow speed in the gastric tube using ICG fluorescence imaging may help predict anastomotic leakage after esophageal cancer surgery.
- Further studies are needed to identify clinical factors affecting blood perfusion in the gastric wall and also to establish surgical procedures and postoperative management according to the outcomes of fluorescence imaging.

1 Introduction

The purpose of blood flow assessment in upper gastrointestinal surgery is to decrease the incidence of postoperative anastomotic leakage by making anastomosis at a site with sufficient blood perfusion. In particular, leakage from cervical anastomosis after esophageal cancer surgery is more likely to occur and is associated with mortality. Among various factors affecting the outcomes of anastomosis, decreased blood perfusion in the reconstructed organs is thought to be the major cause of anastomotic leakage.

Recently, intraoperative fluorescence imaging using ICG has been introduced to gastrointestinal cancer surgery with the advancement of imaging equipment [1], following applications to the identification of liver cancers/hepatic segments and sentinel lymph nodes in breast and gastric cancer surgery. We have also reported the efficacy of this technique in evaluating blood perfusion in the gastric tube during resection for esophageal cancer. In this chapter, we describe the current status and technical details of fluorescence navigation in esophageal cancer surgery.

2 Conventional Techniques for Evaluation of Blood Perfusion in the Reconstructed Organ After Esophagectomy

After the resection of esophageal cancer, the stomach, small intestine, and colon can be used for reconstruction. In any case, a sufficient length of the digestive tract should be elevated from the abdominal cavity to the neck, but securing this elevation may cause ischemia at the tip of the elevated organ. In the gastric tube, for example, blood is mainly supplied from the right gastroepiploic artery, but communications with the left gastroepiploic artery and short gastric artery are often not observed. Therefore, blood perfusion around the anastomotic site mainly depends on the vessel communications in the gastric wall. In order to determine the optimal anastomotic site in esophageal cancer surgery, various techniques for perfusion assessment of the gastric tube have been proposed.

Conventionally, blood perfusion in the reconstructed organs was estimated by the color tone of the organ surfaces and the pulsation of the feeding arteries, based on surgeons' experience. However, these methods are subjective, and it is not easy to distinguish the demarcation line between areas with good and poor blood perfusion. Thermography and laser Doppler were also used as objective methods to evaluate blood perfusion. The thermographic method was an

Supplementary Information The online version contains supplementary material available at https://doi.org/10.1007/978-981-19-7372-7_11.

K. Koyanagi (✉) · S. Ozawa · K. Yatabe · I. Higuchi · M. Yamamoto
Department of Gastroenterological Surgery, Tokai University School of Medicine, Isehara, Kanagawa, Japan
e-mail: kkoyanagi@tsc.u-tokai.ac.jp

Y. Ninomiya
Department of Gastroenterological Surgery, Tokai University Hachioji Hospital, Hachioji, Tokyo, Japan

T. Ishizawa (ed.), *Fluorescence-Guided Surgery*, https://doi.org/10.1007/978-981-19-7372-7_11

attempt to evaluate blood flow by temperature changes in the gastric tube, but only a few reports were published, and the results were not related to postoperative anastomotic leakage. The laser Doppler method can detect microcirculation in any part of the digestive gastric tube, as reported in esophageal cancer surgery in the late 1990s and early 2000s [2]. However, despite the advantages in point-by-point observation of the small spot, laser Doppler was not adequate for the evaluation of a large area. Because of the limitations in reproducibility and predictability for postoperative anastomotic leakage, these techniques were not widely used in actual clinical practice.

3 History of Perfusion Assessment of the Upper Gastrointestinal Tracts by Indocyanine Green Fluorescence Imaging

In esophageal cancer surgery, ICG fluorescence imaging was initially used to evaluate the patency of the vessel anastomosis after supercharging the gastric tube. Later, ICG fluorescence imaging was used to evaluate the blood perfusion in the gastric tube, and the relationship with postoperative anastomotic leakage was evaluated by several studies. While the evaluation of blood perfusion by fluorescence imaging has shifted from the qualitative method to the quantitative way, standardization of perfusion assessment with fluorescence imaging remains the future challenge in the field of upper gastrointestinal surgery.

3.1 Qualitative Assessment of Blood Perfusion by Indocyanine Green Fluorescence Imaging

Zehetner et al. reported that the incidence of anastomotic leakage was 45% when anastomosis was performed at a site where ICG fluorescence was not observed at all during esophageal cancer surgery, whereas the leakage rate was 2% when anastomosis was performed at a site with sufficient fluorescence signals [3]. On the other hand, Shimada et al. reported that postoperative anastomotic leakage developed in 3 of 40 cases of gastric tube reconstruction for esophageal cancer despite fluorescence signals detected in the gastric conduit in all cases, suggesting that intraoperative evaluation of blood flow by ICG fluorescence imaging alone may be insufficient [4]. Although qualitative evaluation of blood perfusion based only on the presence or absence of fluorescence signals is the simplest method, a more objective assessment based on quantitative measurement of fluorescence signals is needed for improving the reliability of ICG fluorescence imaging.

3.2 Development of Quantitative Measurement

Recently, Yukaya et al. measured the trend of ICG fluorescence intensities between two points in the gastric tube and quantitatively demonstrated the impairment of blood inflow and outflow in the target organ [5]. Kumagai et al. measured the arrival time of ICG fluorescence signals from intravenous ICG injection to the planned anastomotic site and suggested that blood perfusion was sufficient if the interval was within 90 seconds [6]. Kamiya et al. also demonstrated the efficacy of quantitative perfusion assessment on the free jejunal graft to be used for reconstruction after esophageal cancer surgery in confirmation of the patency of the anastomosed vessels [7]. However, none of these reports demonstrated a direct relationship between outcomes of fluorescence imaging and the incidence of postoperative anastomotic leakage, suggesting that further studies are needed to develop ICG fluorescence imaging into an essential navigation tool during upper gastrointestinal surgery.

Point

- Indocyanine green fluorescence imaging is rapidly becoming a common method for evaluating intraoperative blood perfusion in the upper gastrointestinal tract.
- Recently, ICG fluorescence imaging enables quantitative assessment of blood perfusion in the gastric tube during esophageal cancer surgery, although its association with the incidence of postoperative anastomotic leakage remains to be clarified.

4 Measurement of Blood Flow Speed in the Gastric Wall Using Indocyanine Green Fluorescence Imaging

In our institute, we have developed navigation surgery focusing on the temporal and spatial changes of ICG fluorescence signals in the reconstructed gastric tube after esophagectomy [8].

4.1 Preparation of the Gastric Tube

A 3.5-cm wide gastric tube was made, preserving the right gastroepiploic artery (Fig. 11.1). We simulate the anastomosis by extending the prepared gastric tube to the neck on the chest wall, and evaluate the color tone of the tube and the pulsation of the gastroepiploic artery by palpation. In the meantime, ICG (1.25 mg/body) for intravenous injection and saline for boosting are prepared.

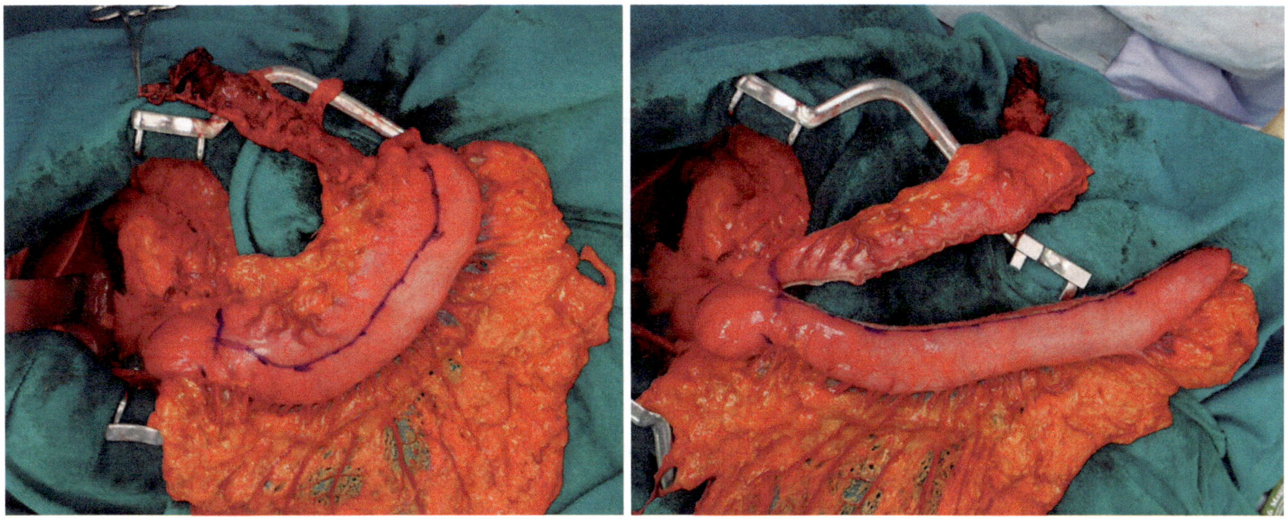

Fig. 11.1 Creation of a narrow gastric tube on the side of the greater curvature

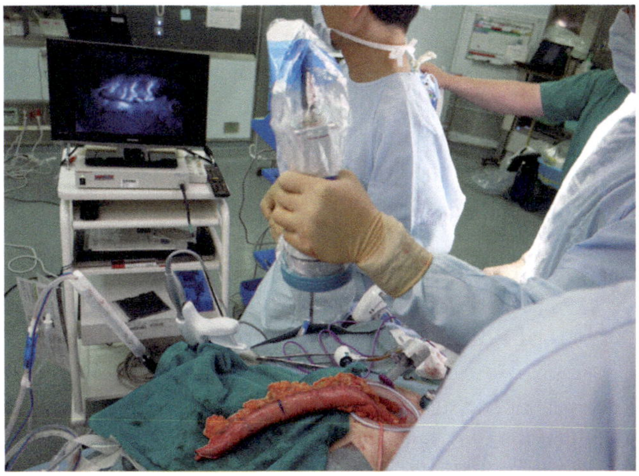

Fig. 11.2 Measurement of ICG fluorescence imaging

4.2 Equipment for Indocyanine Green Fluorescence Imaging

We use pde-neo® (Hamamatsu Photonics) for ICG fluorescence imaging (Fig. 11.2). During fluorescence imaging with this device, it is necessary to turn off the surgical lights in the operating room (the ceiling light can be turned on). Fluorescence signals are usually observed in the root of the right gastroepiploic artery within 30 seconds after intravenous injection of ICG.

4.3 Measurement of Fluorescence Imaging of Gastric Tube

Indocyanine green (1.25 mg/body) is administered intravenously through a central venous catheter and boosted with saline. Normally, ICG fluorescence signals can be observed in the root of the right gastroepiploic artery within 30 seconds after intravenous injection. We focused on the transition speed of ICG fluorescence and set the following four observation points on the gastric tube (Fig. 11.3). The distance between point "a" and the other points is measured. In addition, by measuring the transition time of the fluorescence signals, the transition speed can be determined (Movie 11.1). After reconstruction, ICG fluorescence imaging is measured in the same way. When reconstruction is performed by the posterior sternum root, continuous measurement is not possible. Even in that situation, however, the transition speed can be calculated by confirming fluorescence signals at the root of the right gastroepiploic artery and cervical anastomosis.

4.4 Transition Speed of Indocyanine Green Fluorescence Imaging in the Gastric Tube

The results of ICG fluorescence imaging measurements in 109 esophageal cancer surgery patients who underwent three-field lymph node dissection and gastric tube reconstruction via the posterior sternal route between 2014 and 2017 are as follows: The average length of the prepared gastric tube before anastomosis was 34.6 cm, and the distance between "a" and "b" was 22.9 cm. ICG fluorescence imaging showed that the blood flow between "a" and "c" was 32.3 cm, and that between "a" and "d" was 26.2 cm, both of which were more distal than the blood flow observed by palpation. The transition time of ICG fluorescence was 12.2 seconds on average between "a"–"c" and 7.4 seconds between "a"–"d," and the transient speed of ICG fluorescence (velocity of blood flow evaluated by ICG fluorescence imaging) was 2.9 cm/sec in the gastric tube wall and 4.0 cm/sec in the gastroepiploic artery. Anastomoses should be made on the site where ICG fluorescence is observed in the gastric tube wall.

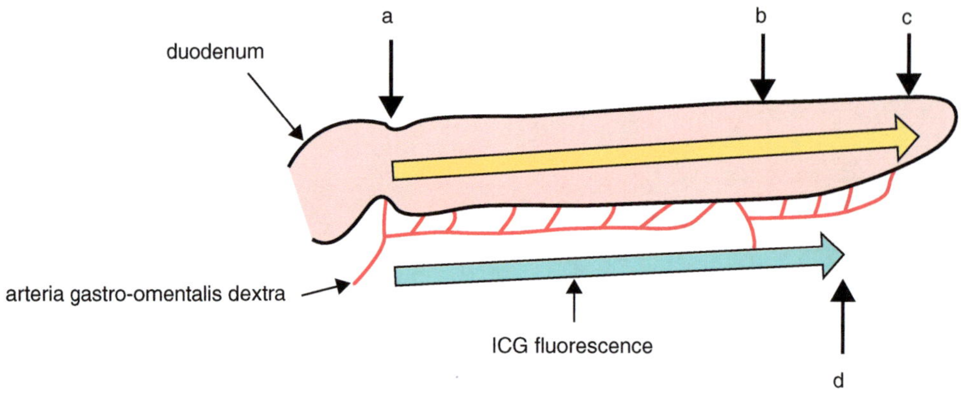

Fig. 11.3 Measurement of the transient speed in the gastric tube using ICG fluorescence imaging. **(a)** Pylorus. **(b)** Advanced beating area by visual inspection and palpation of the right gastroepiploic artery. **(c)** Advanced ICG blood flow in the gastric tube wall. **(d)** Advanced ICG blood flow in the gastroepiploic artery

Point

- We apply a quantitative evaluation method focusing on the transient speed of ICG fluorescence.
- The transient speed can be calculated by measuring the distance and time from the pylorus to the advanced part of ICG fluorescence in the gastric tube wall and the gastroepiploic artery.
- The transient speed of ICG fluorescence in the gastric tube wall would reflect the microcirculation.

5 Association Between Indocyanine Green Fluorescence Imaging and Anastomotic Leakage After Esophageal Cancer Surgery

The expected role of ICG fluorescence imaging as a blood flow evaluation method is to predict the risk of postoperative anastomotic leakage. The results of our quantitative evaluation method focusing on the transient speed of ICG fluorescence and its relationship to postoperative leakage are demonstrated below.

5.1 Association Between Indocyanine Green Fluorescence Imaging and Postoperative Anastomotic Leakage

In the 109 patients described above, postoperative anastomotic leakage developed in 15 cases, although all gastric anastomoses were made at the site where ICG fluorescence was observed in the gastric tube wall. There was no association between the development of anastomotic leakage and clinical, oncological, or operative factors such as intraoperative blood loss and operative time. On the other hand, the transient speed of ICG fluorescence in the gastric tube wall ("a"–"c") was significantly lower in the group with anastomotic leakage than in the group without leakage (1.91 cm/sec vs. 2.78 cm/sec, $p < 0.0001$). The transient speed in the right gastroepiploic artery ("a"–"d") was not significantly different between the groups.

5.2 The Use of Indocyanine Green Fluorescence Imaging as a Predictor of Postoperative Anastomotic Leakage

Receiver operating characteristics (ROC) curve analysis identified that the transient speed of ICG fluorescence in the gastric tube wall had a predictive value for anastomotic leakage after esophageal cancer surgery with an AUC of 0.92 (cutoff value, 2.07 cm/sec; Fig. 11.4). Multivariate analysis revealed that that the transient speed of blood flow in the gastric tube evaluated by ICG fluorescence imaging was an independent predictor of anastomotic leakage ($p < 0.0001$).

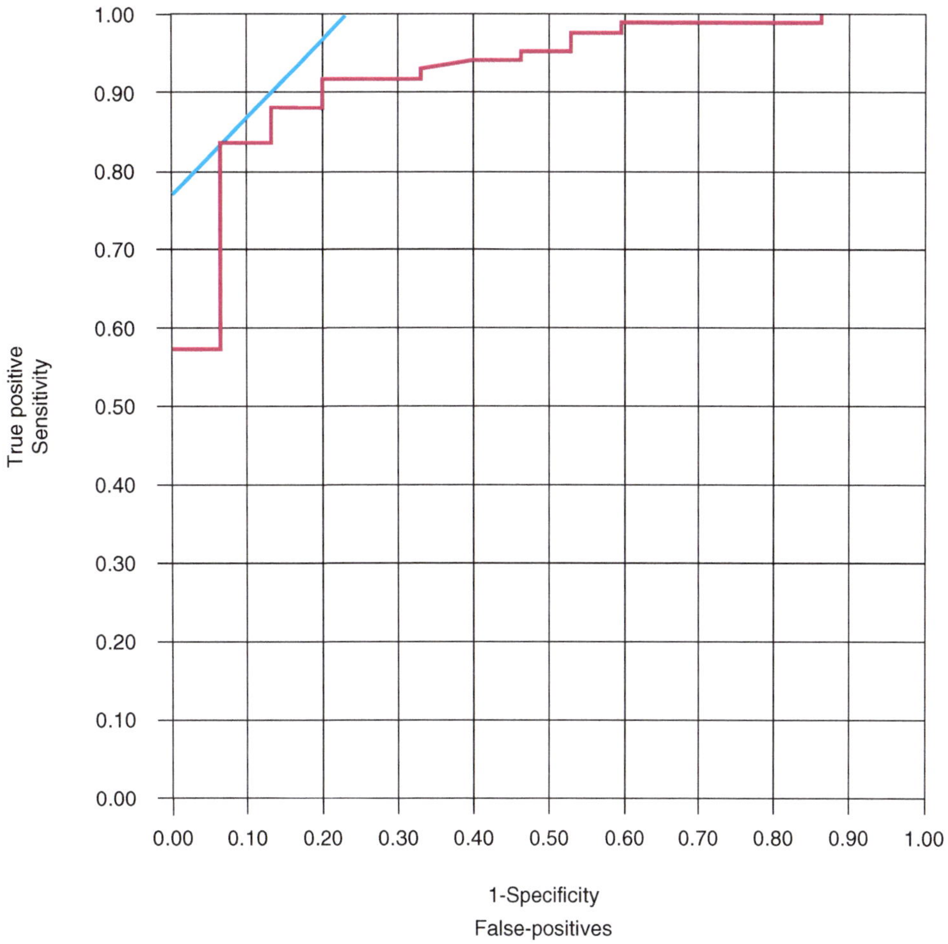

Fig. 11.4 ROC curve: Transient speed of blood flow in the gastric tube wall for a prediction of anastomotic leakage sensitivity and specificity (cut-off value 2.07 cm/sec)

6 Precautions for Blood Flow Measurement by Indocyanine Green Fluorescence Imaging

In our study, all anastomoses of the gastric tube were made at the site where ICG fluorescence was confirmed in the gastric tube wall. However, anastomotic leakage still occurred in 15 cases. This suggests the limitation of the current techniques for qualitative evaluation of blood perfusion by fluorescence imaging. In addition to blood perfusion, there can be other factors associated with anastomotic leakage other than blood perfusion.

In the development of postoperative anastomotic leakage, not only the inflow of arterial blood (inflow) but also venous return (outflow) would play an important role. Yukaya et al. tried to quantify disorders disturbance of both inflow and outflow in reconstructed gastric vessels, but unfortunately, they could not reveal a significant association with postoperative anastomotic leakage [4]. Our evaluation method based on the transient speed of fluorescence signals showed an association with anastomotic leakage. However, our method would be based on comprehensive information on blood perfusion, because it is still difficult to discriminate between inflow and outflow, and we believe that our method is a comprehensive blood flow evaluation.

Point

- Quantitative evaluation is important in perfusion assessment of the gastric tube using ICG fluorescence imaging.
- Current quantitative measurement by ICG fluorescence imaging includes comprehensive information on blood perfusion in the gastric tube, without discriminating disorders in inflow and outflow.

7 Prospects for Blood Flow Measurement by Indocyanine Green Fluorescence Imaging in the Upper Gastrointestinal Surgery

For the further development of ICG fluorescence imaging for perfusion assessment in upper gastrointestinal surgery, we still need studies based on quantitative evaluation of blood flow [9, 10]. Quantitative evaluation of fluorescence signals will discriminate perfusion disorders caused by inflow problems and those by outflow problems, which will enable surgeons to revise surgical procedures more appropriately. For this purpose, advances in imaging equipment and analysis software are highly awaited. With the current development of endoscopic systems with near-infrared light imaging, applications of ICG fluorescence imaging for perfusion assessment will be expanded to laparoscopic upper gastrointestinal surgery.

References

1. Miwa M. ICG Keikouhou no Genri to Kikikaihatsu. (ICG fluorescence method principles and instrument development). Geka. 2009;71:913–7. (in Japanese)
2. Ikeda Y, Niimi M, Kan S, et al. Clinical significance of tissue blood flow during esophagectomy by laser doppler flowmetry. J Thorac Cardiovasc Surg. 2001;122:1101–6.
3. Zehetner J, DeMeester SR, Alicuben ET, et al. Intraoperative assessment of perfusion of the gastric graft and correlation with anastomotic leaks after esophagectomy. Ann Surg. 2015;262:74–8.
4. Shimada Y, Okumura T, Nagata T, et al. Usefulness of blood supply visualization by indocyanine green fluorescence for reconstruction during esophagectomy. Esophagus. 2011;8:259–66.
5. Yukaya T, Saeki H, Kasagi Y, et al. Indocyanine green fluorescence angiography for quantitative evaluation of gastric tube perfusion in patients undergoing esophagectomy. Am Coll Surg. 2015;221:e37–42.
6. Kumagai Y, Hatano S, Sobajima J, et al. Indocyanine green fluorescence angiography of the reconstructed gastric tube during esophagectomy: efficacy of the 90-second rule. Dis Esophagus. 2018;31:1–4.
7. Kamiya K, Unno N, Miyazaki S, et al. Quantitative assessment of the free jejunal graft perfusion. J Surg Res. 2015;194:394–9.
8. Koyanagi K, Ozawa S, Oguma J, et al. Blood flow speed of the gastric conduit assessed by indocyanine green fluorescence: new predictive evaluation of anastomotic leakage after esophagectomy. Medicine. 2016;95:30(e4386).
9. Daele EV, van Nieuwenhove Y, Ceelen W, et al. Near-infrared fluorescence guided esophageal reconstructive surgery: a systematic review. World J Gastrointest Oncol. 2019;11:250–63.
10. Ladak F, Dang JT, Switzer N, et al. Indocyanine green for the prevention of anastomotic leaks following esophagectomy: a meta-analysis. Surg Endosc. 2019;33:384–94.

Evaluation of Blood Perfusion in Colorectal Surgery

12

Hiro Hasegawa, Yuichiro Tsukada, and Masaaki Ito

Summary

- Indocyanine green fluorescence imaging is the most popular technique in colorectal surgery, enabling objective real-time evaluation of blood perfusion in the reconstructed intestine during surgery.
- Indocyanine green fluorescence imaging may help reduce the incidence of anastomotic leakage, which is a serious postoperative complication.

1 Introduction

Recently, indocyanine green (ICG) fluorescence imaging has been attracting attention as an intraoperative diagnostic tool to evaluate blood perfusion in the gastrointestinal tract in real time. In this chapter, we have reviewed ICG fluorescence imaging in the field of colorectal surgery.

2 Application of Indocyanine Green Fluorescence Imaging in Colorectal Surgery

Despite advances in surgical techniques and equipment, anastomotic leakage (AL) is one of the most serious postoperative complications in colorectal surgery. The incidence of AL in rectal surgery is particularly high, with a reported incidence of 9–14% [1, 2]. AL increases postoperative mortality [1, 2], prolongs hospital stay, and increases medical costs [1]. In rectal resection, AL not only impairs defecation function and quality of life [3] but also negatively affects the recurrence rate of cancer [4].

Various factors, such as patient, oncological, and surgical parameters, are involved in the development of AL [5]. Among them, an adequate blood supply to the reconstructed intestine is considered the most important factor for avoiding AL [5].

Blood perfusion in the reconstructed intestine is evaluated intraoperatively by surgeons based on the color tone of the intestinal wall, presence or absence of intestinal peristalsis, presence or absence of bleeding from the dissection margin of the intestine, and beating of the mesenteric artery. However, these conventional methods are subjective and unreliable because their reliability depends on the experience and skill of the surgeon [6]. Doppler methods and near-infrared spectroscopy have been utilized to objectively evaluate intraoperative blood perfusion in the gastrointestinal tract [7]. They have not been accepted in clinical practice because they are difficult to perform and lack reproducible results. Recently, ICG fluorescence imaging has been attracting attention as a simple, minimally invasive, and objective method for real-time evaluation of blood perfusion in the gastrointestinal tract during surgery. In the field of colorectal surgery, ICG fluorescence imaging may help reduce the incidence of AL, and its clinical application has been expanding worldwide [8–19].

3 Methods of Indocyanine Green Fluorescence Imaging for Perfusion Assessment During Colorectal Surgery

The reported methods of perfusion assessment by ICG fluorescence imaging during colorectal surgery [8–18] are summarized in Table 12.1. Varying dosages of ICG have been reported in the literature [8, 10, 12–18]. Inasmuch as the timing of perfusion assessment is concerned, fluorescence imaging is often used before and/or after anastomosis [8–18]. Evaluation can be performed from both the serosal and mucosal surfaces of the intestine.

We introduced the ICG fluorescence method in 2016 to evaluate blood perfusion in the bowel reconstructed mainly

H. Hasegawa (✉) · Y. Tsukada · M. Ito
Department of Colorectal Surgery, National Cancer Center Hospital East, Kashiwa, Chiba, Japan
e-mail: hirhaseg@east.ncc.go.jp

T. Ishizawa (ed.), *Fluorescence-Guided Surgery*, https://doi.org/10.1007/978-981-19-7372-7_12

Table 12.1 Evaluation of blood perfusion in colorectal anastomoses using ICG fluorescence imaging

Authors	Publication (Year of publication)	Study design	Operative procedure	Number of cases	Evaluation of blood perfusion	Imaging system	ICG dose	Imaging success rate (%)	Perfusion assessment time (ICG fluorescence imaging; s)	Length of procedure (min)	Change of surgical plan (%)	Incidence of anastomotic leakage (%)
Kudzus [8]	Langenbecks Arch Surg (2010)	Retrospective	Colorectal resection	402 ICG fluorescence imaging group: 201 Control group: 201	Before and after anastomosis	IC-View®	0.2–0.5 mg/kg	n.r.	n.r.	6.8	13.9	5.5 ICG group: 3.5 Control group: 7.5
Ris [9]	Surg Endosc (2014)	Prospective	Colorectal resection	30	After anastomosis	PINPOINT system	n.r.	96.7	35	4.5	n.r.	0
Jafari [10]	J Am Coll Surg (2015)	Prospective	Left-sided colectomy Anterior resection	139	Before and after anastomosis	PINPOINT system	3.75–7.5 mg	98.6	n.r.	n.r.	7.9	1.4
Kin [11]	Dis Colon Rectum (2015)	Retrospective	Left-sided colectomy Rectal resection	346 ICG fluorescence imaging group: 173 Control group: 173	Before anastomosis	SPY Imaging System™	n.r.	n.r.	n.r.	n.r.	4.6	6.9 ICG group: 7.5 Control group: 6.4
Boni [12]	Surg Endosc (2016)	Prospective	Colorectal resection	107	Before anastomosis	IMAGE1 S™ system	0.2 mg/kg	100	57	n.r.	3.7	0.9
Boni [13]	Surg Endosc (2017)	Retrospective	Low anterior resection	80 ICG fluorescence imaging group: 42 Control group: 38	Before anastomosis	IMAGE1 S™ system	0.2 mg/kg	100	n.r.	n.r.	4.7	2.5 ICG group: 0 Control group: 5.3
Kim [14]	Dis Colon Rectum (2017)	Retrospective	Sphincter-sparing surgery	657 ICG fluorescence imaging group: 310 Control group: 347	Before and after anastomosis	Firefly™	10 mg	n.r.	44	n.r.	n.r.	3.0 ICG group: 0.6 Control group: 5.2
Ris [15]	Br J Surg (2018)	Prospective	Colorectal resection	504	Before and after anastomosis	PINPOINT system	7.5 mg	100	29	4	5.8	2.4

Hasegawa [16]	Int J Colorectal Dis (2019)	Retrospective	Sphincter-sparing surgery	844 ICG fluorescence imaging group: 141 Control group: 703	Before anastomosis	IMAGE1 S™ system 1588 AIM Platform and SPY Fluorescence technology HyperEye Medical System Handy	5 mg	100	36	n.r.	17.0	10.0 ICG group: 2.8 Control group: 12.4
Alekseev [17]	Colorectal Dis (2020)	Randomized controlled trial	Sigmoidectomy Rectal resection	377 ICG fluorescence imaging group: 187 Control group: 190	Before and after anastomosis	IMAGE1 S™ system	0.2 mg/kg	100	n.r.	n.r.	19.2	12.7 ICG group: 9.1 Control group: 16.3
Nardi [18]	Surg Endosc (2020)	Randomized controlled trial	Left-sided colectomy Rectal resection	240 ICG fluorescence imaging group: 118 Control group: 122	Before and after anastomosis	IMAGE1 S™ system	0.3 mg/kg	100	n.r.	2–4	11.0	7.1 ICG group: 5.1 Control group: 9.0

Notes: *n.r.* not reported, *ICG* indocyanine green

during rectal resections with a high risk of AL. In our institution, ICG fluorescence imaging is performed before anastomosis, prior to the dissection of the proximal bowel. The planned resection line is determined by visual assessment, and ICG (5 mg) is administered intravenously to confirm the presence of blood perfusion by ICG fluorescence imaging (Fig. 12.1). The median time from the administration of ICG to the appearance of fluorescence in the intestinal wall is 29–57 s [9, 12, 14–16]. If the proximal bowel is not visually enhanced at the planned resection line, it is resected up to the point where the bowel shows fluorescent signals, and anastomosis is performed at that point. For post-anastomosis evaluation, ICG fluorescence imaging is used on the serosal or mucosal surface around the suture line, depending on the location of the anastomosis (Figs. 12.2 and 12.3). If the blood perfusion is judged to be inadequate, revision of the surgical procedure, including re-anastomosis, is considered.

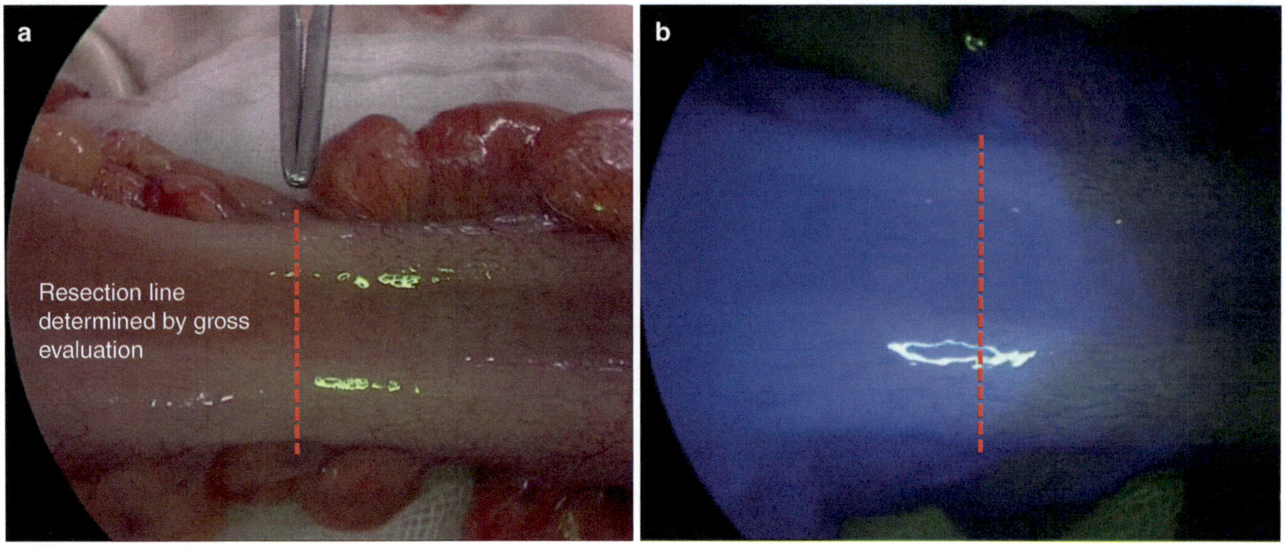

Fig. 12.1 Evaluation of blood perfusion before anastomosis from the serosal surface. (**a**) White-light image after dissection of the mesentery. The planned resection line in the proximal bowel was determined by gross evaluation. (**b**) ICG fluorescent image. ICG fluorescence (*blue*) is present up to the planned resection line, indicated by the dotted line. ICG indocyanine green

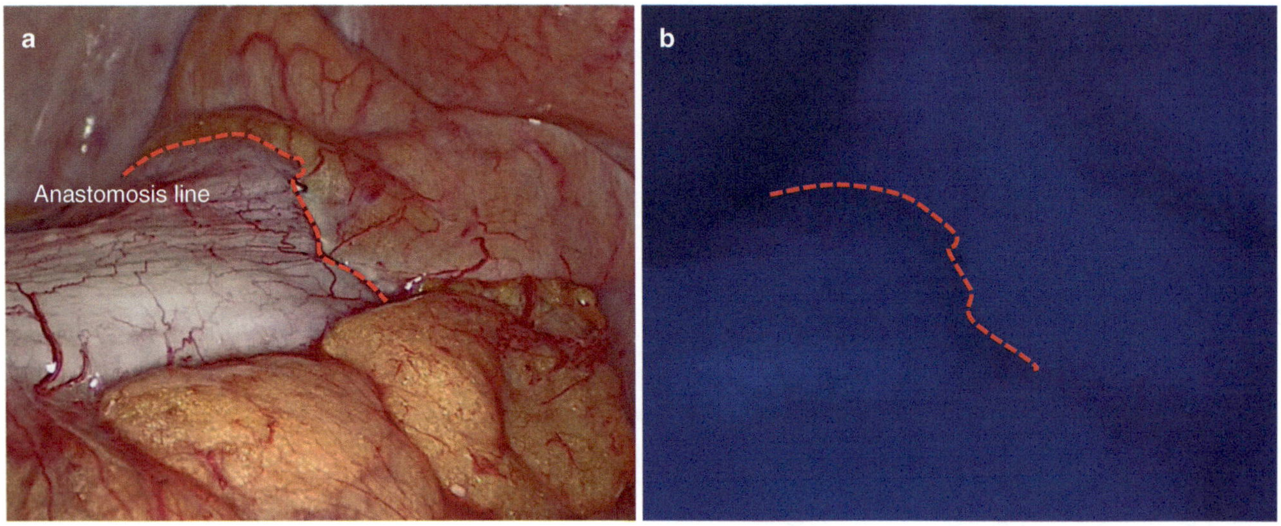

Fig. 12.2 Evaluation of blood perfusion after anastomosis from the serosal surface. (**a**) White-light image after completion of anastomosis using a stapler. (**b**) ICG fluorescent image. ICG fluorescence (*blue*) is present up to the anastomotic site, indicated by the dotted line. ICG indocyanine green

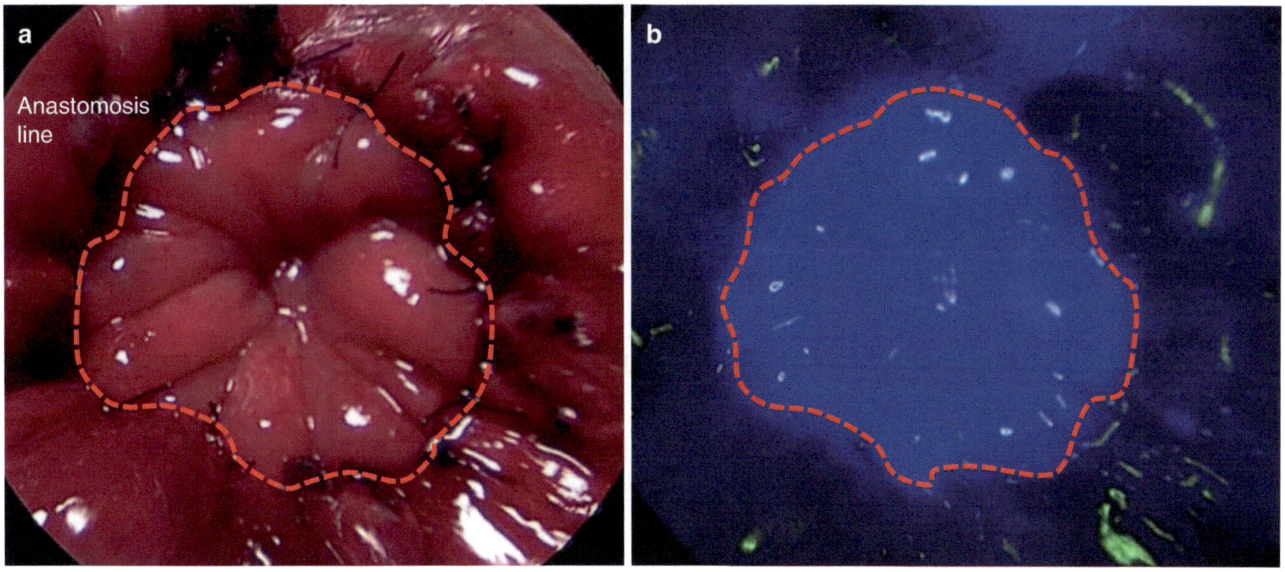

Fig. 12.3 Evaluation of blood perfusion after anastomosis from the mucosal surface. (**a**) White-light image after completion of hand-sewn anastomosis. (**b**) ICG fluorescent image. ICG fluorescence (*blue*) is present up to the anastomotic site, indicated by the dotted line. ICG indocyanine green

4 Outcomes of Indocyanine Green Fluorescence Imaging for Perfusion Assessment During Colorectal Surgery

The reported results of intraoperative intestinal blood perfusion evaluation using ICG fluorescence in colorectal surgery [8–18] are shown in Table 12.1.

In previous studies, ICG fluorescence imaging enabled blood perfusion evaluation in most patients [9, 10, 12, 13, 15–18], and the time required for the operation was short (2–7 min) [8, 9, 15, 18]. The major causes of unsuccessful fluorescence imaging were device malfunction and lack of rapid intravenous ICG infusion. No cases of adverse effects caused by ICG administration have been reported [8–18]. ICG fluorescence imaging is considered a safe, simple, and minimally invasive method for intraoperative evaluation of intestinal blood perfusion. Its use leads to the revision of surgical planning, such as additional resection of the proximal bowel, in 3.7–17.0% of cases [8, 10–13, 15–18]. Congenital defects of the marginal arteries (Sudeck or Griffith points) or damage to the marginal arteries due to intraoperative manipulation is believed to be the major factor causing insufficient blood perfusion at the site for colorectal anastomosis.

In 2010, Kudszus et al. [9] first reported the safety and usefulness of ICG fluorescence imaging in colorectal resection. In 13.9% (28/201) of patients in the ICG fluorescence imaging group, the planned resection line of the proximal bowel was changed, and the incidence of AL was lower in the ICG fluorescence imaging group than in the control group (3.5% vs. 7.5%). In 2015, Jafari et al. [10] reported the results of a multicenter phase II study (PILLAR-II) of left-sided colectomy and anterior resection. The overall incidence of AL was 1.4% (2/139). ICG fluorescence imaging was performed before and after anastomosis, and the surgical plan was changed according to the imaging results in 7.9% (11/139) of the patients. In 9 of these 11 patients, the planned resection line of the proximal bowel was changed. Among the remaining two patients, post-anastomosis evaluation led to re-anastomosis in one patient and avoidance of temporary stoma in one patient. However, none of the patients developed AL. In 2016, Boni et al. [13] reported the results of rectal resection using the ICG fluorescence technique. ICG fluorescence imaging was performed before and after anastomosis, and the results showed a change in the surgical plan in 4.7% (2/42) patients; none of these patients developed AL. In 2017, Kim et al. [14] reported the usefulness of ICG fluorescence imaging in robot-assisted rectal resection. The ICG fluorescence imaging group had a significantly lower incidence of AL than the control group (0.6% vs. 5.2%). In 2020, Hasegawa et al. [16] reported the usefulness of ICG fluorescence imaging in rectal resection. Before propensity score matching ($n = 844$), the overall incidence of AL in the ICG fluorescence imaging and control groups was 2.8% (4/141) and 12.4% (87/703), respectively. After 1:2 matching using propensity scores ($n = 420$) to balance patient backgrounds between the two groups, the incidence of AL in the ICG fluorescence imaging and control groups was 2.8% (4/141) and 13.6% (38/279), respectively. Logistic regression analysis using propensity scores showed that the odds ratio of AL was significantly lower in the ICG fluorescence imaging group than in the control group. In 2020, Alekseev et al. [17] reported the results of a single-center randomized controlled trial of sigmoid and rectal resections. The incidence of AL was significantly lower in the ICG fluorescence imaging group than in the control group (9.1% vs. 16.3%; $p = 0.04$). ICG fluorescence imaging did not decrease the incidence of AL of high anastomoses (1.3% vs. 4.6%; $p = 0.37$). However, ICG fluorescence imaging decreased the incidence of AL of low anastomoses (14.4% vs. 25.7%; $p = 0.04$).

In contrast, Kin et al. [11] reported in 2015 that there was no difference in the incidence of AL between the ICG fluorescence imaging group and control group in colorectal resection (7.5% vs. 6.4%; $p = 0.67$), although this study was potentially limited by the presence of selection bias and a small sample size. De Nardi et al. [18] reported the results of a multicenter randomized controlled trial of left-sided colon and rectal resection. The incidence of AL was lower in the ICG fluorescence imaging group than in the control group, although the difference was not statistically significant (5.1% vs. 9.0%).

A systematic review has also suggested that the ICG fluorescence technique is useful in reducing the incidence of AL in colorectal resection [19]. Taken together, ICG fluorescence imaging can help reduce the incidence of AL in colorectal surgery.

Take Home Points

- Indocyanine green fluorescence imaging is a safe, simple, and minimally invasive method for intraoperative evaluation of intestinal blood perfusion.
- The use of ICG fluorescence imaging may help reduce the incidence of AL after colorectal surgery.

5 Challenges of Indocyanine Green Fluorescence Imaging in Colorectal Surgery

Currently, the major limitation of ICG fluorescence imaging for perfusion assessment is its objectivity based on the qualitative assessment of fluorescence signals. Wada et al. [20] retrospectively analyzed the fluorescence intensity of the colorectal wall used for anastomosis in 112 patients who underwent ICG fluorescence imaging and reported that patients with low fluorescence intensity could develop AL, despite sufficient blood perfusion on qualitative assessment. Hence, further large prospective studies are warranted to evaluate the efficacy of quantitative analysis of blood perfusion by ICG fluorescence imaging for preventing AL after colorectal surgery.

6 Conclusions

Indocyanine green fluorescence imaging in colorectal surgery is a safe, simple, and minimally invasive tool for evaluating blood perfusion during colorectal surgery. Such in vivo fluorescence imaging techniques are expected to help improve the clinical outcomes of patients with colorectal cancer.

References

1. Kang CY, Halabi WJ, Chaudhry OO, et al. Risk factors for anastomotic leakage after anterior resection for rectal cancer. JAMA Surg. 2013;148:65–71.
2. Snijders HS, Wouters MW, van Leersum NJ, et al. Meta-analysis of the risk for anastomotic leakage, the postoperative mortality caused by leakage in relation to the overall postoperative mortality. Eur J Surg Oncol. 2012;38:1013–9.
3. Mongin, Maggiori L, Agostini J. Does anastomotic leakage impair functional results and quality of life after laparoscopic sphincter-saving total mesorectal excision for rectal cancer? A case-matched study. Int J Color Dis. 2014;29:459–67.
4. Mirnezami A, Mirnezami R, Chandrakumaran K, et al. Increased local recurrence and reduced survival from colorectal cancer following anastomotic leak: systematic review and meta-analysis. Ann Surg. 2011;253:890–9.
5. Kingham TP, Pachter HL. Colonic anastomotic leak: risk factors, diagnosis, and treatment. J Am Coll Surg. 2009;208:269–78.
6. Karliczek A, Harlaar NJ, Zeebregts CJ, et al. Surgeons lack predictive accuracy for anastomotic leakage in gastrointestinal surgery. Int J Color Dis. 2009;24:569–76.
7. Nachiappan S, Askari A, Currie A, et al. Intraoperative assessment of colorectal anastomotic integrity: a systematic review. Surg Endosc. 2014;28:2513–30.
8. Kudszus S, Roesel C, Schachtrupp A, et al. Intraoperative laser fluorescence angiography in colorectal surgery: a noninvasive analysis to reduce the rate of anastomotic leakage. Langenbeck's Arch Surg. 2010;395:1025–30.
9. Ris F, Hompes R, Cunningham C, et al. Near-infrared (NIR) perfusion angiography in minimally invasive colorectal surgery. Surg Endosc. 2014;28:2221–6.
10. Jafari MD, Wexner SD, Martz JE, et al. Perfusion assessment in laparoscopic left-sided/anterior resection (PILLAR II): a multi-institutional study. J Am Coll Surg. 2015;220:82–92.
11. Kin C, Vo H, Welton L, et al. Equivocal effect of intraoperative fluorescence angiography on colorectal anastomotic leaks. Dis Colon Rectum. 2015;58:582–7.
12. Boni L, David G, Dionigi G, et al. Indocyanine green-enhanced fluorescence to assess bowel perfusion during laparoscopic colorectal resection. Surg Endosc. 2016;30:2736–42.
13. Boni L, Fingerhut A, Marzorati A, et al. Indocyanine green fluorescence angiography during laparoscopic low anterior resection: results of a case-matched study. Surg Endosc. 2017;31:1836–40.
14. Kim JC, Lee JL, Park SH. Interpretative guidelines and possible indications for indocyanine green fluorescence imaging in robot-assisted sphincter-saving operations. Dis Colon Rectum. 2017;60:376–84.
15. Ris F, Liot E, Buchs NC, et al. Multicentre phase II trial of near-infrared imaging in elective colorectal surgery. Br J Surg. 2018;105:1359–67.

16. Hasegawa H, Tsukada Y, Wakabayashi M, et al. Impact of intraoperative indocyanine green fluorescence angiography on anastomotic leakage after laparoscopic sphincter-sparing surgery for malignant rectal tumors. Int J Color Dis. 2020;35:471–80.
17. Alekseev M, Rybakov E, Shelygin Y, et al. A study investigating the perfusion of colorectal anastomoses using fluorescence angiography: results of the FLAG randomized trial. Color Dis. 2020;22:1147–53.
18. De Nardi P, Elmore U, Maggi G, et al. Intraoperative angiography with indocyanine green to assess anastomosis perfusion in patients undergoing laparoscopic colorectal resection: results of a multicenter randomized controlled trial. Surg Endosc. 2020;34:53–60.
19. Degett TH, Andersen HS, Gögenur I. Indocyanine green fluorescence angiography for intraoperative assessment of gastrointestinal anastomotic perfusion: a systematic review of clinical trials. Langenbeck's Arch Surg. 2016;401:767–75.
20. Wada T, Kawada K, Takahashi R, et al. ICG fluorescence imaging for quantitative evaluation of colonic perfusion in laparoscopic colorectal surgery. Surg Endosc. 2017;31:4184–93.

Perfusion Assessment in HBP Surgery and Liver Transplantation

13

Satoru Seo

Summary

- In the surgery for liver and gallbladder cancers, indocyanine green (ICG) fluorescence imaging enables real-time navigation of the hepatic transection line.
- Perfusion assessment by ICG fluorescence imaging may also be used to determine the necessity of concomitant resection of gastrointestinal tracts in pancreatic cancer surgery.
- In liver transplantation, visualization of blood perfusion status in the transplanted graft is expected to improve surgical outcomes.

1 Introduction

Because hepatobiliary pancreatic (HBP) surgery and liver transplantation are technically difficult and associated with fatal complications, surgeons usually prepare for the surgeries by understanding the anatomy of each patient, sometimes by drawing sketches based on preoperative imaging studies. During surgery, however, only visual inspection and ultrasonography are available for understanding the patient's anatomy. Therefore, surgeons have long wished to develop an intraoperative navigation system. In recent years, the usefulness of ICG fluorescence imaging as an intraoperative navigation tool has been reported in various applications. We herein describe the usefulness of ICG fluorescence imaging in the evaluation of blood perfusion in HBP surgery and liver transplantation.

Supplementary Information The online version contains supplementary material available at https://doi.org/10.1007/978-981-19-7372-7_13.

S. Seo (✉)
Department of Surgery,
Kochi Medical School, Nankoku, Kochi, Japan
e-mail: rutosa@kuhp.kyoto-u.ac.jp

2 Limitations of Conventional Techniques for Perfusion Assessment

2.1 In HBP Surgery

2.1.1 Identification of Hepatic Segmental Boundaries for Anatomic Hepatectomy

In order to perform safe and precise anatomic hepatectomy, two major methods have been used. One is a dye-staining technique first reported by Makuuchi et al. in which the portal vein branch corresponding to the tumor-bearing hepatic segment is punctured and injected with a dye (indigocarmine) to delineate the demarcation line on the liver surface [1]. The other method, reported by Takasaki, is to close the Glissonian sheath supplying the tumor-bearing hepatic segment to delineate the demarcation line as boundaries of hepatic ischemic regions [2]. The ideal procedures for anatomic hepatectomies are to start hepatic dissection based on the demarcation line identified by these methods and to expose the landmark hepatic veins running along the intersegmental planes. The major problem in the conventional techniques lies in the fact that surgeons cannot identify accurate intersegmental planes connecting the starting lines set on the liver surfaces and ending points usually set on the root of the corresponding Glissonian sheath during hepatic transection. Despite the meticulous use of intraoperative ultrasonography for understanding the intrahepatic vessel anatomy, it is not always easy to reach the endpoints of hepatic transection without hesitation. In addition, in cases of cirrhosis or re-hepatectomy with severe adhesions, it is often difficult to see the demarcation line on hepatic surfaces. Thus, intraoperative real-time navigation is highly awaited for safe and precise anatomic hepatectomy.

2.1.2 Identification of the Optimal Extent of Hepatectomy for Gallbladder Cancer

The aim of hepatectomy for gallbladder cancer is to secure a surgical margin and also to resect possible micrometastases. The extended cholecystectomy (resection of hepatic paren-

T. Ishizawa (ed.), *Fluorescence-Guided Surgery*, https://doi.org/10.1007/978-981-19-7372-7_13

chyma around the gallbladder bed) for the former purpose and hepatectomy of S4a + 5 for the latter have been performed, although the significance of these procedures has not yet been clarified [3]. Identification of the optimal extent of hepatectomy by injecting blue dye into the stump of the cystic artery is a possible alternative, although staining of the hepatic regions is not always successful.

2.1.3 Evaluation of Gastrointestinal Blood Perfusion During Pancreatic Surgery

In resection for advanced pancreatic cancer, all major arteries feeding the gastrointestinal tracts (e.g., the colic arteries in the case of pancreatic head cancer and the gastric arteries in distal pancreatectomy following previous distal gastrectomy) can be resected. In such situations, concomitant resection of the adjacent organs should be considered. However, it is often difficult to decide the need for the additional resection, without using a novel intraoperative navigation enabling clear visualization of blood perfusion in the target organs.

2.2 In Liver Transplantation

In living donor liver transplantation using the right liver graft, reconstruction of middle hepatic vein draining S5 and S8 (V5 and V8, respectively) and inferior right hepatic vein is considered to prevent congestion of hepatic parenchyma other than the drainage areas by right hepatic vein, which can cause postoperative graft dysfunction. In our department, the need for venous reconstruction is determined based on the estimation of future congested liver volume for each branch of the vein using 3D simulation images constructed from preoperative CT. In the conventional techniques based on preoperative imaging data, however, the real extent and severity of congestion in the liver graft cannot be measured, making it difficult to evaluate the accuracy of preoperative simulation.

3 Development History of Indocyanine Green Fluorescence Imaging in HBP and Transplantation Surgery

3.1 Application to HBP Surgery

In the field of HBP surgery, the efficacy of ICG fluorescence imaging has been reported for the visualization of hepatic segmental boundaries [4], identification of liver tumors [5], and bile ducts [6]. In particular, visualization of hepatic segments by ICG fluorescence imaging has been used for anatomic hepatectomy since it was first reported by Aoki et al. in 2008 [4]. ICG is taken up into hepatocytes via organic anion transmembrane polypeptide (OATP) and sodium taurocholate cotransporting polypeptide (NTCP) expressed on hepatocyte membranes, and accumulates in hepatocytes until it is excreted into bile through MDR3 expressed in the bile canaliculi, which expanded applications of ICG fluorescence imaging, especially in hepatobiliary surgery. Although there have been only a few reports on its application to pancreatic surgery, some authors have demonstrated the efficacy of ICG fluorescence imaging for real-time evaluation of blood perfusion in the stomach in distal pancreatectomy with resection of the celiac axis [7].

3.2 Application to Liver Transplantation

Although previous publications on this topic are still limited, we have used fluorescence imaging for visualization of regional perfusion in the liver graft, which is usually difficult to recognize by the naked eye, by mixing ICG into the perfusion fluid used for the explanted liver graft in bench surgery.

4 Clinical Practice of Indocyanine Green Fluorescence Imaging

4.1 Evaluation of Blood Perfusion in HBP Surgery

4.1.1 Real-Time Navigation Surgery for Anatomic Hepatectomy

We have developed the Medical Imaging Projection System (MIPS), a system for directly projecting ICG fluorescence images onto a patient's organs, as an industry-academia collaborative project [8] by applying projection mapping technology that has been widely used in entertainment (Fig. 13.1a). With this system, we were able to reduce the misalignment to less than 2 mm and the time lag to projection to less than 0.2 seconds. This enabled ICG fluorescence images to follow the movement and deformation of the liver in real time. In addition, by using an algorithm that converts fluorescing areas to specific colors and non-fluorescing areas to white, MIPS can illuminate surgical fields brightly even when the surgical lights in the OR are turned off for fluorescence imaging (Fig. 13.1b).

With the use of MIPS, the boundaries between the resected liver and the remnant liver are continuously projected as the presence or absence of ICG fluorescence throughout hepatic parenchymal transections. In other words, real-time navigation of hepatectomy has been realized by the introduction of MIPS (Fig. 13.2a, b, Movie 13.1).

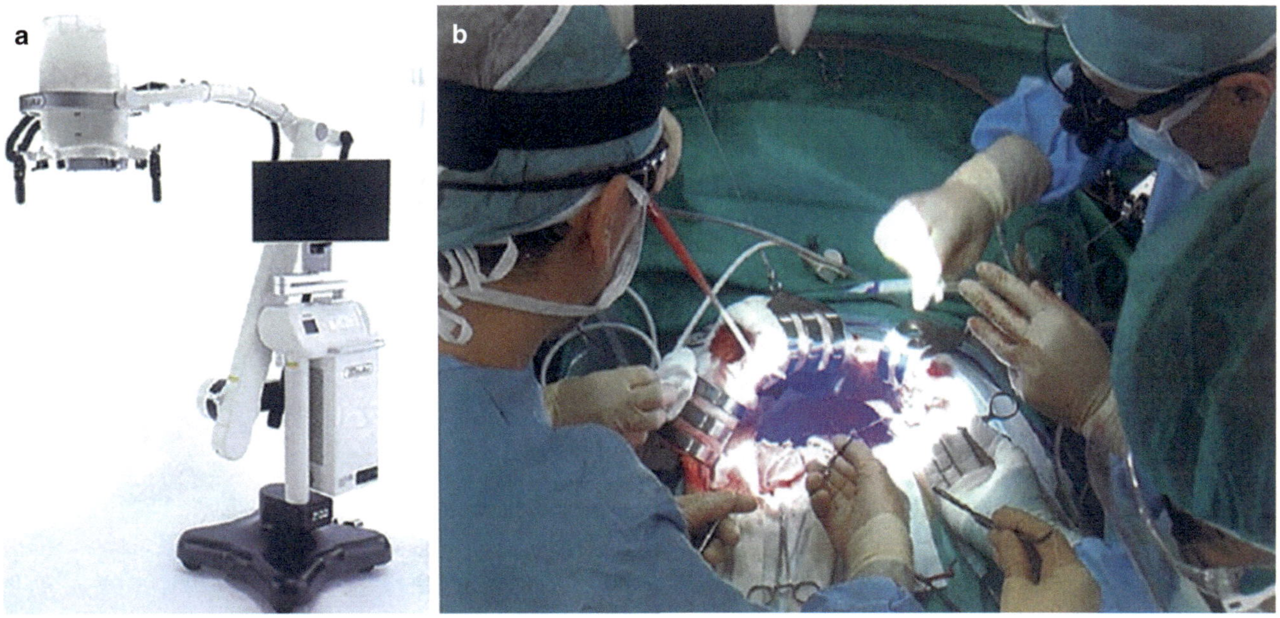

Fig. 13.1 Appearance of the Medical Imaging Projection System (MIPS). (**a**) Appearance of MIPS. (**b**) Surgical view during hepatectomy obtained with MIPS

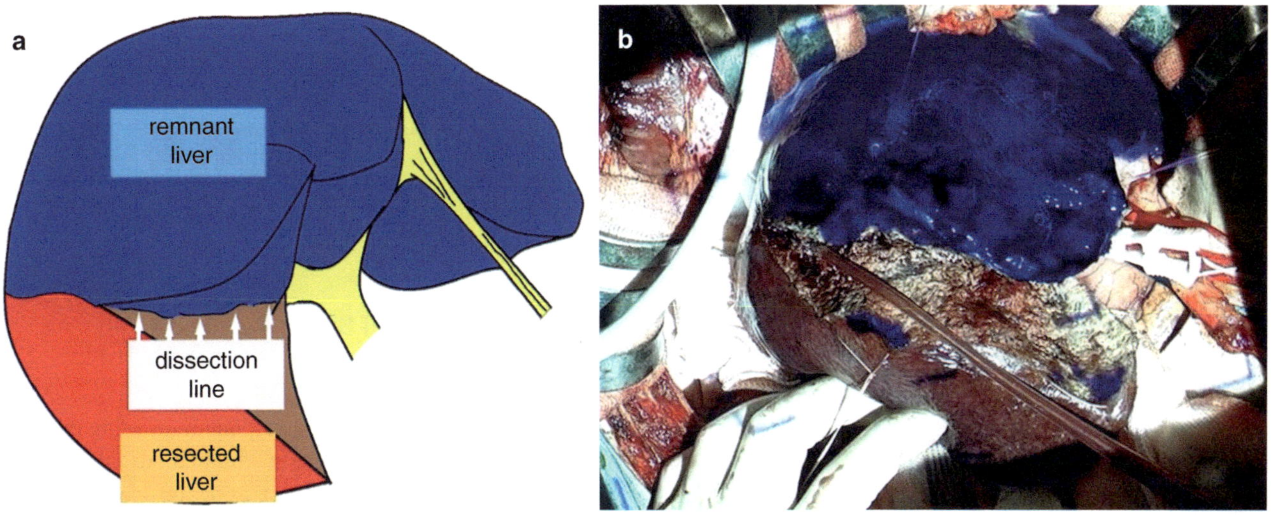

Fig. 13.2 Real-time navigation hepatectomy using MIPS. (**a**) Schematic view of hepatic raw surfaces after resection. (**b**) Actual hepatic cutting plane

4.1.2 Determination of the Optimal Extent of Hepatectomy for Gallbladder Cancer

We performed extended cholecystectomy using the ICG fluorescence imaging technique in four patients from January 2012 to August 2016, determining the perfusion areas by the cystic artery as the extent of hepatectomy. After the intraoperative injection of ICG (2.5 mg) through the cystic artery, the hepatic perfusion areas were delineated using a PDE camera, which can be used as hepatic transection lines (Movie 13.2). In the four patients, these fluorescence areas did not coincide with possible hepatic transection lines based on surgical margins 2 cm distant from the gallbladder bed or S4a + 5 regions (Fig. 13.3). The mean operative time was 439.8 min, blood loss was 504.5 mL, and the postoperative hospital stay was 16.5 days in this series. The cancer depth was mp (n = 1), ss (n = 2), and se (n = 1), with lymph node metastasis (N1) in one patient. External surgical margins were all negative. All patients are alive without evidence of recurrence except for one case of postoperative lymph node metastasis. MIPS may be useful for real-time navigation in the resection of gallbladder cancer as well as hepatectomy for liver cancers.

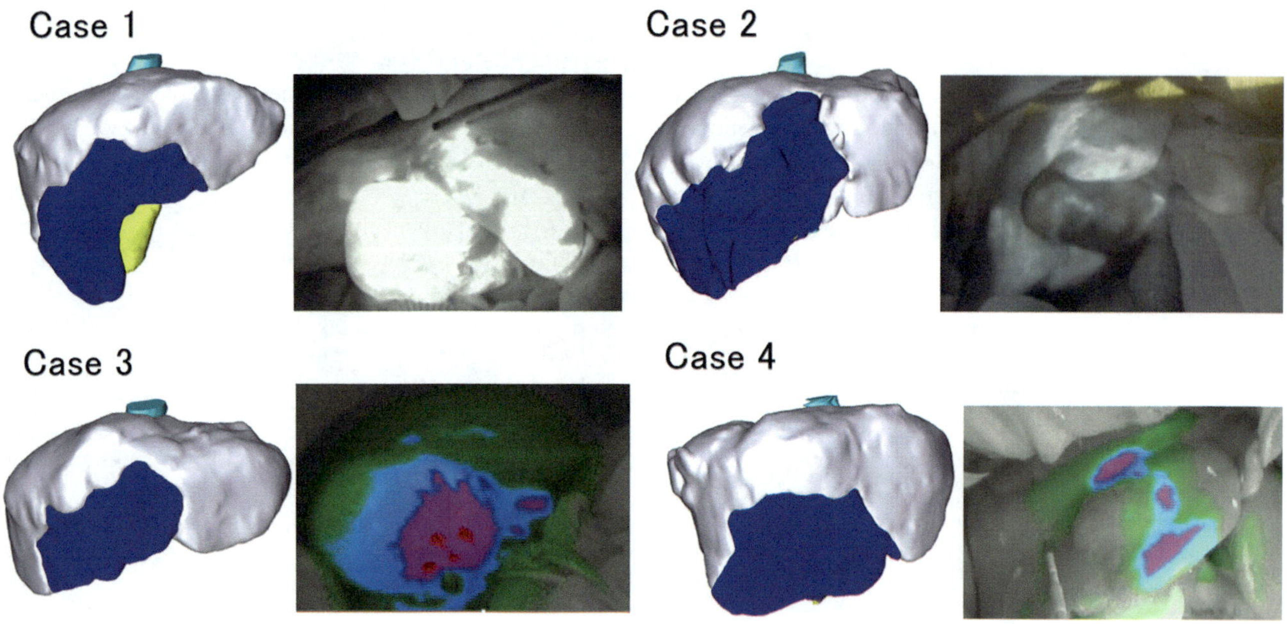

Fig. 13.3 Comparison of preoperative simulation of S4a + 5 resection and intraoperative ICG fluorescence imaging

4.1.3 Perfusion Assessment of the Gastrointestinal Tract During Pancreatic Surgery

In pancreatectomy, perfusion assessment is needed when the mesocolon is removed due to direct invasion of pancreatic cancer, without concomitant resection of the colon itself. In the case presented in Fig. 13.4a, b (Movie 13.3), the beating of the marginal artery was not palpated after the resection, but the color of the colon wall was not so ischemic, making it difficult to determine the need for the additional colectomy for avoiding ischemic complications. Then, fluorescence imaging was performed by injecting ICG (2.5 mg) intravenously with the use of a PDE camera. As a result, fluorescence signals in the corresponding colon wall were obviously insufficient, enabling us to conclude that the patient had intestinal ischemia and to perform an additional partial resection of the transverse colon. No postoperative complications developed in this case.

In distal pancreatectomy after a previous gastrectomy, all major arteries feeding the remnant stomach can be dissected as a result of resection for pancreatic cancer, which can cause postoperative necrosis of the gastric wall. On the other hand, it is also important to consider the decrease in the patient's QOL by total resection of the remaining stomach to prevent a potential risk of postoperative necrosis. The elderly patient presented in Fig. 13.4c, d (Movie 13.4) was in his 80 s and had undergone distal gastrectomy for gastric cancer 7 years ago. During distal pancreatectomy for pancreatic cancer, ICG there were no color changes suggesting ischemia, and considering his age, we tried to avoid total removal of the remaining stomach. Then, fluorescence imaging was used following intravenous administration of ICG (2.5 mg) and a PDE camera, which visualized sufficient blood perfusion in the remnant stomach. The patient followed an uneventful postoperative course.

4.2 Evaluation of Blood Perfusion in the Liver Graft

In a case of living donor liver transplantation using the right liver graft, we used perfusion fluid mixed with ICG (2.5 mg/L) in bench surgery after graft harvest in order to visualize the extent of graft perfusion by fluorescence imaging. In the case shown in Fig. 13.5 (Movie 13.5), when perfusion was started with V5 and V8 clamped, no fluorescence signals were observed in the congested areas estimated by the preoperative simulation. Then, the clamp was released and perfusion was continued. At this time, fluorescence imaging identified a gradual increase of fluorescence signals in the regions drained by V5 and V8, and finally, the whole graft fluoresced. With the information obtained by fluorescence imaging, we were able to transplant the liver with confirmation of sufficient blood perfusion in the entire graft.

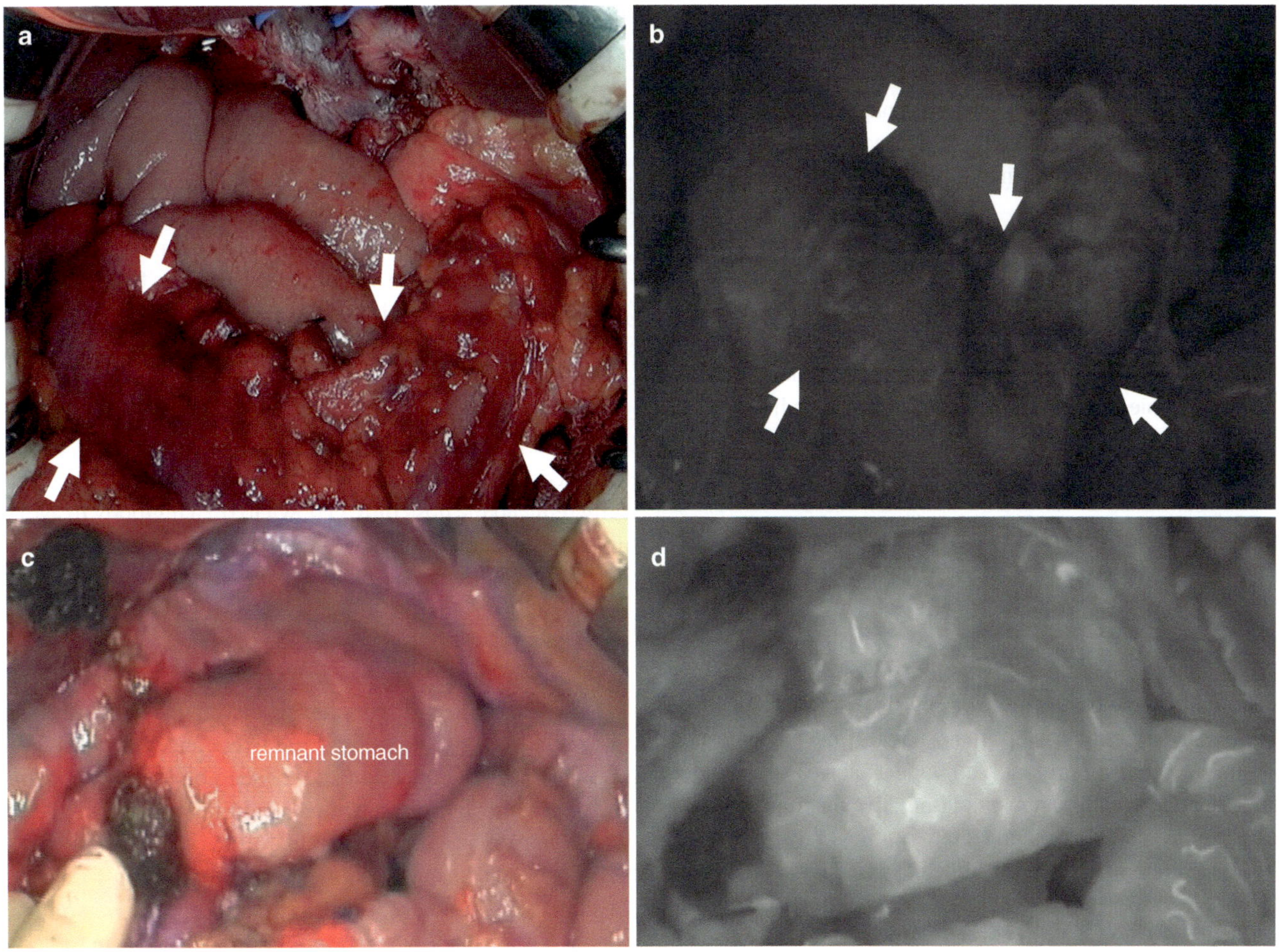

Fig. 13.4 Visualization of gastrointestinal ischemic regions using ICG fluorescence imaging. (**a**) The color of the colon wall was not so ischemic (*arrow*). (**b**) Significant decrease in fluorescence signals was detected in a part of the transverse colon (*arrow*). (**c**) The color of remnant stomach was good. (**d**) The fluorescence imaging was visualized sufficient blood perfusion in the remnant stomach

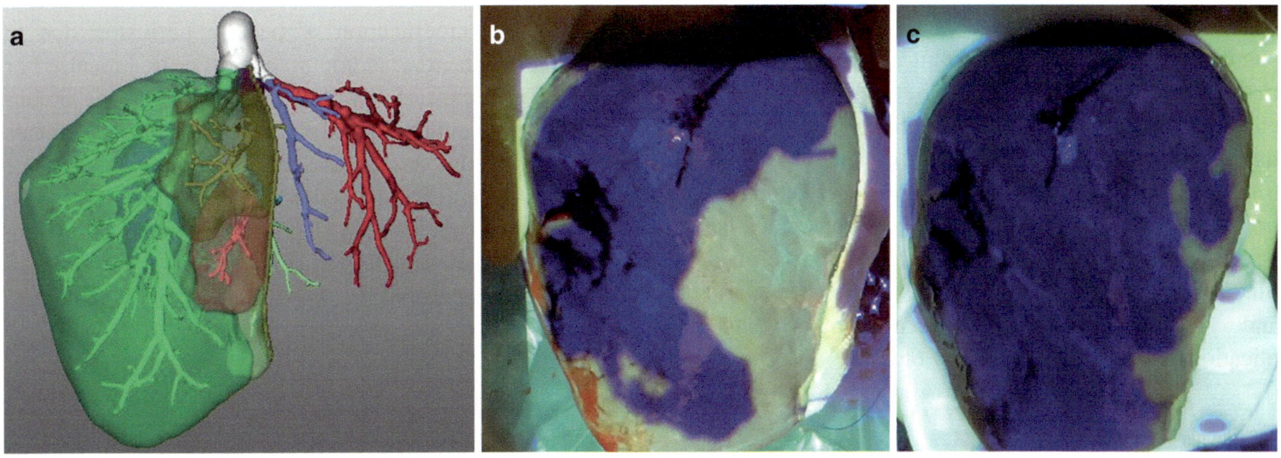

Fig. 13.5 Visualization of the graft perfusion status using ICG fluorescence imaging. (**a**) Preoperative simulation. (**b**) After V5 and V8 are clamped. (**c**) After V5 and V8 clamps are released

Point

- The Medical Imaging Projection System enables real-time navigation throughout hepatectomy procedures.
- Indocyanine green fluorescence imaging enables visualization of the hepatic regions perfused by the cystic artery, which may provide a new approach in the treatment of gallbladder cancer.
- In pancreatic cancer surgery, perfusion assessment by ICG fluorescence imaging can be used to avoid unnecessary resections of the adjacent gastrointestinal tracts.
- Indocyanine green fluorescence imaging visualizes the perfusion status of the graft before transplantation and allows objective assessment of the need for hepatic vein reconstructions.

5 Expected Roles of Fluorescence Imaging in HBP Surgery and Liver Transplantation

Fluorescence imaging is characterized by minimally invasive, real-time visualization of invisible or difficult-to-see biological structures. This makes real-time navigation of anatomic hepatectomy and gallbladder cancer surgery possible, leading to improvement of operative outcomes in terms of accuracy, curability, and operation time. In addition, since ICG is excreted into bile, the bile duct anatomy in the hepatic hilum can also be visualized in a certain period of time after intravenous administration of ICG. If ICG fluorescence cholangiography can be an alternative to conventional radiographic cholecystectomy, it will not only shorten the operation time but also avoid radiation exposure of patients and medical staff. In the highly invasive surgery for pancreatic cancer, the ability to evaluate gastrointestinal blood perfusion intraoperatively and to determine the necessity of combined resection of the gastrointestinal tracts is expected to contribute to shortening the operation time, reducing surgical complications, and improving postoperative QOL. ICG fluorescence imaging may also be used in living donor liver transplantation for easy and non-invasive detection of poor perfusion of the liver graft due to congestion of major hepatic veins, which would contribute to shortening the operation time by omitting unnecessary hepatic vein reconstruction and to improving the operative outcomes by avoiding complications associated with outflow block.

6 Precautions and Challenges in Perfusion Assessment by Indocyanine Green Fluorescence Imaging

In perfusion assessment by fluorescence imaging, the optimal route and dose of ICG administration have not yet been established completely. We initially standardized the dose of ICG at 2.5 mg/body for intravenous injection. However, in a case in which the hepatic artery and portal vein were divided individually through hilar dissection, fluorescence signals in hepatic segments to be removed gradually increased probably because of the arterial inflow via the connective tissues in the hepatic hilum bridging the hepatic segments to be removed and the remnant liver. Since then, we have successfully reduced the dose of ICG down to 0.25 mg/body to avoid unexpected enhancement of fluorescence signals. Since the dose of ICG should be further decreased in the case of transportal and transarterial administration, we need to establish the optimal dose of ICG according to applications and routes of administration.

At present, it is impossible to distinguish between the fluorescence signals emitted from liver cancer as a result of preoperative ICG injection and those to be used for hepatic segmentation by intraoperative administration of ICG, which may mislead surgeons to follow accurate intersegmental planes during anatomic hepatectomy (Fig. 13.6a, b). In such cases, it is essential to refer to preoperative simulation and intraoperative ultrasonography in addition to fluorescence imaging. Future development of novel cancer-specific fluorophores and imaging devices may enable discrimination of fluorescence signals used for cancer localization and hepatic segmentation (Fig. 13.6c). Discrimination of fluorescence signals is also needed in patients with modification of hepatic perfusion due to tumor-related factors and/or preoperative treatments. For example, in a case of right hepatectomy following preoperative portal vein embolization, weak fluorescence signals appearing on the surface of the right paramedian sector could be distinguished from demarcation between the right and left hemi-liver by adjusting intensities of illumination lights and camera sensitivities (Fig. 13.7).

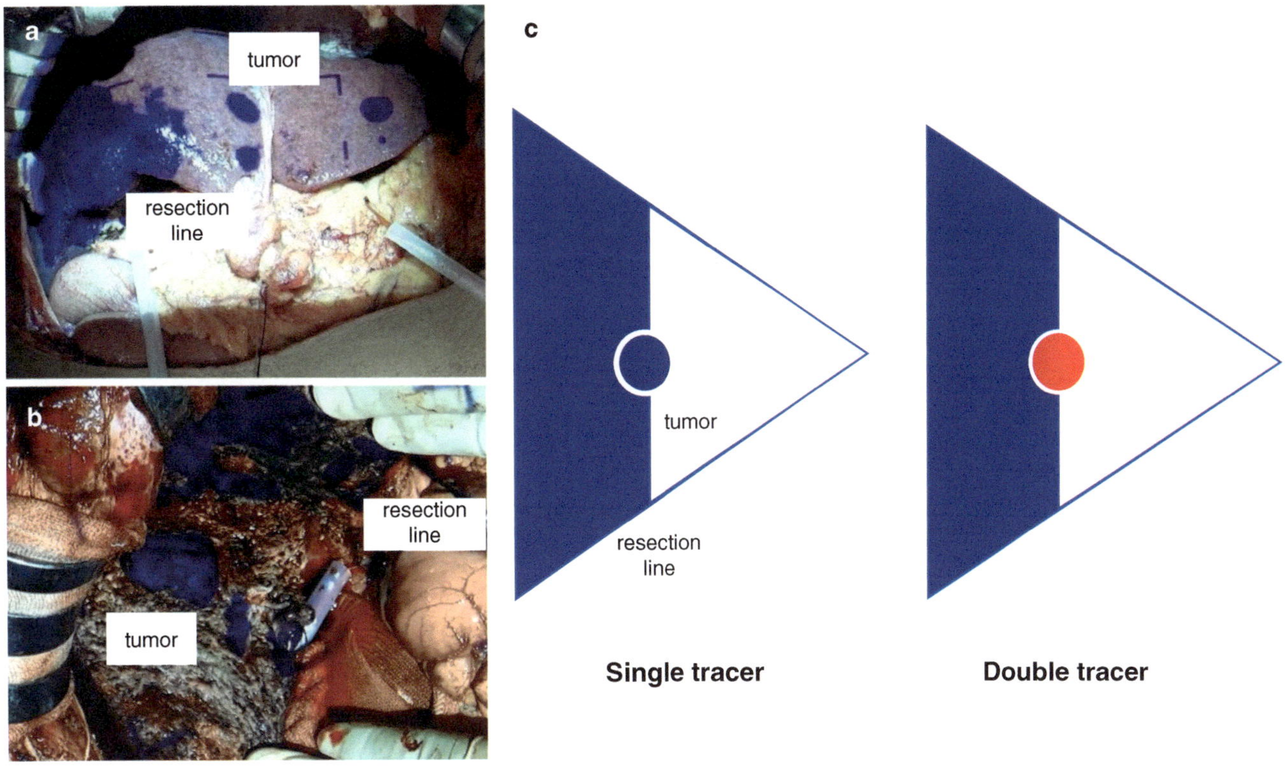

Fig. 13.6 Fluorescence imaging of liver cancer and hepatic segment. (**a** and **b**) It is impossible to distinguish between the fluorescence of cancer and those to be used for hepatic segmentation. (**c**) The development of novel cancer-specific fluorophores and imaging devices enable discrimination of fluorescence of cancer and hepatic segmentation

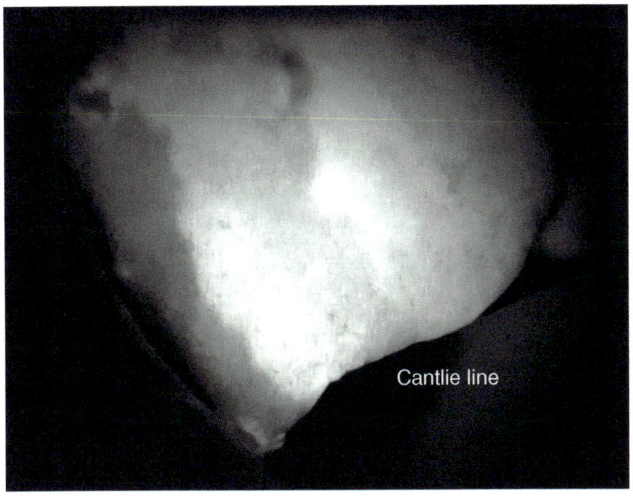

Fig. 13.7 Indocyanine green fluorescence imaging during right hepatectomy following preoperative portal vein embolization

7 Conclusion

With the advent of fluorescence imaging systems to be used in open, laparoscopic, and robot-assisted surgery, ICG fluorescence imaging would develop into a gold standard in HBP surgery and liver transplantation. We believe that, when the imaging techniques could be standardized through the accumulation of clinical experiences, real-time navigation of HBP surgery and liver transplantation would be realized.

References

1. Makuuchi M, Hasegawa H, Yamazaki S. Ultrasonically guided subsegmentectomy. Surg Gynecol Obstet. 1986;161:346–50.
2. Takasaki K. Glissonean pedicle transection method for hepatic resection: a new concept of liver segmentation. J Hepato-Biliary-Pancreat Surg. 1998;5:286–91.
3. Horiguchi A, Miyakawa S, Ishihara S, et al. Gallbladder bed resection or hepatectomy of segments 4a and 5 for pT2 gallbladder carcinoma: analysis of Japanese registration cases by the study group for biliary surgery of the Japanese Society of Hepato-Biliary-Pancreatic Surgery. J Hepatobiliary Pancreat Sci. 2013;20:518–24.
4. Aoki T, Yasuda D, Shimizu Y, et al. Image-guided liver mapping using fluorescence navigation system with indocyanine green for anatomical hepatic. World J Surg. 2008;32:1763–7.
5. Ishizawa T, Fukushima N, Shibahara J, et al. Real-time identification of liver cancers by using indocyanine green fluorescent imaging. Cancer. 2009;115:2491–504.
6. Ishizawa T, Bandai Y, Ijichi M, et al. Fluorescent cholangiography illuminating the biliary tree during laparoscopic cholecystectomy. Br J Surg. 2010;97:1369–77.
7. Oba A, Inoue Y, Sato T, et al. Impact of indocyanine green-fluorescence imaging on distal pancreatectomy with celiac axis resection combined with reconstruction of the left gastric artery. HPB (Oxford). 2019;21:619–25.
8. Nishino H, Hatano E, Seo S, et al. Real-time navigation for liver surgery using projection mapping with indocyanine green fluorescence: development of the novel Medical Imaging Projection System. Ann Surg. 2018;267:1134–40.

Column 2: Establishment and Activities of the International Society for Fluorescence Guided Surgery (ISFGS)

Takeaki Ishizawa

In April 2010, the author presented a video on "Laparoscopic Cholecystectomy using Fluorescence Cholangiography with intravenous injection of ICG" at the IHPBA (International Hepato-Biliary-Pancreatic Association) congress in Buenos Aires. One day, 1 year after I finished a kind of never-ending journey to Argentina from Japan that almost gave me a bedsore, I received an email from an Argentinean surgeon, named Dr. Fernando Dip. When I opened the mail with some suspicion, I found that he introduced himself as "the first person in Argentina to perform laparoscopic cholecystectomy using fluorescence imaging" and asked me why the fluorescence cholangiography that I had presented at the IHPBA had not spread in Japan since the original report in 1992 (Fig. 1). The fact was that the report by Araki et al. (Kumamoto National Hospital) [1] that he pointed out was not a fluorescence imaging technique, but a technique using the "coloration" of the bile ducts observed after intravenous injection of a large dose of ICG (50 mg).

In 2013, I receive an abrupt email from him again. He then moved to the Cleveland Clinic Florida in the USA and asked me if I would like to join him in forming an international group to expand fluorescence imaging to other areas. The content of the article is very ambitious (Fig. 1). I was puzzled by such a sudden offer, but wanted to see him once during the SAGES annual meeting in Baltimore. Although the weather at the Chicago airport used for transit was very rough and almost all flights were delayed or canceled, forcing me to overnight at the airport, I finally met him during the congress and successfully settled on the purpose of the study group and the activity policies of (1) making a homepage, (2) listing core members, (3) editing textbooks, and (4) holding scientific meetings (Fig. 2).

The author's homework was to select speakers and make a program for our first annual meeting. Since I have few connections outside of hepatobiliary and pancreatic surgery, I searched the PubMed base and sent e-mails to all the authors of major papers outside of my specialty. I imagine that a sudden invitation e-mail from an unknown Japanese HBP surgeon was a kind of fishing e-mail for most surgeons and researchers. However, the response was more than expected, probably because many fluorescence imaging researchers felt the need for such a research organization, and eventually we were able to hold the first annual meeting in February 2014 with about 40 valuable abstracts from many fields. Since then, we have planned and held seven annual meetings including my term as the President (2018–2019). I would like to take this opportunity to express my sincere gratitude to the many surgeons and researchers from Japan and abroad who have participated in these meetings. Especially, without the participation of Japanese surgeons and researchers, we would not have been able to hold even the first meeting (Fig. 3).

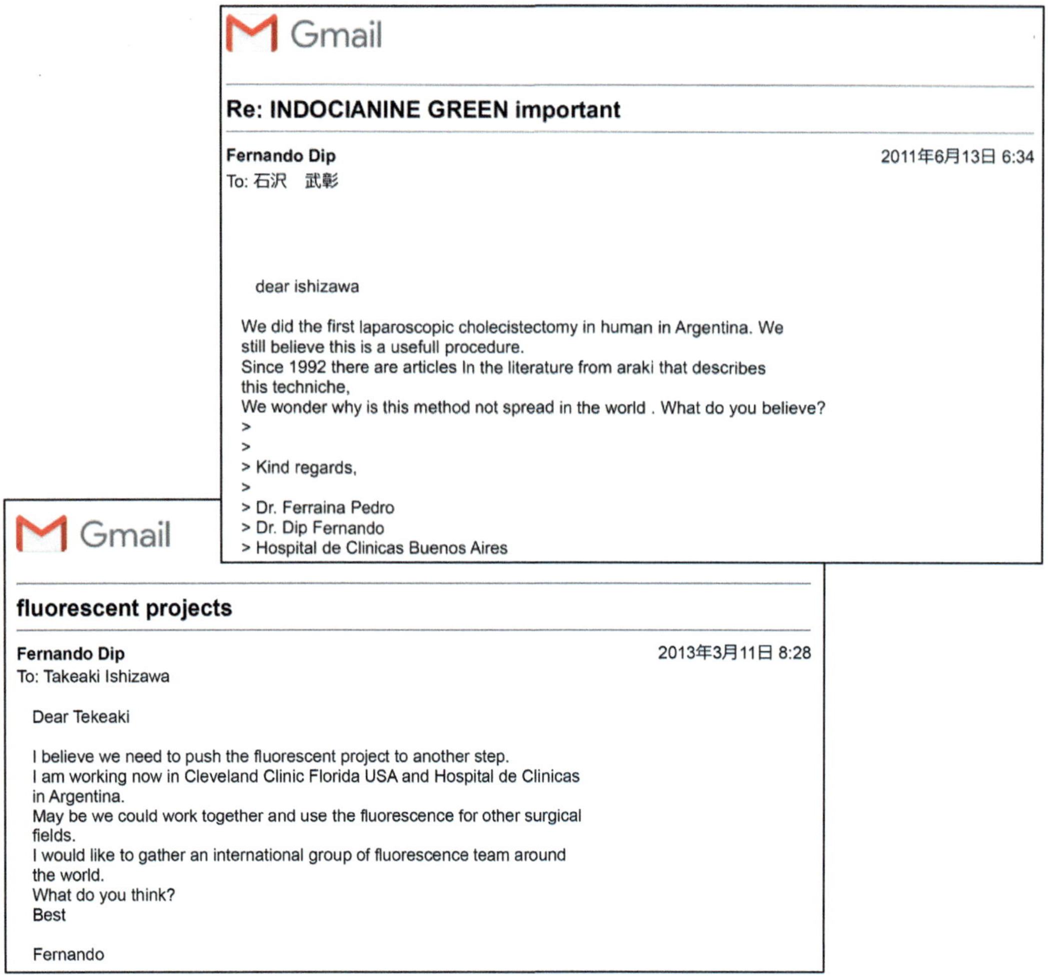

Gmail

Re: INDOCIANINE GREEN important

Fernando Dip 2011年6月13日 6:34
To: 石沢　武彰

dear ishizawa

We did the first laparoscopic cholecistectomy in human in Argentina. We
still believe this is a usefull procedure.
Since 1992 there are articles In the literature from araki that describes
this techniche,
We wonder why is this method not spread in the world . What do you believe?
>
>
> Kind regards,
>
> Dr. Ferraina Pedro
> Dr. Dip Fernando
> Hospital de Clinicas Buenos Aires

Gmail

fluorescent projects

Fernando Dip 2013年3月11日 8:28
To: Takeaki Ishizawa

Dear Tekeaki

I believe we need to push the fluorescent project to another step.
I am working now in Cleveland Clinic Florida USA and Hospital de Clinicas
in Argentina.
May be we could work together and use the fluorescence for other surgical
fields.
I would like to gather an international group of fluorescence team around
the world.
What do you think?
Best

Fernando

Fig. 1 Initial email from Dr. Dip (above) and subsequent somewhat ambitious email (below)

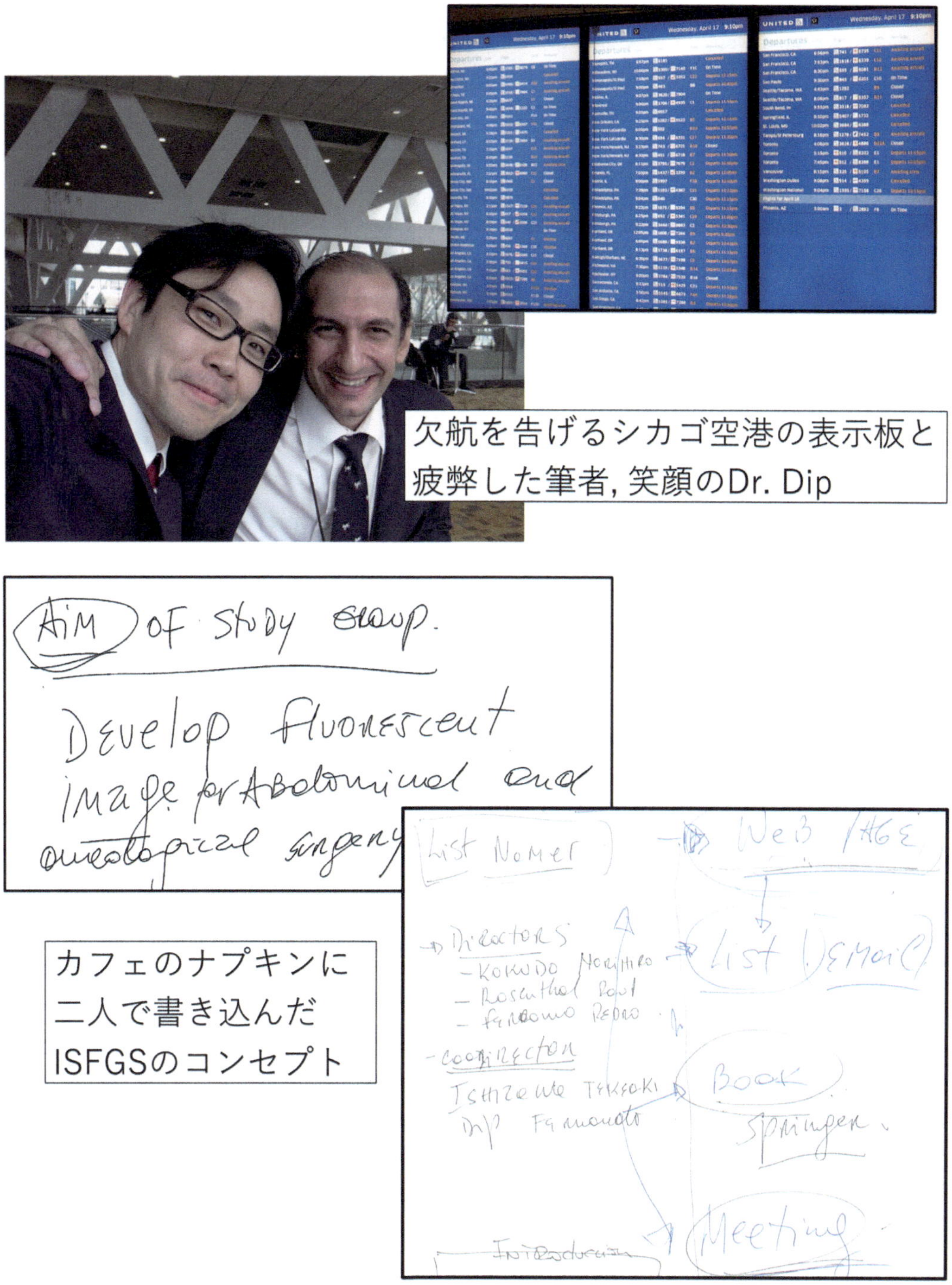

Fig. 2 Kickoff meeting with Dr. Dip (Baltimore)
A lot of flights were canceled due to storming. The author seems to be tired, with smiling Ferando!
Concepts of ISFGS, written on a napkin of the cafeteria

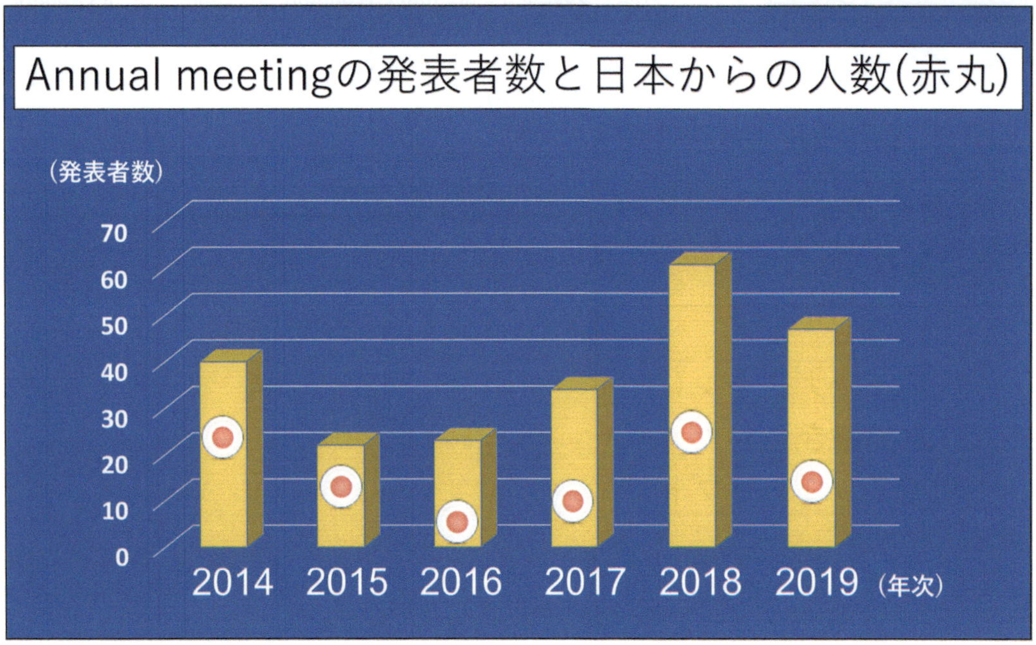

Fig. 3 The first ISFGS annual meeting (top) and the number of presenters (bottom)
Dr. Norihiro Kokudo, Dr. Fernando Dip, I (Ishizawa), and Dr. Raul Rosenthal
The number of presenters at the annual meetings and Japanese surgeons/researchers (red circle)

The textbook, which was another homework with Dr. Dip, has already been published in 2015 [2]. We are also working on the formation of a fluorescence imaging consensus using the Delphi method and the evaluation of the development stage using the IDEAL framework [3] in each field, which will be published in due course. Various webinars are also available. Please refer to the website of the society (https://www.isfgs.org/).

References

1. Araki K, Namikawa K, Mizutani J, et al. Indocyanine green staining for visualization of the biliary system during laparoscopic cholecystectomy. Endoscopy. 1992;24:803.
2. Dip FD, Ishizawa T, Kokudo N, Rosenthal R. Fluorescence imaging for surgeons. Springer International Publishing Switzerland; 2015.
3. McCulloch P, Altman DG, Campbell WB, et al. No surgical innovation without evaluation: the IDEAL recommendations. Lancet. 2009;374:1105–12.

Part III

Intraoperative Fluorescence Imaging [Practice] – Imaging of Cancer

Takeaki Ishizawa

The most expected role of intraoperative fluorescence imaging is to visualize the exact localization of cancer tissues for improving long-term outcomes after resection. Despite recent advances in preoperative diagnosis, it is still difficult to identify tiny cancers and actual spread of cancer invasions on site during surgery. Real-time and accurate identification of cancerous tissues by intraoperative fluorescence imaging would enable surgeons to remove cancerous tissues completely and also to avoid non-curative extensive resection in a patient with advancing cancer beyond preoperative estimation. In fact, fluorescence-guided cancer surgery has already been realized in some surgical fields, although diagnostic sensitivity and specificity of fluorescence imaging should be improved for further development.

Liver Cancer (Primary Liver Cancer, Metastatic Liver Cancer)

14

Yoshiharu Kono, Takeaki Ishizawa, and Kiyoshi Hasegawa

Summary
In this chapter, among the main applications of ICG fluorescence imaging in the field of hepatobiliary surgery (1) fluorescence cholangiography, (2) identification of cancer, and (3) delineation of the hepatic region, we focus on "identification of cancer" and outline the history of its development and its actual application.

1 Introduction

The location of hepatocellular carcinoma and metastatic liver cancer can be determined by preoperative intravenous indocyanine green (ICG) and infrared observation can be identified as a fluorescent region.

In laparoscopic hepatectomy in particular, it is expected to be a substitute for palpation in identifying tumors under the liver capsule and confirming the surgical margin from a detached section.

The sensitivity is high, but the false positive rate is relatively high.

Supplementary Information The online version contains supplementary material available at https://doi.org/10.1007/978-981-19-7372-7_14.

Y. Kono · K. Hasegawa (✉)
Department of Hepatobiliary and Pancreatic Surgery, Artificial Organ and Transplantation Surgery, The University of Tokyo, Bunkyo-ku, Tokyo, Japan
e-mail: kihase-tky@umin.ac.jp

T. Ishizawa
Department of Hepatobiliary-Pancreatic Surgery, Graduate School of Medicine, Osaka Metropolitan University, Osaka, Japan

Japanese Society for Fluorescence Guided Surgery (JSFGS), Tokyo, Japan

2 Methods and Problems in the Identification of Liver Cancer During Surgery

In hepatectomy, the surgeon confirms the location of the liver tumor (primary liver cancer, metastatic liver cancer, etc.) by visual palpation and intraoperative ultrasonography. However, it is not always easy to identify a tumor "on the spot" during surgery that was depicted by CT or MRI before surgery because the liver is not transparent and its location changes constantly during the operation. In particular, in laparoscopic hepatectomy, which has recently become widespread, the surgeon cannot directly touch the surface of the liver, so it is sometimes difficult to recognize the location of a tumor that is easily palpable in open surgery.

In 2009, a surgeon in Japan reported a technique for confirming the location of a tumor by fluorescence imaging during laparotomy, using the property that ICG injected intravenously for liver function tests before surgery stays in and around the liver cancer. This technique attracted attention as a forerunner of navigation surgery, in which cancer is fluorescently labeled and resected. However, it was not widely used in clinical practice because liver function tests using ICG were not as common overseas as in Japan, and the available imaging equipment was limited at that time. Later, however, when laparoscopic instruments with near-infrared fluorescence observation became commercially available from various companies, many surgeons began to experience that liver cancer could be easily visualized by fluorescence imaging, and this technique is now spreading both in Japan and overseas.

T. Ishizawa (ed.), *Fluorescence-Guided Surgery*, https://doi.org/10.1007/978-981-19-7372-7_14

3 History of the Development and Clinical Application of Liver Cancer Identification Using Indocyanine Green Fluorescence Imaging

In 2008, Ishizawa et al. attempted to perform biliary angiography using ICG during a hepatectomy procedure and found that hepatocellular carcinoma was strongly fluorescent even before ICG was administered for biliary angiography. Although the reason for this was initially unclear, we realized that patients undergoing hepatectomy were given intravenous ICG for liver function tests before surgery, and we expected that the ICG administered at that time must have been trapped in the cancerous tissue. In fact, fluorescence imaging of the circumferential surface of the resected specimen revealed that ICG was distributed inside the cancerous tissue in highly differentiated hepatocellular carcinoma and in a ring-like distribution in the surrounding non-cancerous liver parenchyma in poorly differentiated hepatocellular carcinoma and liver metastasis from colorectal cancer (Fig. 14.1). In 2009, we demonstrated the applicability of this mechanism to intraoperative fluorescence imaging of hepatocellular carcinoma with fluorescence patterns in 63 nodules of hepatocellular carcinoma and 28 nodules of hepatic metastasis of colorectal carcinoma [1]. In the same year, Gotoh et al. [2] also reported that ICG fluorescence imaging was performed in 10 cases of hepatic resection for hepatocellular carcinoma and that tumors undetectable by intraoperative ultrasound were depicted in four cases.

It was published in 2012 [3].

The mechanism of preoperative intravenous (IV) ICG retention in and around hepatocellular carcinoma has been investigated by fluorescence microscopy and immunohistochemistry, and Ishizawa et al. [4] found that a transporter involved in ICG uptake (NTCP, Na+/taurocholate cotransporting polypeptide, and OATP8, organic aniontransporting polypeptide-8) expression is relatively preserved, while the bile excretion process is impaired, resulting in prolonged ICG retention in the cancer tissue. We found that ICG remains in cancer tissues for a long time. In this study, we hypothesized that ICG would not be taken up by cancer tissues because the expression of the above transporters is poor in poorly differentiated hepatocellular carcinoma and metastatic hepatocellular carcinoma and that ICG would show ring-shaped fluorescence reflecting bile stasis in the surrounding tissues. A research team from the Netherlands [5] concluded that immature hepatocytes with low bile excretion capacity exist around liver metastases of colorectal cancer and that this is the cause of the ring-shaped fluorescence. Such differences in fluorescence patterns may reflect differences in the biological properties of cancer tissues. For example, Shibasaki et al. [6] found that the expression of a transporter (MDR3; multidrug resistance p-glycoprotein-3) involved not only in ICG uptake but also in ICG excretion was maintained in hepatocellular carcinoma, a type of cancer in which ICG is retained in cancer tissue, and that MDR3-negative hepatocellular carcinoma showed long-term growth after resection. The company reports poor performance.

However, in Japan, ICG is often administered intravenously for preoperative liver function tests, and it is difficult to optimize the dose and timing for fluorescence imaging. However, in Japan, ICG is often administered intravenously for preoperative liver function tests, making it difficult to optimize the dose and timing for fluorescence imaging [1, 4]. If surgery is performed more than 7–10 days after the ICG test, additional administration of 0.2 mg/kg [7] of ICG 24–48 hours before surgery or 2.5 mg/body [8] the day before surgery may improve the identification rate of liver cancer. It has also been reported that additional administration of 0.2 mg/kg [7] or 2.5 mg/body [8] of ICG 24–48 hours before surgery may improve the identification rate of liver cancer.

The application of laparoscopic hepatectomy was reported for the first time from Japan in 2010 [9] and for the resection of 32 nodules of liver tumors in 2014 [10], and recently there have been increasing opportunities for presentations from overseas institutions. In laparoscopic hepatectomy, not only can the tumor be visualized from the liver surface before hepatic resection, but it can also be used to confirm the resection margin during hepatic dissection (resection so that the fluorescence around the cancer does not remain in the liver on the sparing side) [11, 12], which is thought to be one of the reasons why the clinical application is expanding.

Point

The diagnosis of liver cancer using ICG fluorescence imaging was developed in Japan and has been applied to laparoscopic surgery.

It is spreading both domestically and internationally.

Indocyanine green administered intravenously for preoperative liver function tests can be used to visualize the tumor.

Therefore, it is very easy to use.

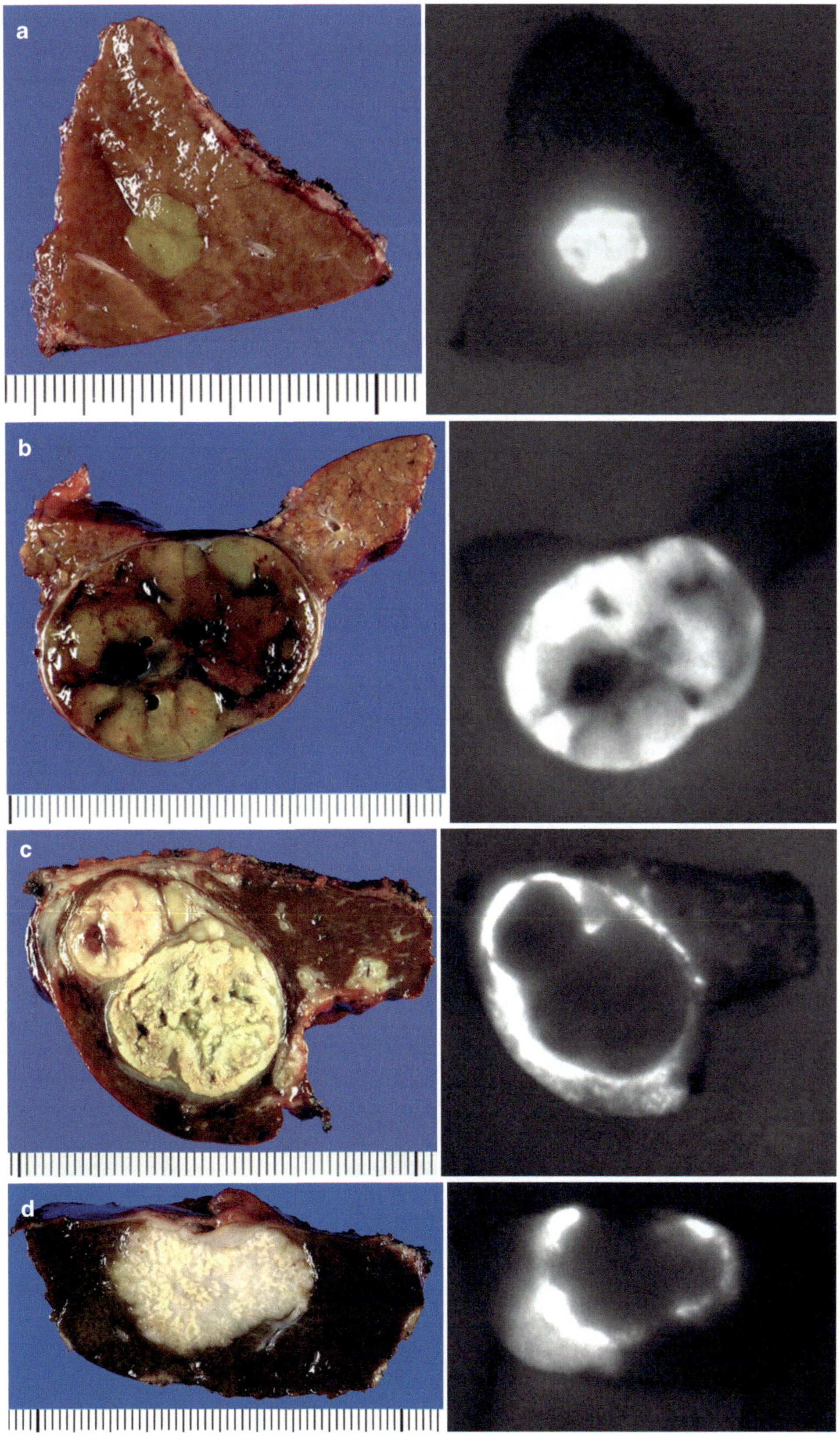

Fig. 14.1 Fluorescence pattern of hepatocellular carcinoma sections. (**a**) Highly differentiated hepatocellular carcinoma (a pattern of fluorescence of the entire tumor). (**b**) Moderately differentiated hepatocellular carcinoma (a pattern of heterogeneous fluorescence observed within the tumor). (**c**) Poorly differentiated hepatocellular carcinoma (a pattern of ring-shaped fluorescence observed in the surrounding non-cancerous liver parenchyma rather than inside the tumor). (**d**) Colorectal cancer liver metastasis (ring-shaped fluorescence pattern observed in the surrounding non-cancerous liver parenchyma, not inside the tumor)

4 Practical Methods for Identifying Liver Cancer Using Indocyanine Green Fluorescence Imaging

We present two cases of laparoscopic liver resections performed at our institution using the ICG fluorescence imaging method for liver cancer identification.

Case 1: Enucleation for hepatocellular carcinoma (S2) (Fig. 14.2 and Movie 14.1)

- Indocyanine green (0.5 mg/kg) was administered 6 days before surgery. Fifteen-minute ICG stagnation rate was normal (9.8%), but portal hypertension was prominent. The tumor had a simple nodular morphology with a firm coating. Therefore, in order to preserve liver function to the maximum extent, laparoscopic tumor enucleation was applied.
- PINPOINT (Stryker) was used. The operation time was 2 hours and 12 minutes, and the blood loss was 20 mL.

1. Fluorescence imaging clearly shows hepatocellular carcinoma (arrow) protruding from the lateral surface of liver S2. Note that non-cancerous lesions such as regenerative nodules are also fluorescent (arrowheads).
2. Initially, the hepatic detachment is started with a small amount of liver parenchyma attached to the tumor.

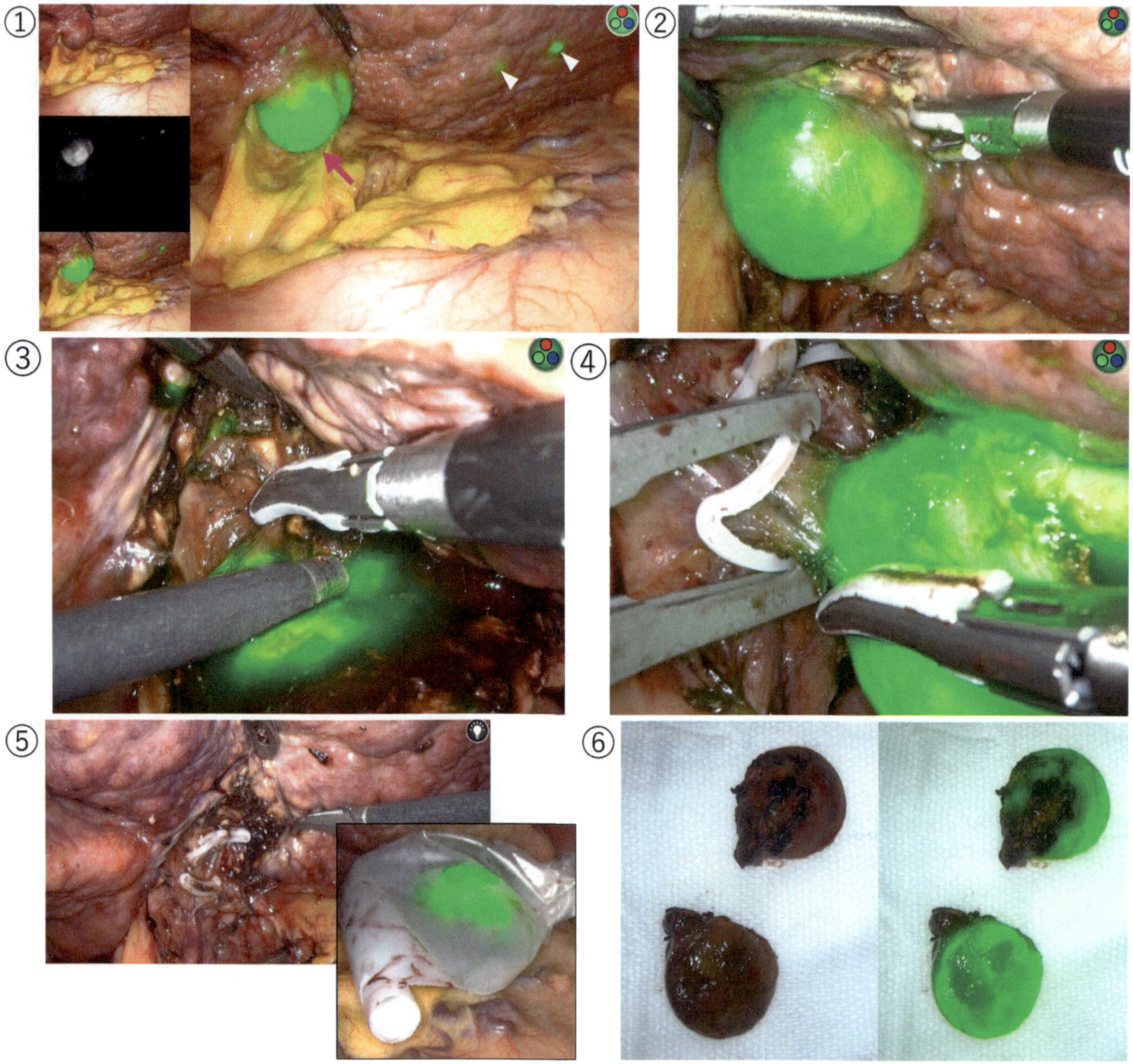

Fig. 14.2 Enucleation for hepatocellular carcinoma (S2)

3. In the vicinity of the Glisson's sheath, the liver parenchyma is gently sprained with forceps while recognizing the tumor border with fluorescence imaging, and the remaining small vessels are dissected with a vascular sealing system.
4. Large vessels are dissected after clipping.
5. Controlled section after enucleation. Fluorescence imaging (lower right) still shows the presence of the tumor contained in the retrieval bag.
6. Split section of the resection specimen. The entire tumor shows fluorescence of preoperatively administered ICG. Histopathologically, it was diagnosed as highly to moderately differentiated hepatocellular carcinoma.

Case 2: Partial hepatectomy for intrahepatic cholangiocarcinoma (S7) (Fig. 14.3 and Movie 14.2)

- Indocyanine green (0.5 mg/kg) was administered 2 days before surgery, and the ICG 15-minute stagnation rate was normal (7.3%).
- AIM 1688 (Stryker) was used. The operation time was 3 hours and 11 minutes, and the blood loss was 100 mL.

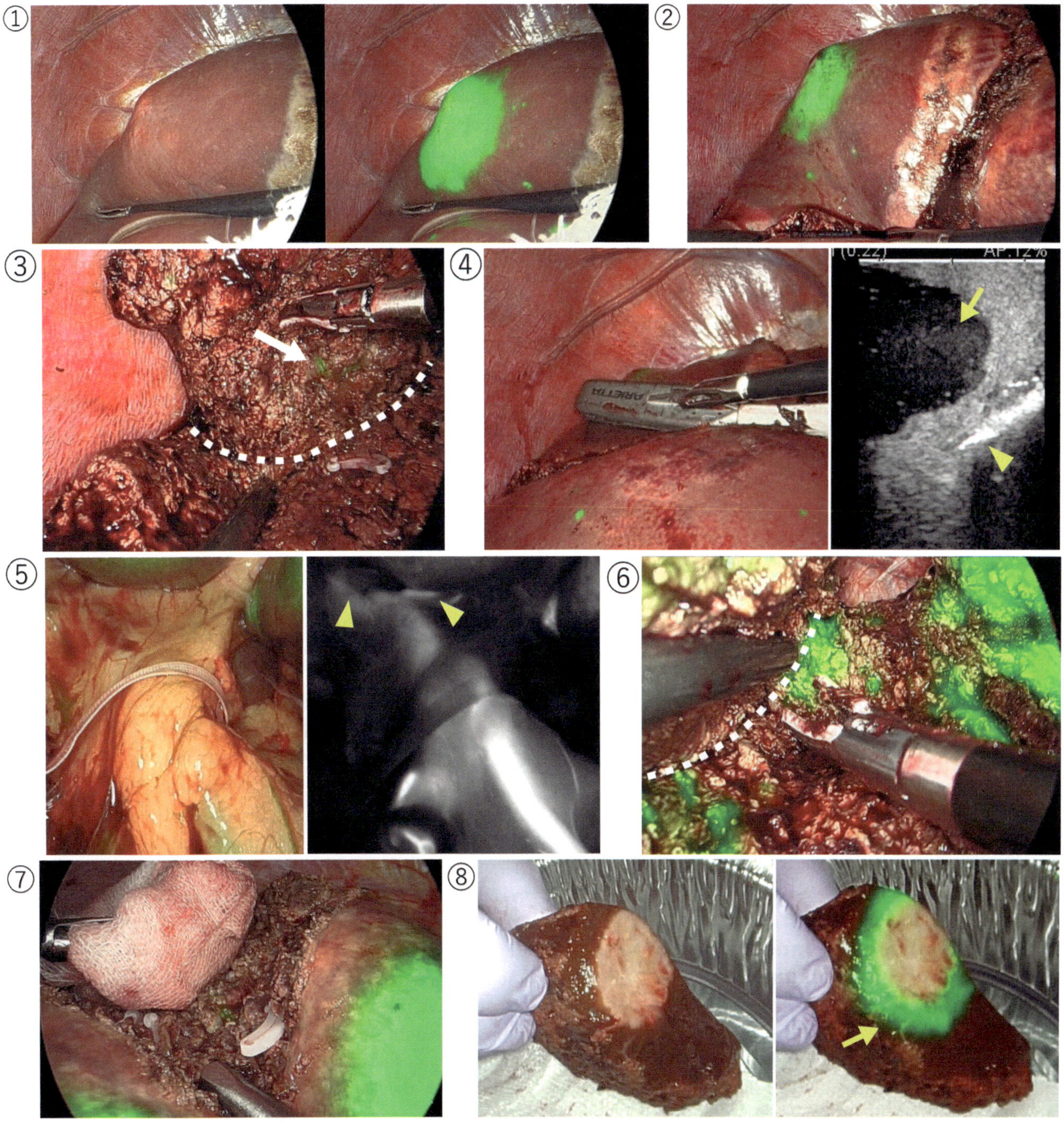

Fig. 14.3 Partial hepatectomy for intrahepatic bile duct cancer (S7)

1. Fluorescence imaging shows a tumor slightly protruding from the diaphragmatic surface of liver S7. Because of the 4 K image quality, the fine structures of the diaphragm and liver surface are finely depicted.
2. Use fluorescence imaging and intraoperative ultrasound to set the hepatic resection line.
3. In principle, it is desirable to perform the liver dissection so that the fluorescent area emitted from the periphery of the tumor is not visible in order to secure the resection margin, but fluorescence may be observed from the area where the liver parenchyma is thinned (arrow). However, fluorescence may be observed from areas of thinner liver parenchyma (arrows). In such cases, the line of liver dissection should be modified slightly deeper to avoid exposure of the tumor (dashed line).
4. Intraoperative ultrasound is also used to confirm the positional relationship between the tumor (arrow) and the detached section (arrowhead) at the depth of the detachment.
5. In this case, ICG 2.5 mg was additionally administered intravenously intraoperatively to confirm the blood flow in the surrounding liver parenchyma after the responsible vessel of the tumor was dissected. In the arterial phase, the right and left hepatic arteries running through the hilar region of the liver were depicted (arrowheads).
6. Using this method, it is also possible to proceed with liver dissection by recognizing the boundary (dashed line) between the ischemic area to be resected and the surrounding liver parenchyma where blood flow is maintained.
7. Using ICG, which is excreted as bile after intravenous injection, we check for bile leak after hepatectomy using fluorescence imaging (in this case, there was no adherence of the fluorescent area to the gauze, and we can conclude that there was no bile leak).
8. Section of the resection specimen. In this case, the fluorescence area of the non-cancerous liver parenchyma around the tumor was thicker (about 5 mm) than that of the liver metastasis of colorectal cancer, probably because the pathological diagnosis was intrahepatic cholangiocarcinoma. The use of fluorescence imaging in the resection of liver metastases from colorectal cancer is basically the same as in this case.

5 Effectiveness of Indocyanine Green Fluorescence Imaging for Liver Cancer Identification

Since the fluorescence of ICG penetrates biological tissues as small as 8 mm, the sensitivity of fluorescence imaging for tumors close to the liver surface is good (around 90%). As mentioned earlier, it is expected to be a good substitute for palpation, especially in laparoscopic hepatectomy, and to be useful for quickly setting the hepatic resection line and confirming the resection margin during liver dissection. In addition, fluorescence imaging may be effective in cases where there are strong adhesions on the hepatic surface due to previous surgery, or for tumors that are difficult to identify by conventional intraoperative diagnostic methods (palpation and ultrasonography), such as hepatocellular carcinoma of indistinct type or liver metastases from colorectal cancer that have responded well to preoperative chemotherapy. For example, Terasawa et al. [11] reported that 3 nodules (6%) of 53 malignant tumors resected by laparoscopic hepatectomy could not be identified without ICG fluorescence imaging.

In hepatocellular carcinoma with Gleason infiltration, not only the location of the tumor but also the area of bile stasis associated with the tumor infiltration is depicted on the liver surface as a fluorescent image, which may be useful in determining the resection line [13]. According to Satou et al. [14], ICG fluorescence imaging of distant metastases (lung metastases, lymph node metastases, peritoneal dissemination, adrenal metastases, etc.) of hepatocellular carcinoma has a sensitivity of 92% for distant metastasis of hepatocellular carcinoma (lung metastasis, lymph node metastasis, peritoneal dissemination, adrenal metastasis, etc.). Similarly, ICG fluorescence imaging may be useful for achieving curative resection in hepatic resection of hepatoblastoma and extrahepatic metastases that maintain bile production capacity [15, 16].

Point

This method is highly sensitive for depicting liver tumors directly under the liver capsule.

Observation from the liver detachment section is also useful.

On the other hand, it should be noted that fluorescence may occur even if the lesion is not cancerous.

6 Cautions and Issues in Liver Cancer Identification Methods Using Indocyanine Green Fluorescence Imaging

The first caveat of this technique is what to do when a fluorescent area without preoperative diagnosis is visualized. This technique visualizes "bile stasis inside or around liver cancer" and is not truly cancer cell-specific imaging. In fact, the false-positive rate of this method for new lesions without preoperative diagnosis is about 40% [1, 3], and it has been reported that the smaller the diameter of the fluorescent area and the lower the fluorescence intensity, the higher the possibility that the lesion is non-cancerous [3]. When a new fluorescent area is visualized during surgery, palpation, ultrasonography, and preoperative imaging should be thoroughly reviewed, and additional resection should be considered if malignancy is suggested by methods other than fluorescence imaging. There is a technical limit to the depth of observation (up to approximately 8 mm), and intraoperative ultrasonography should always be performed at least once to confirm the depth of information, even if the tumor is depicted by fluorescence imaging.

Regarding insurance indications, in the 2020 revision, the scope of "K939-2 Additional fee for intraoperative imaging of blood vessels, etc.: Calculated when blood vessels and tumors are confirmed by fluorescence measurement, etc. using indocyanine green or aminolevulinic acid hydrochloride..." was expanded to include liver resection including laparoscopic surgery. It is thought that there is no problem in the act itself of observing liver cancer using fluorescence imaging. However, it should be noted that ICG (diagno green), which is a reagent, is not yet indicated for intraoperative diagnosis of liver cancer.

7 Future Prospects

In recent years, the sensitivity and image quality of near-infrared observation systems have been remarkably improved, and the contrast between liver (red) and tumor (often displayed in pseudo-color green) in superimposed imaging is excellent. It is expected that the use of ICG fluorescence imaging for liver cancer identification will increase in the future.

Kaibori et al. [17] showed that the combination of 5-aminolevulinic acid (5-ALA), which has already been clinically applied to detect brain tumors and bladder cancer, improved the specificity of ICG fluorescence imaging compared to ICG fluorescence imaging alone for liver cancer imaging. Our research group developed a novel fluorescent probe (gGlu-HMRG) that is hydrolyzed by γ-glutamyltranspeptidase (GGT), which is overexpressed in cancer tissues, to rapidly produce fluorescence in the visible region. We reported that ICG fluorescence imaging can label adenocarcinoma tissue that does not itself fluoresce, and that the stronger the fluorescence (i.e., the degree of GGT expression) of colorectal cancer liver metastases, the higher the risk of recurrence after resection [18]. Overseas, for example, clinical trials of a technique to delineate primary and metastatic foci of colorectal cancer using an anti-CEA monoclonal antibody (SGM-101) are underway [19]. If such a new technology is introduced in the future, it will not only improve the accuracy of intraoperative diagnosis of hepatocellular carcinoma and contribute to the achievement of R0 resection, but it will also be useful for predicting the biological behavior of tumors (risk of recurrence and resistance to anticancer drugs).

References

1. Ishizawa T, Fukushima N, Shibahara J, et al. Real-time identification of liver cancers by using indocyanine green fluorescent imaging. Cancer. 2009;115:2491–504.
2. Gotoh K, Yamada T, Ishikawa O, et al. A novel image-guided surgery of hepatocellular carcinoma by indocyanine green fluorescence imaging navigation. J Surg Oncol. 2009;100:75–9.
3. Yokoyama N, Otani T, Hashidate H, et al. Real-time detection of hepatic micrometastases from pancreatic cancer by intraoperative fluorescence. Cancer. 2012;118:2813–9.
4. Ishizawa T, Masuda K, Urano Y, et al. Mechanistic background and clinical applications of indocyanine green fluorescence imaging of hepatocellular carcinoma. Ann Surg Oncol. 2014;21:440–8.
5. van der Vorst JR, Schaafsma BE, Hutteman M, et al. Near-infrared fluorescence-guided resection of colorectal liver metastases. Cancer. 2013;119:3411–8.
6. Shibasaki Y, Sakaguchi T, Hiraide T, et al. Expression of indocyanine green-related transporters in hepatocellular carcinoma. J Surg Res. 2015;193:567–76.
7. Alfano MS, Molfino S, Benedicenti S, et al. Intraoperative ICG-based imaging of liver neoplasms: a simple yet powerful tool: preliminary results. Surg Endosc. 2019;33:126–34.
8. Kobahashi K, Kawaguchi Y, Kobayashi Y, et al. Identification of liver lesions using fluorescence imaging: comparison of methods for administering indocyanine green. HPB. 2020; (in press)
9. Ishizawa T, Bandai Y, Harada N, et al. Indocyanine green-fluorescent imaging of hepatocellular carcinoma during laparoscopic hepatectomy: an initial experience. Asian J Endosc Surg. 2010;3:42–5.
10. Kudo H, Ishizawa T, Tani K, et al. Visualization of subcapsular hepatic malignancy by indocyanine-green fluorescence imaging during laparoscopic. Surg Endosc. 2014;28:2504–8.
11. Terasawa M, Ishizawa T, Saiura A, et al. Applications of fusion fluorescence imaging using indocyanine green in laparoscopic hepatectomy. Surg Endosc. 2017;31:5111–8.
12. Aoki T, Murakami M, Koizumi T, et al. Determination of the surgical margin in laparoscopic liver resections using infrared indocyanine green fluorescence. Langenbecks Arch Surg. 2018;403:671–80.
13. Harada N, Ishizawa T, Muraoka A, et al. Fluorescence navigation hepatectomy by visualization of localized cholestasis from bile duct tumor. J Am Coll Surg. 2010;210:e2–6.
14. Satou S, Ishizawa T, Masuda K, et al. Indocyanine green fluorescent imaging for detecting extrahepatic metastasis of hepatocellular carcinoma. Gastroenterol. 2013;48:1136–43.
15. Yamamichi T, Oue T, Yonekura T, et al. Clinical application of indocyanine green (ICG) fluorescent imaging of hepatoblastoma. J Pediatr Surg. 2015;50:833–6.
16. Kitagawa N, Shinkai M, Mochizuki K, et al. Navigation using indocyanine green fluorescence imaging for hepatoblastoma pulmonary metastases surgery. Pediatr Surg Int. 2015;31:407–11.
17. Kaibori M, Matsui K, Ishizaki M, et al. Intraoperative detection of superficial liver tumors by fluorescence imaging using indocyanine green and 5- aminolevulinic acid. Anticancer Res. 2016;36:1841–9.
18. Miyata Y, Ishizawa T, Kamiya M, et al. Intraoperative imaging of hepatic cancers using γ-glutamyltranspeptidase-specific fluorophore enabling real-time identification and estimation of recurrence. Sci Rep. 2017;7:3542.
19. Boogerd LSF, Hoogstins CES, Schaap DP, et al. Safety and effectiveness of SGM101, a fluorescent antibody targeting carcinoembryonic antigen, for intraoperative detection of colorectal cancer: a dose-escalation pilot study. Lancet Gastroenterol Hepatol. 2018;3:181–91.

Lung Cancer (Marking the Tumor Site) 15

Toyofumi Fengshi Chen-Yoshikawa

Summary

- Although direct fluorescence labeling of nodular lesions including lung cancer has not yet been established, VAL-MAP (Virtual-Assisted Lung MAPping), in which dye is injected transbronchially, is widely used.
- The use of indocyanine green (ICG) and radiographic contrast material in VAL-MAP (ICG-VAL-MAP) improves the preoperative and intraoperative visibility of the marking site.

1 Introduction

Current advances in diagnostic imaging such as CT have led to an increase in the number of cases in which micro lesions are found in the peripheral lung. These micro lesions should be removed surgically because it is often difficult to make an accurate diagnosis of malignancy and benign lesions by CT-guided biopsy or bronchoscopic biopsy. However, even in surgery, these tiny lesions are difficult to identify by visual inspections or palpations by surgeons. Therefore, when surgical resection of these tiny lesions is needed from the oncological perspective, invasive surgical procedures such as lung lobectomy or segmentectomy may be indicated to secure surgical margins even for small lesions.

For the intraoperative identification of the tiny lesions, preoperative marking, such as CT-guided placement of a hook wire or coil or injection of a dye/lipiodol, can be performed in some institutions with good cooperation with the radiology department. However, the conventional percutaneous marking technique is associated with a risk of hemoptysis, pneumothorax, and more critical complications, such as myocardial and cerebral infarction due to air embolization [1, 2]. Furthermore, depending on the site of the lesion (e.g., the apex of the lung, the dorsal side behind the scapula, the mediastinum, and the lung near the diaphragm), marking itself may be technically difficult or impossible. Therefore, the development of safer preoperative marking methods is essential for dealing with the increasing number of peripheral microvascular lesions in the lung field. In this chapter, we detail the ICG-VAL-MAP method, which has been developed based on the VAL-MAP (Virtual-Assisted Lung MAPping) method (Fig. 15.1), a current standard of a transbronchial marking method.

Supplementary Information The online version contains supplementary material available at https://doi.org/10.1007/978-981-19-7372-7_15.

T. F. Chen-Yoshikawa (✉)
Department of Thoracic Surgery, Nagoya University Graduate School of Medicine, Nagoya, Aichi, Japan
e-mail: tyoshikawa@med.nagoya-u.ac.jp

T. Ishizawa (ed.), *Fluorescence-Guided Surgery*, https://doi.org/10.1007/978-981-19-7372-7_15

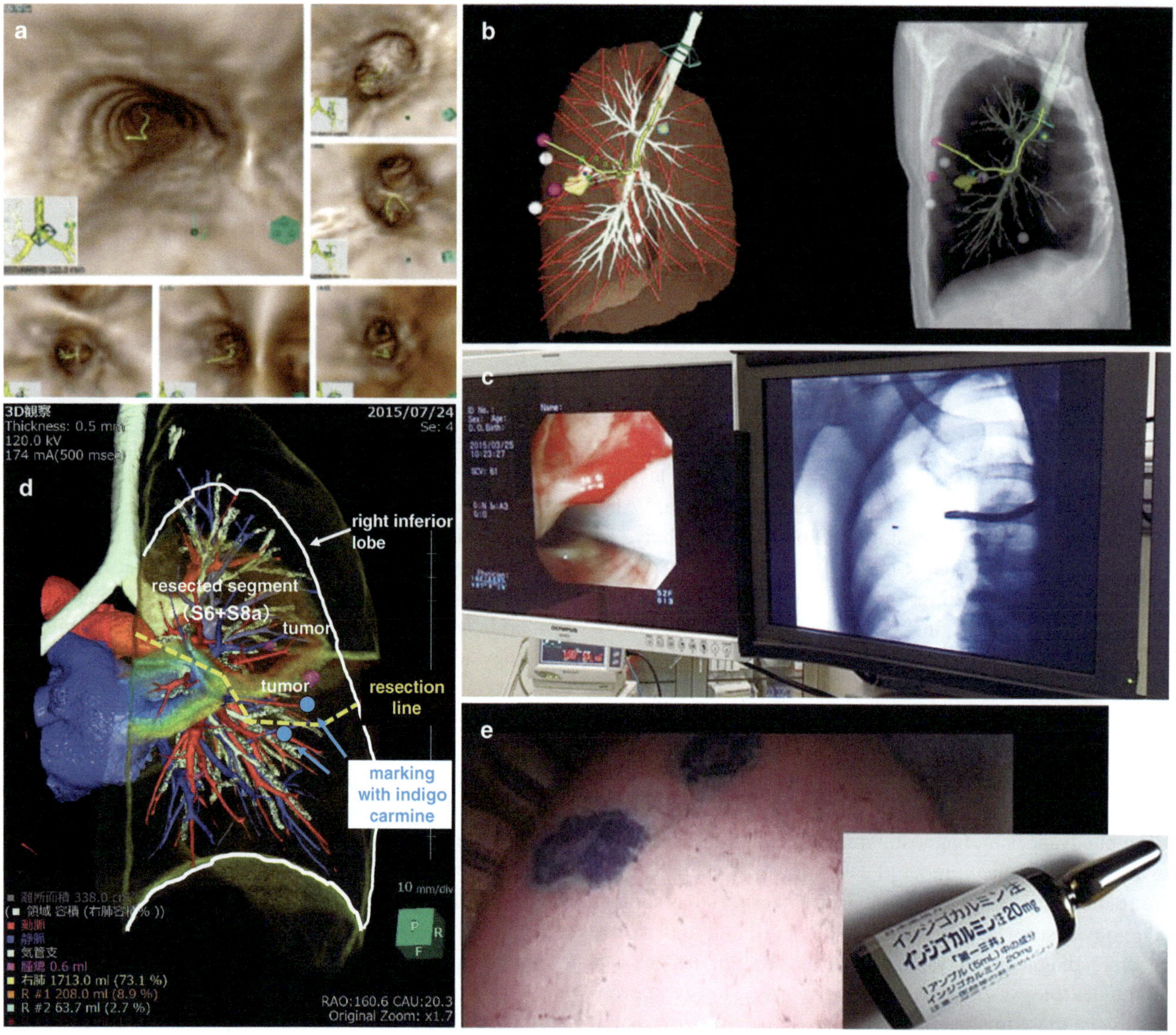

Fig. 15.1 Overview of Virtual Assisted Lung MAPping (VAL-MAP). **(a)** A virtual bronchoscope is created based on preoperative CT to determine the tumor-bearing bronchial branch. **(b)** 3D simulation images are created to understand spatial relationships between the nodule and the planned marking site. **(c)** Marking by injecting dye just below the pleura using a bronchoscope. **(d)** 3D-CT is taken to confirm the location of the nodule and the marking. **(e)** Perform surgery using the marking site as a guide

2 Conventional Techniques and Limitations

With the advancement of imaging techniques, the incidence of micronodular lung lesions has been increasing. In the surgical diagnosis and treatment of such micropulmonary lesions, a diagnostic partial lung resection is often attempted (Fig. 15.2). However, intraoperative diagnosis of "invisible" and/or "impalpable" nodules is extremely difficult. Even if the resection areas are extended to the level of a lung segment, it is still difficult to assure complete resection of such a tiny lesion by intraoperative examinations on the resected specimen. In fact, there have been many reports of cases of reoperation due to missed lesions. Furthermore, percutaneous marking using a hook wire is associated with a risk of complications such as myocardial infarction, cerebral infarction, hemorrhage, and pneumothorax due to air embolization, and its application is limited according to the location of the lesions [3].

As an alternative method, VAL-MAP (staining of the pleural surface with a dye [indigo carmine] using a bronchoscope guided by a virtual bronchoscope reconstructed from CT images) has been developed [4, 5]. Since multiple markings can be easily made with VAL-MAP, this technique can

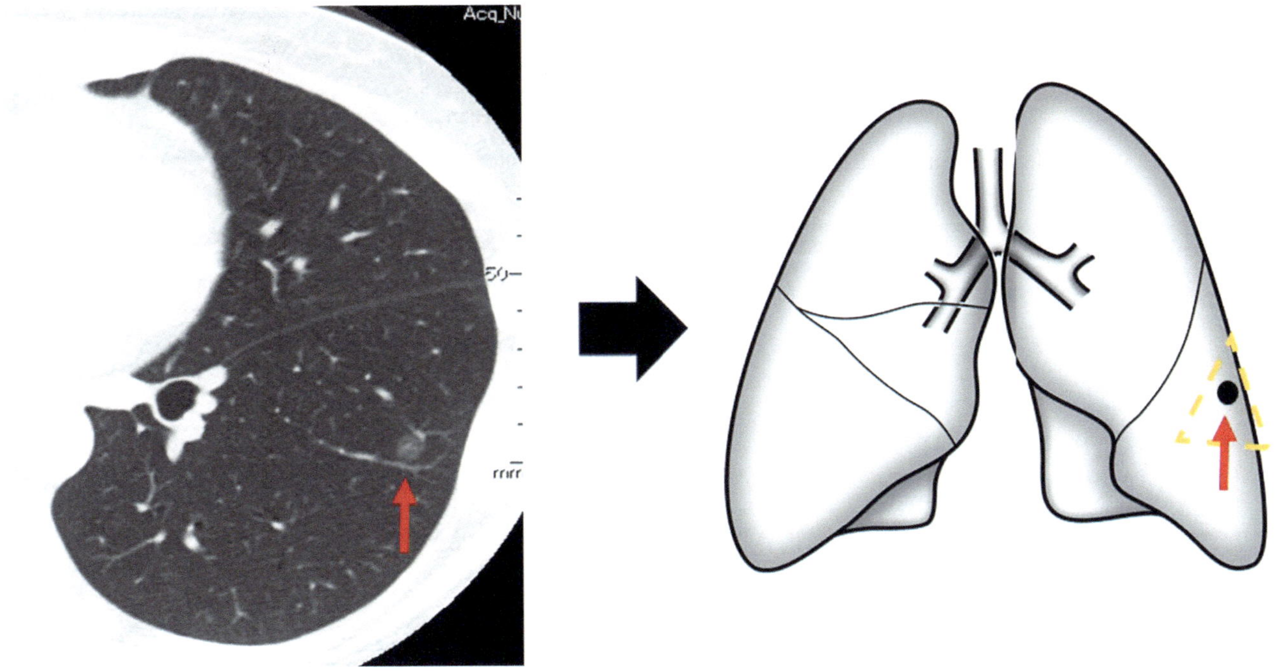

Fig. 15.2 Schema of partial lung resection for micropulmonary nodules

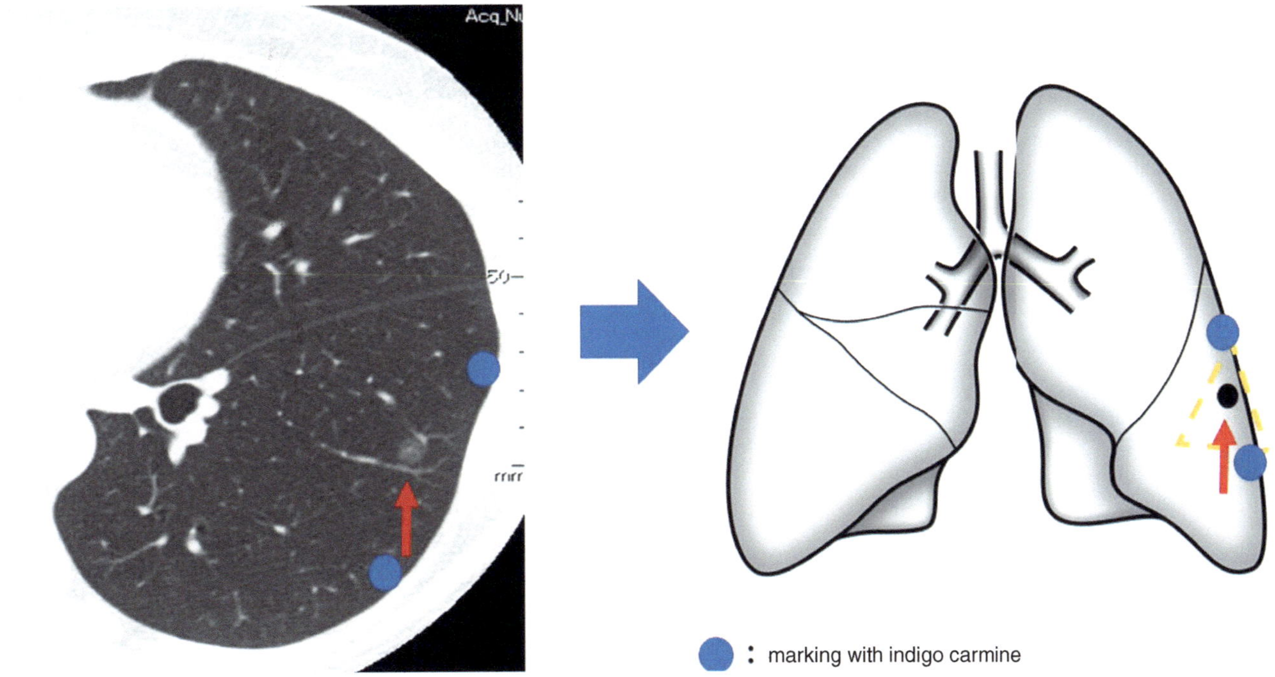

Fig. 15.3 Schema of partial lung resection in VAL-MAP

be used as a surgical navigation providing information on the tumor location and resection line (Fig. 15.3). However, VAL-MAP has technical difficulties in marking just beneath the pleura and confirming the location of the marked area on post-marking CT (Fig. 15.4) or in the lung of heavy smokers with a high degree of charcoal dust deposition during surgery. In order to solve these difficulties, a new method using ICG and a radiographic contrast agent instead of indigo carmine (ICG-VAL-MAP) has been developed (Fig. 15.5). ICG-VAL-MAP has been reported to be easier than VAL-MAP in confirming the target lesion by post-marking CT and during surgery.

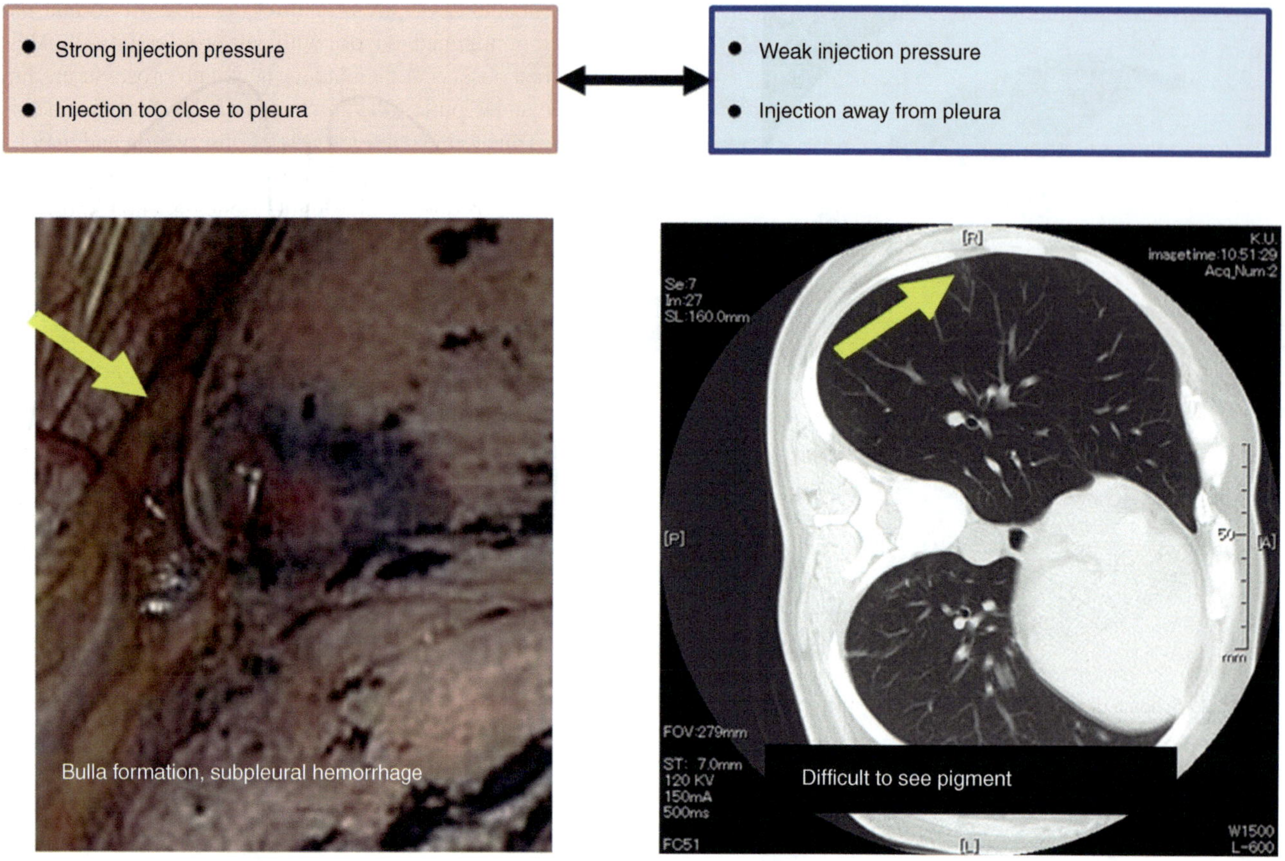

Fig. 15.4 Intraoperative images and CT for confirmation after VAL-MAP. Intraoperatively, bulla formation at the injection site is seen probably because the dye was injected just below the pleura or the injection pressure was too high. On the other hand, CT fails to identify the injection site clearly because the marking was too far from the pleura or the injection pressure was too weak

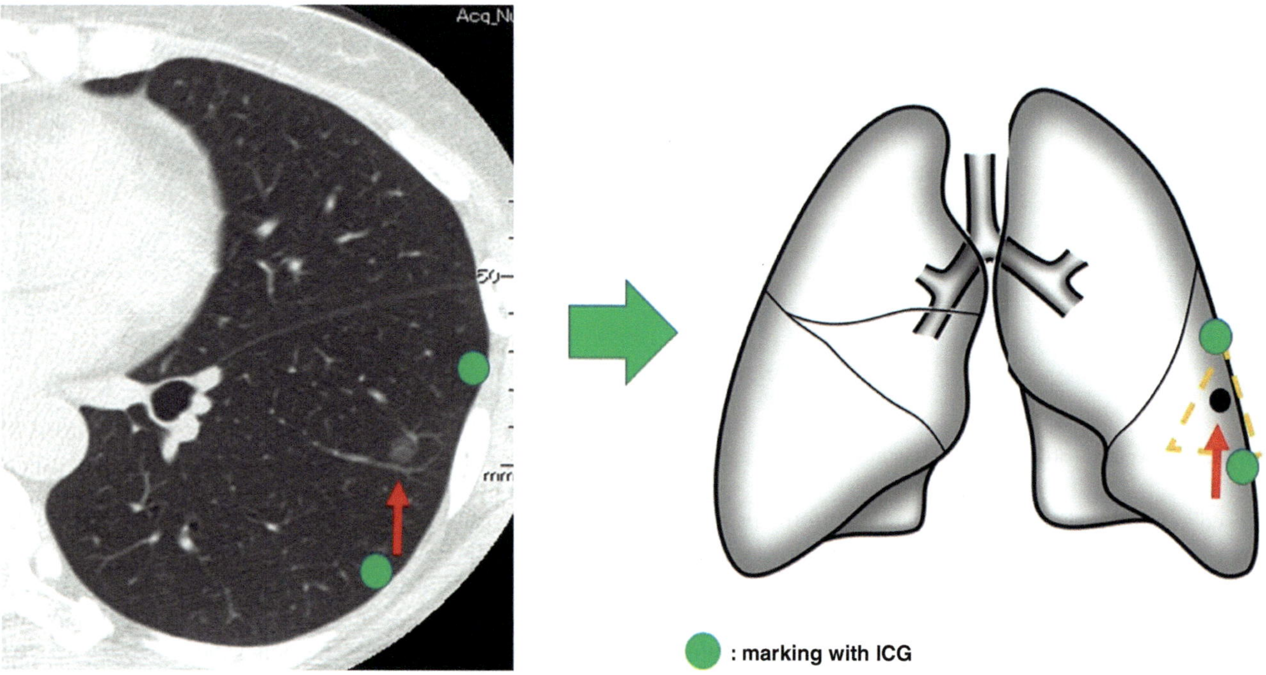

Fig. 15.5 Schema of partial lung resection using ICG-VAL-MAP

3 Development History of Fluorescence Imaging in Pulmonary Surgery

Recently, ICG fluorescence imaging has been applied actively to pulmonary surgery [6]. The main focus of this technique in pulmonary surgery has been the identification of sentinel lymph nodes and visualization of intersegmental planes [7], but in recent years, attempts have been made to use ICG for tumor localization [8]. For example, some authors reported clinical applications of ICG fluorescence imaging for intraoperative identification of lung tumors with ICG uptake function, such as hepatoblastoma [9], although not all tumors in the lung have property of ICG uptake sufficient for intraoperative visualization by fluorescence imaging [10]. In Japan, applications of ICG fluorescence imaging in the field of pulmonary surgery is extending from the identification of intersegmental planes during segmentectomies of the lung to mapping of the target lesions by intratracheal administration of ICG, such as ICG-VAL-MAP [11].

Point
- Indocyanine green fluorescence imaging is now beginning to be widely used in pulmonary surgery, particularly for the identification of intersegmental planes.
- Recently, a clinical application of fluorescence imaging has been performed in the localization of primary and metastatic lung cancers as well as in sentinel node mapping.

4 Clinical Practice of ICG-VAL-MAP

In principle, preoperative marking is performed at least 48 hours before the surgery. In most cases, (i) the marking is performed in the bronchoscopy room in the morning of the surgery day, (ii) the patient is moved directly to the CT room and a chest CT is taken to confirm the marked area, (iii) 3D-CT reconstruction images are created before the surgery, and (iv) the surgery is performed in the afternoon of the same day.

The specific procedures are detailed below:

(i) Bronchoscopy is performed under local anesthesia as usual (Fig. 15.6). The bronchial branch to be marked is selected in advance (Fig. 15.7). Based on this information, marking is performed in the bronchoscopy room. Using a bronchoscopic spray tube as shown in Fig. 15.8, a dye solution containing ICG and radiographic contrast agent is injected into the target area. Currently, 0.5 mL of ICG solution (25 mg/10 mL) is well mixed with 4.5 mL of radiographic contrast medium, and 0.1–0.3 mL of the mixture is injected. The target bronchus is selected, and the spreading tube is advanced just below the pleura using fluoroscopy. Then, the dye is injected after withdrawing the tip of the spreader tube about 1 cm below the pleura. At this time, while feeling the pressure with the 10-mL syringe attached to the hand side of the spreader tube, the dye is injected at the point where the pressure is released (when the pressure is no longer felt) so that the marking can be done at the right place (slightly central to the pleura). If fluoroscopy is used at the same time as the injection, the status of the dye injection can be confirmed on fluoroscopic images (Fig. 15.9). In this case, it is not necessary to inject a large amount of the dye. Because the sensitivity of the fluorescence imaging is high, too much injection can lead to the spreading of ICG solution, making pinpoint

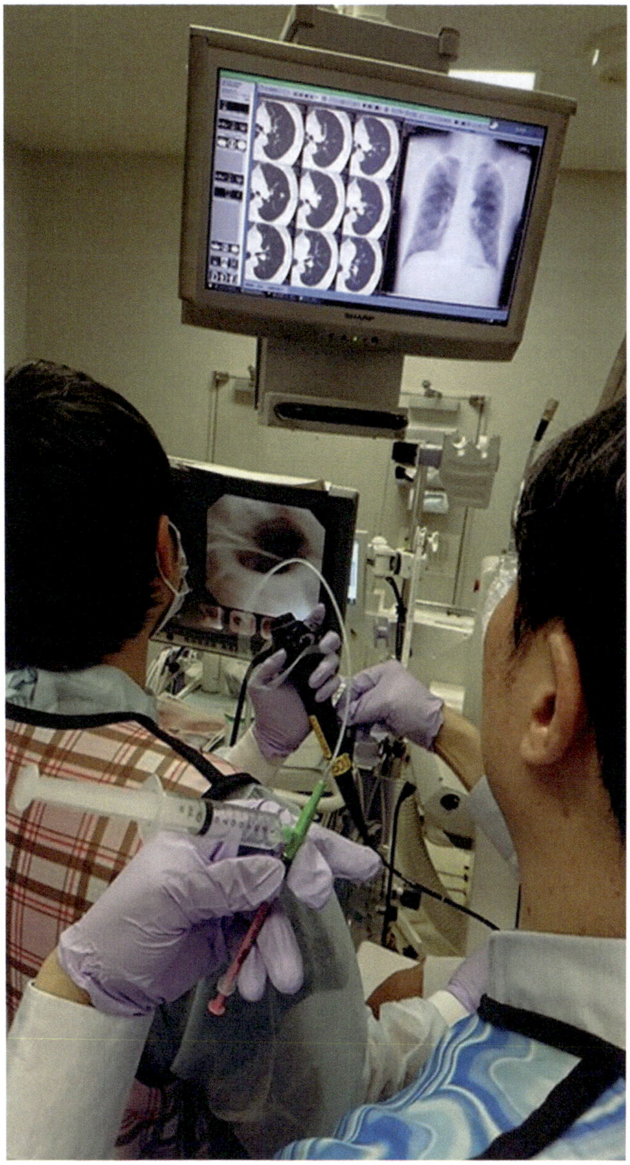

Fig. 15.6 Bronchoscopy under local anesthesia. The spraying tube for dye injection is applied

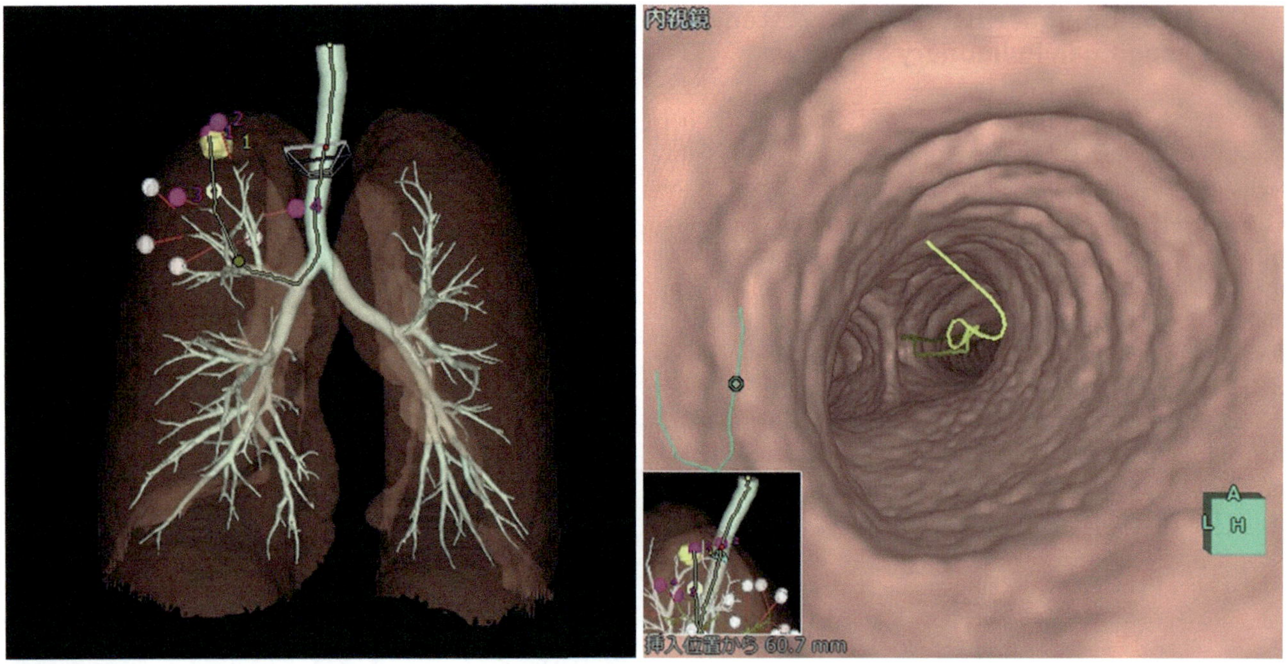

Fig. 15.7 Selection of bronchi to be marked using virtual bronchoscopy software

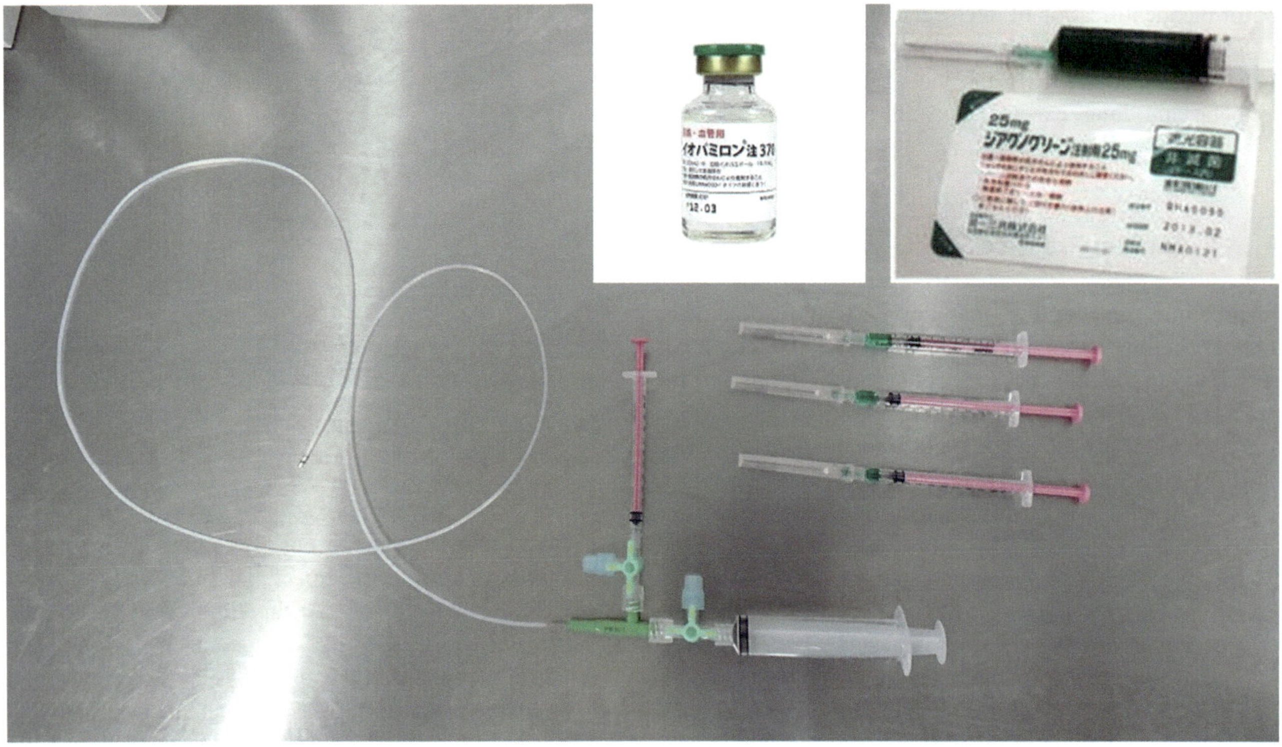

Fig. 15.8 The bronchoscopic spreading tube and preparation of dye (ICG + radiographic contrast agent) for injection

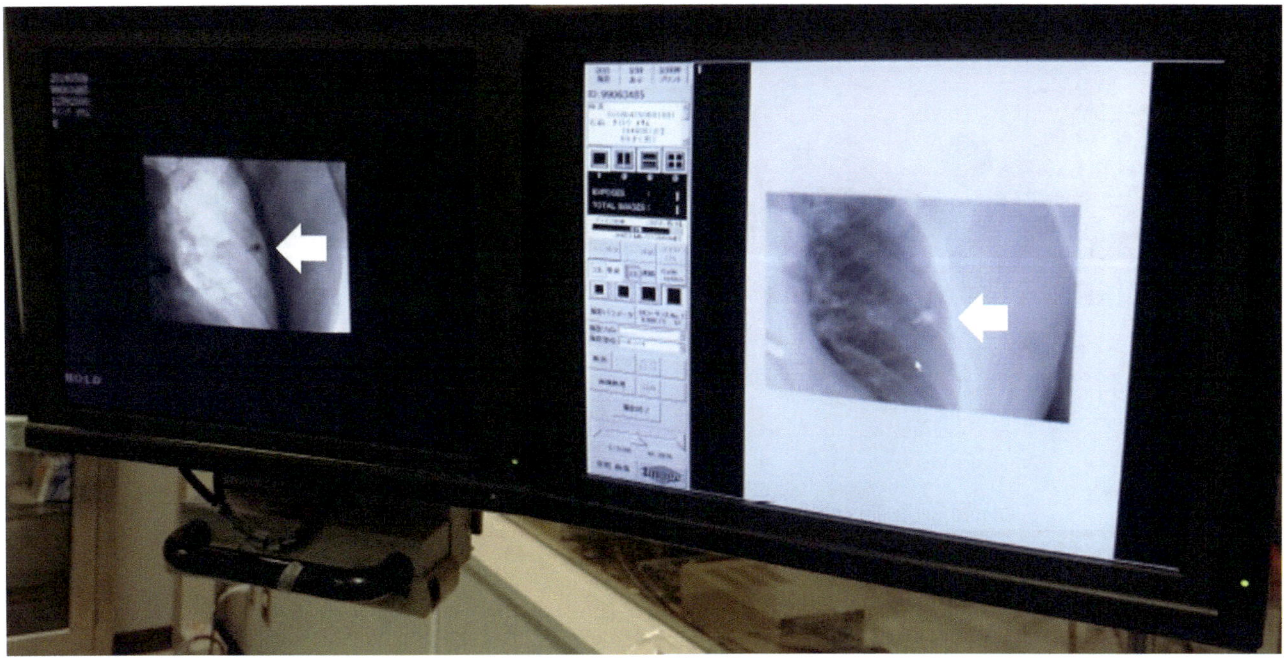

Fig. 15.9 X-ray fluoroscopy after dye (ICG + radiographic contrast agent) injection. Because ICG is injected with a mixture of radiographic contrast material, the injection site can be clearly identified on radiography

marking impossible. In other words, marking can be completed with a smaller amount of dye than with conventional VAL-MAP. The injection is terminated when the dye is visible on fluoroscopy, followed by air injection with a 10 cc syringe as with VAL-MAP. These procedures are repeated for the required number of markings.

(ii) CT is taken to confirm the marked area. In some cases, CT images are obtained in the same position as the intraoperative position so that the 3D-CT reconstruction image is more consistent with the intraoperative findings. Figure 15.10 shows an example of how the lung lesion with or without dye (ICG + radiographic contrast agent or indigo carmine) is demonstrated in the post-marking CT.

(iii) 3D-reconstructed CT images are created with a workstation to understand the spatial positional relationships between the marking site and the tumor before surgery (Fig. 15.11).

(iv) Surgical resection is performed based on the intraoperative images of the surgical field, guided by the 3D-reconstructed images (Movie 15.1).

In VAL-MAP, more than two markings have been performed in case of failure in identifying the target lesion. In contrast, the success rate of ICG-VAL-MAP has been improved, leading to reducing the total number of markings required.

Point

- The amount of ICG injected should be minimum for avoiding diffusion of fluorescence signals.
- Thanks to the high sensitivity of ICG-VAL-MAP, the total number of markings can be reduced as compared with conventional VAL-MAP.
- The secret of successful marking is adequate planning before the procedure, which is the same for both VAL-MAP and ICG-VAL-MAP.

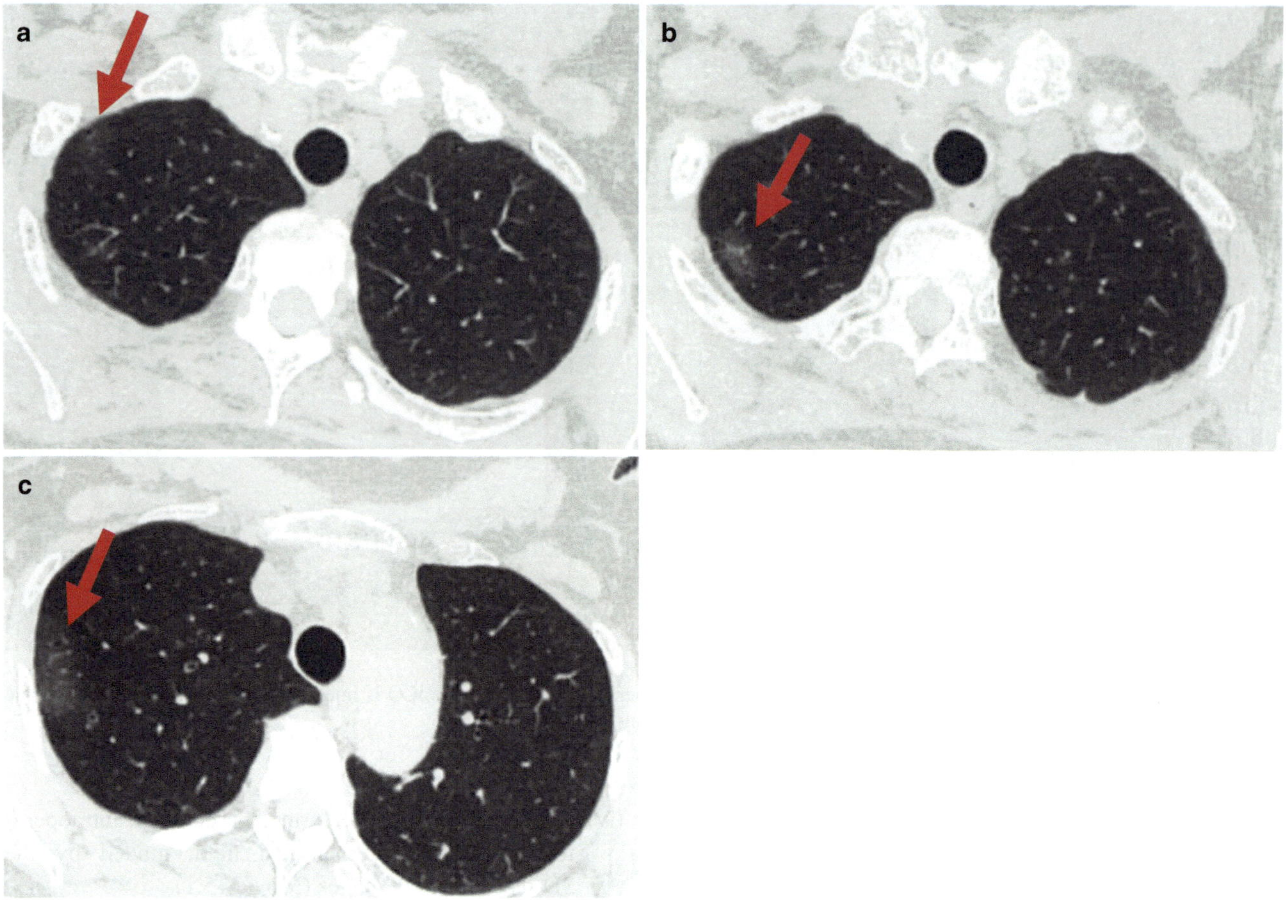

Fig. 15.10 CT images after marking in the case of a combination of VAL-MAP with indigo carmine and ICG-VAL-MAP with ICG and contrast agent. (**a**) VAL-MAP (indigo carmine). (**b**) Plain CT identifying peripheral lung lesion in right S1. (**c**) ICG-VAL-MAP (ICG + contrast agent)

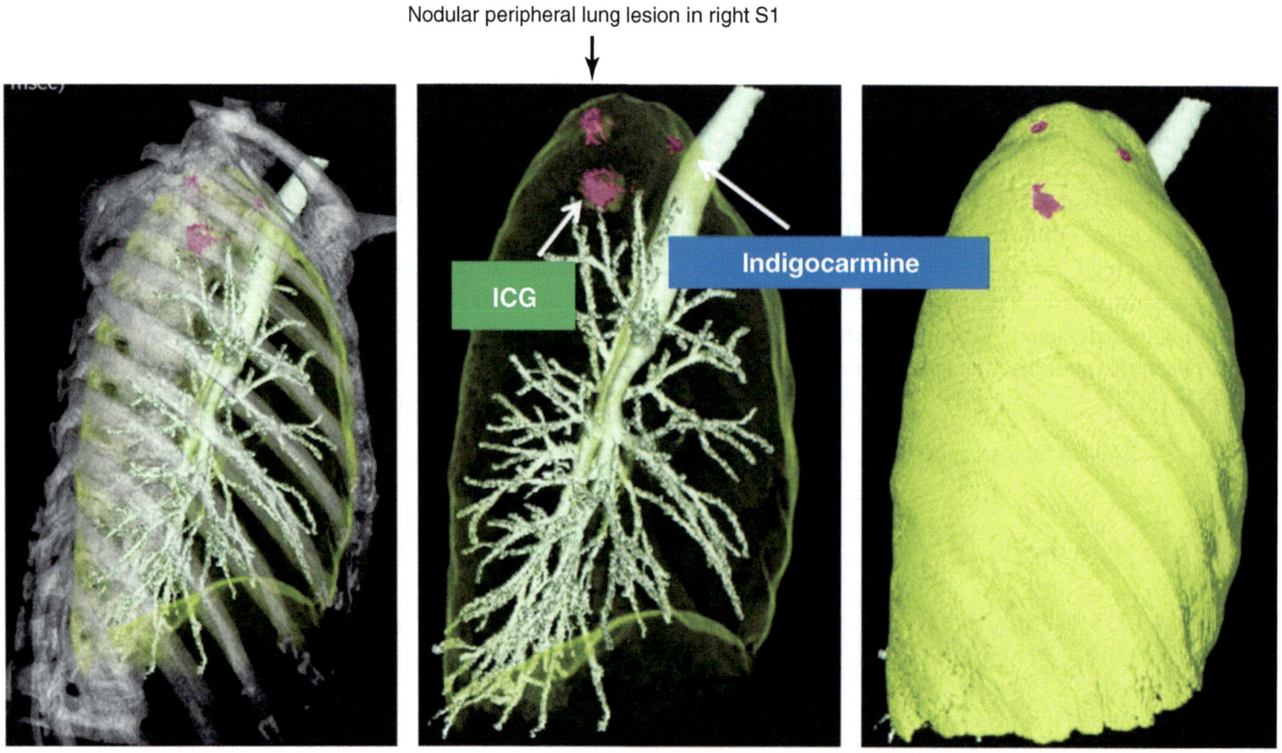

Fig. 15.11 Preoperative 3D-reconstructed CT images

5 Expected Effects of Fluorescence Imaging

ICG-VAL-MAP improves the visibility of pulmonary lesions, which is a weak point of conventional VAL-MAP, by using ICG for sensitive fluorescence imaging (Fig. 15.12 and Movie 15.2). At present, ICG-VAL-MAP needs the bronchoscopy technique, but it will play a significant role in marking where surgeons want to mark in a less invasive manner in comparison with a conventional percutaneous marking. Future development of cancer-specific fluorophores enables intraoperative navigation of cancerous tissues in the true sense.

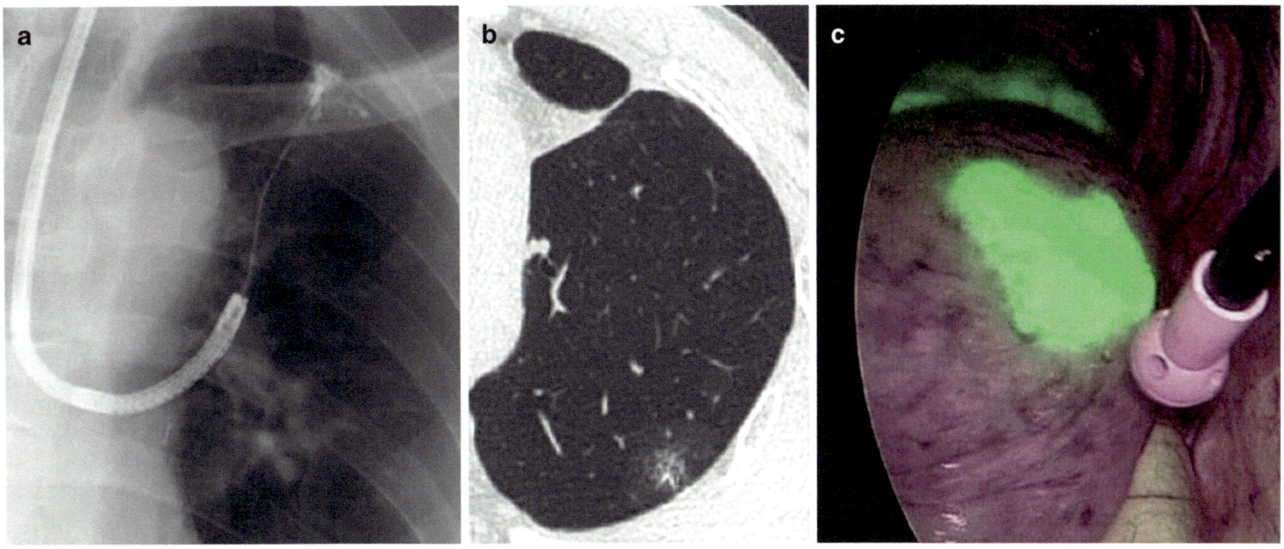

Fig. 15.12 Visibility in ICG-VAL-MAP. Sufficient visibility of the lesion at the time of marking (**a**), CT after marking (**b**), and during surgery (**c**)

6 Limitations and Challenges of ICG-VAL-MAP

Although VAL-MAP is covered by the national insurance system in Japan, it should be noted that the bronchial administration of ICG and radiographic contrast media is not approved and should be performed as a clinical trial at present. It should also be noted that this technique is basically contraindicated in patients who are allergic to ICG/contrast media or asthmatic, although there have been no reports of serious complications with ICG-VAL-MAP so far.

7 Conclusions

The ICG-VAL-MAP was developed as a tool for pulmonary surgeons to perform diagnostic resections of peripheral microlesions accurately. Currently, this technique can be used not only for the identification of micropulmonary nodular lesions but also to mark intersegmental boundaries of the lung (Movie 15.3). In addition, by combining different dyes such as indigo carmine and ICG in case of marking failures, more accurate identification of the lesions can be performed.

References

1. Oikawa T, Nomoto Y, Kinoshita K. A case of residual left hemiparesis due to cerebral air embolization during CT-guided marking of a small peripheral lung cancer. J JPN Assoc Chest Surg. 2008;22:914–9.
2. Matsuura Y, Watari M. Two cases of intracardiac air emboli after CT-guided pulmonary puncture. J JPN Assoc Chest Surg. 2010;24:967–71.
3. Chen F, Tatsumi A. Clinical study of preoperative CT-guided marking of peripheral mass lesions in the lung field. J JPN Assoc Chest Surg. 2001;15:87–91.
4. Sato M, Omasa M, Chen F, et al. Use of virtual assisted lung mapping (VAL-MAP), a bronchoscopic multispot dye-marking technique using virtual images, for precise navigation of thoracoscopic sublobar lung resection. J Thorac Cardiovasc Surg. 2014;147:1813–9.
5. Sato M, Yamada T, Menju T, et al. Virtual-assisted lung mapping: outcome of 100 consecutive cases in a single institute. Eur J Cardiothorac Surg. 2015;47:e131–9.
6. Kasai Y, Tarumi S, Chang SS, et al. Clinical trial of new methods for identifying lung intersegmental borders using infrared thoracoscopy with indocyanine green: comparative analysis of 2- and 1-wavelength methods. Eur J Cardiothorac Surg. 2013;44:1103–7.
7. Okusanya OT, Hess NR, Luketich JD, et al. Infrared intraoperative fluorescence imaging using indocyanine green in thoracic surgery. Eur J Cardiothorac Surg. 2018;53:512–8.
8. Rho J, Lee JW, Quan YH, et al. Fluorescent and iodized emulsion for preoperative localization of pulmonary nodules. Ann Surg. 2019. [Epub ahead of print].
9. Chen-Yoshikawa TF, Hatano E, Yoshizawa A, et al. Clinical application of projection mapping technology for surgical resection of lung metastasis. Interact Cardiovasc Thorac Surg. 2017;25:1010–1.
10. Hamaji M, Chen-Yoshikawa TF, Minami M, et al. Near-infrared imaging using intravenous indocyanine green at a conventional dose to locate pulmonary metastases: a pilot study. Thorac Cardiovasc Surg. 2019;67:688–91.
11. Sekine Y, Itoh T, Toyoda T, et al. Precise anatomical sublobar resection using a 3D medical image analyzer and fluorescence-guided surgery with transbronchial instillation of indocyanine green. Semin Thorac Cardiovasc Surg. 2019;31:595–602.

Gastric Cancer (Primary Tumor, Peritoneal Dissemination)

16

Tsuyoshi Takahashi, Yukinori Kurokawa,
Makoto Yamasaki, Hidetoshi Eguchi, and Yuichiro Doki

Summary

- Photodynamic diagnosis using 5-aminolevulinic acid (5-ALA), a natural amino acid, has been applied clinically in the case of brain tumors and bladder cancer.
- We are investigating the possibility of this technique for intraoperative diagnosis of peritoneal dissemination of gastric cancer.

1 Introduction

Peritoneal dissemination as well as hematogenous and lymphatic metastasis is an important clinical factor in the treatment of gastrointestinal cancers. In particular, tiny nodules of peritoneal metastasis are difficult to detect with current diagnostic imaging modalities such as CT and MRI. Therefore, staging laparoscopy is often indicated prior to making a decision for the treatment strategy in cases of advanced gastric cancer and pancreatic cancer with a high risk of peritoneal dissemination [1]. While staging laparoscopy is minimally invasive and has a high positive diagnosis rate of 89–100%, the insufficient false-negative rate (5–17.2%) [2–4] remains a problem in this strategy. We need a higher sensitive diagnostic method for peritoneal dissemination.

2 Current Status of Photodynamic Diagnosis Using 5-Aminolevulinic Acid

5-aminolevulinic acid (5-ALA), a natural amino acid, plays an important role in energy metabolism as a precursor of heme. 5-ALA is taken up by cells and metabolized into protoporphyrin IX (Pp IX), which is translocated into mitochondria (Fig. 16.1). In normal cells, Pp IX is rapidly metabolized to heme. In cancer cells, however, Pp IX is not converted to heme due to the lack of divalent iron, accumulating specifically in the cells [5]. In cancer cells, the expression of PEPT1, a transporter for 5-ALA uptake, is increased and the expression of ABCG2, a transporter for Pp IX efflux, is decreased, which would be the mechanistic background of Pp IX accumulation in cancer cells [6]. In addition, it has been reported that the increased activity of porphobilinogen deaminase promotes the synthesis of Pp IX [7] and the decreased activity of ferrochelatase inhibits the metabolism of Pp IX [8], resulting in the accumulation of Pp IX in cancer cells. As a photosensitizer, Pp IX is excited by blue light (400–410 nm) and emits red fluorescence when it returns to the ground state. These properties of 5-ALA enable photodynamic diagnosis of cancer cells [9].

T. Takahashi (✉)
Medical Education Center, Osaka University School of Medicine, Suita, Osaka, Japan

Department of Gastroenterological Surgery, Osaka University Graduate School of Medicine, Suita, Osaka, Japan
e-mail: ttakahashi2@gesurg.med.osaka-u.ac.jp

Y. Kurokawa · M. Yamasaki · H. Eguchi · Y. Doki
Department of Gastroenterological Surgery, Osaka University Graduate School of Medicine, Suita, Osaka, Japan

T. Ishizawa (ed.), *Fluorescence-Guided Surgery*, https://doi.org/10.1007/978-981-19-7372-7_16

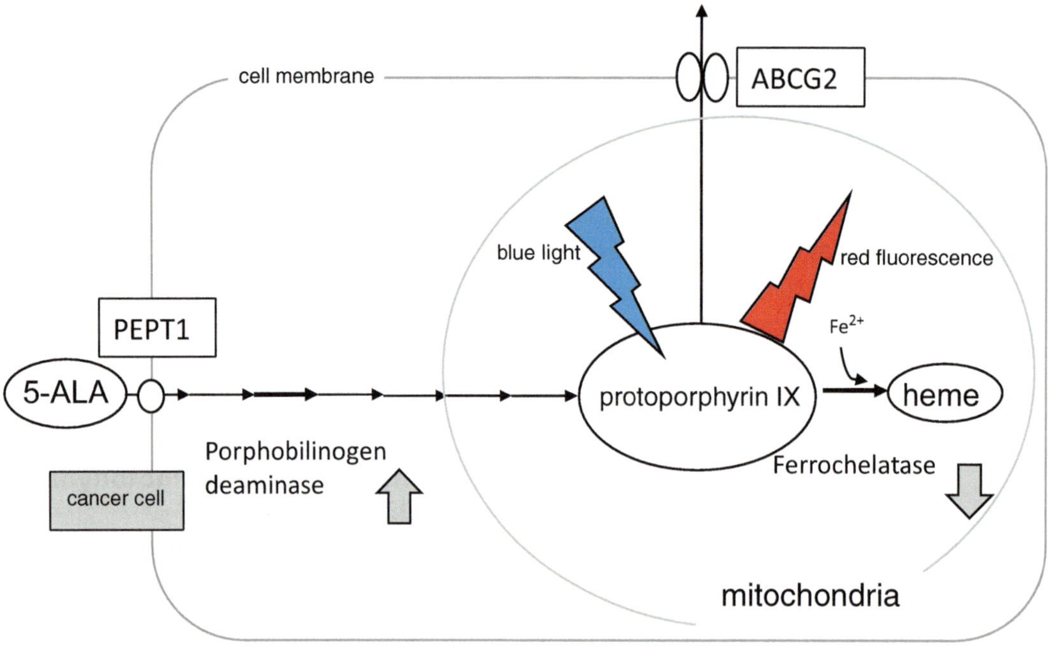

Fig. 16.1 Mechanism of photodynamic diagnosis using 5-ALA. Pp IX shows red fluorescence under blue light illumination

3 Diagnosis of Peritoneal Dissemination of Gastric Cancer Using 5-Aminolevulinic Acid

As an application of photodynamic diagnosis using 5-ALA to gastric cancer, Kishi et al. conducted a clinical trial using this technique for screening laparoscopy in 52 patients with advanced gastric cancer. Among 24 cases in which peritoneal dissemination was not detected by white light, photodynamic diagnosis using 5-ALA detected peritoneal dissemination in five cases (20.8%). The authors concluded that this diagnostic method was effective in increasing the detection sensitivity of peritoneal dissemination from gastric cancer [9].

4 Safety of 5-Aminolevulinic Acid

5-aminolevulinic acid is a naturally occurring amino acid and is relatively safe. Possible adverse reactions include photosensitivity, hypotension, gastrointestinal symptoms, and hepatic dysfunction. Inoue et al. [10] reported increases in LDH (16.2%), AST (16.2%), ALT (16.2%), γ GTP (5.4%), blood bilirubin (2.7%), amylase (10.8%), eosinophils (2.7%), nausea (10.8%), vomiting (8.1%), treatment-induced vomiting (2.7%), and headache (2.7%), and that the severity of these adverse events was mild in all cases.

5 Clinical Application in Our Department

In December 2013, we started a clinical trial of photodynamic diagnosis for advanced gastric cancer patients undergoing screening laparoscopy with the following protocol.

Methods: 5-ALA (1 g) is administered 3–5 hours prior to the start of the laparoscopy. After the administration of 5-ALA, long sleeves, long pants, and sunglasses should be worn to prevent adverse effects due to photosensitivity, and the patient should be shielded from strong light such as sunlight for 48 hours. The examination equipment used is the D-LIGHT system (KARL STORZ: D-LIGHT C PDD light source, IMAGE1-S system camera control unit, IMAGE1-S-FI camera head, HOPKINS® II telescope 30°). In this system, it is possible to easily switch between normal light and blue light using a foot switch.

Laparoscopy procedures: Evaluation with white light is followed by switching to blue light to search for areas with red fluorescence. The white nodule or the area with red fluorescence under blue light is biopsied and confirmed by pathology. If the observation time is prolonged, it becomes difficult to evaluate red fluorescence due to photo breaching; therefore, priority should be given to lesions that cannot be confirmed by white light.

Pathological examination: Pathological diagnosis was made on formalin-fixed specimens because foci of peritoneal dissemination were often too small for intraoperative rapid diagnosis.

Case presentation: In the case of a 50-year-old man, upper gastrointestinal endoscopy was indicated for anorexia and revealed type 3 advanced gastric cancer in the anal part of the gastric body. Abdominal CT identified extraserosal cancer invasion, suggesting peritoneal dissemination. Therefore, we decided to perform a staging laparoscopy. Nodules with red fluorescence on photodynamic diagnosis were found in the pouch of Douglas and below the left diaphragm (Fig. 16.2). Biopsy specimens obtained from these lesions were pathologically diagnosed as peritoneal dissemination of gastric cancer. Red fluorescence was also observed in the primary lesion of the stomach, where cancer tissues were exposed on the gastric serosa (Fig. 16.3).

Point

- In photodynamic diagnosis during staging laparoscopy for gastric cancer, 5-ALA (20 mg/kg) is administered 3–5 hours before laparoscopy.
- Both white-light color imaging and blue light observation shall be conducted.
- Observation of the abdominal cavity should be completed quickly to avoid photo breaching of fluorescence signal.
- 5-aminolevulinic acid photodynamic diagnosis may be useful for the discrimination of malignancy in visible peritoneal nodules and for the detection of new lesions with cancer cells.

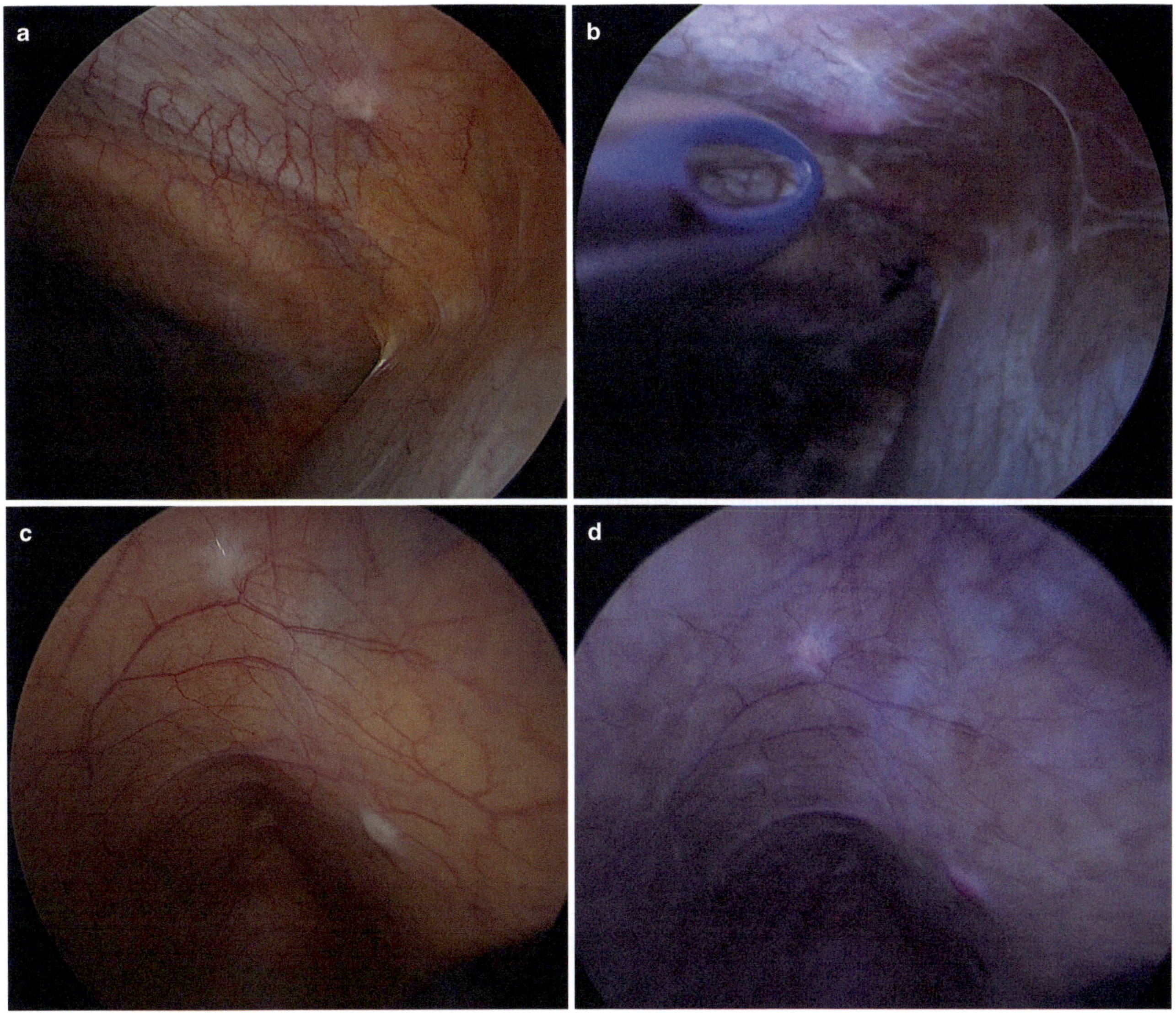

Fig. 16.2 Images in staging laparoscopy. (**a**) White-light color imaging of the right subdiaphragmatic field. (**b**) Fluorescence imaging of the right subdiaphragmatic field. (**c**) White-light color imaging of the pouch of Douglas. (**d**) Fluorescence imaging of the pouch of Douglas

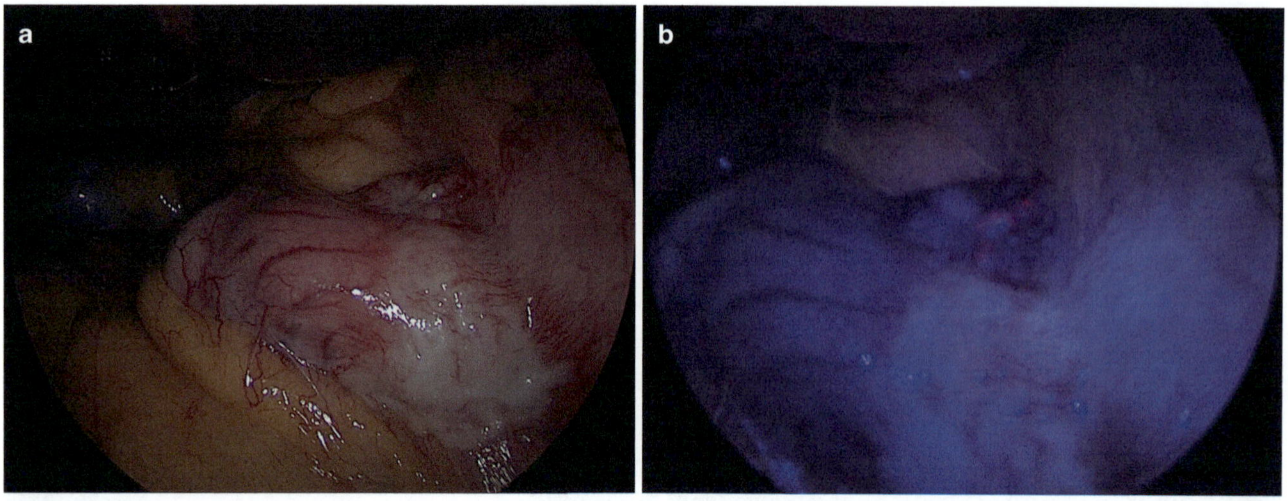

Fig. 16.3 Images of primary lesion of gastric cancer. (**a**) White-light color imaging of the primary lesion. (**b**) Fluorescence imaging of the primary lesion

6 Clinical Significance of 5-Aminolevulinic Acid-Positive Peritoneal Dissemination

Ushimaru et al. reported that, among 113 patients undergoing staging laparoscopy for advanced gastric cancer, 13 patients had microscopic peritoneal dissemination, which could be detected only by 5-ALA photodynamic diagnosis [11]. They reported that the three-year survival rate of these patients was equivalent to that of patients diagnosed with no peritoneal dissemination, probably because of multidisciplinary treatments including chemotherapy enabled by 5-ALA photodynamic diagnosis.

As a result of recent advances in chemotherapy, there have been many reports of successful conversion surgery after confirmation of peritoneal dissemination and chemotherapy. Kishi et al. reported the usefulness of 5-ALA photodynamic diagnosis for the detection of peritoneal dissemination at the time when an indication of curative surgery was evaluated after chemotherapy [12]. In our clinical study, we also experienced a case in which the presence or absence of viable cells after chemotherapy could be determined accurately by photodynamic diagnosis (Fig. 16.4).

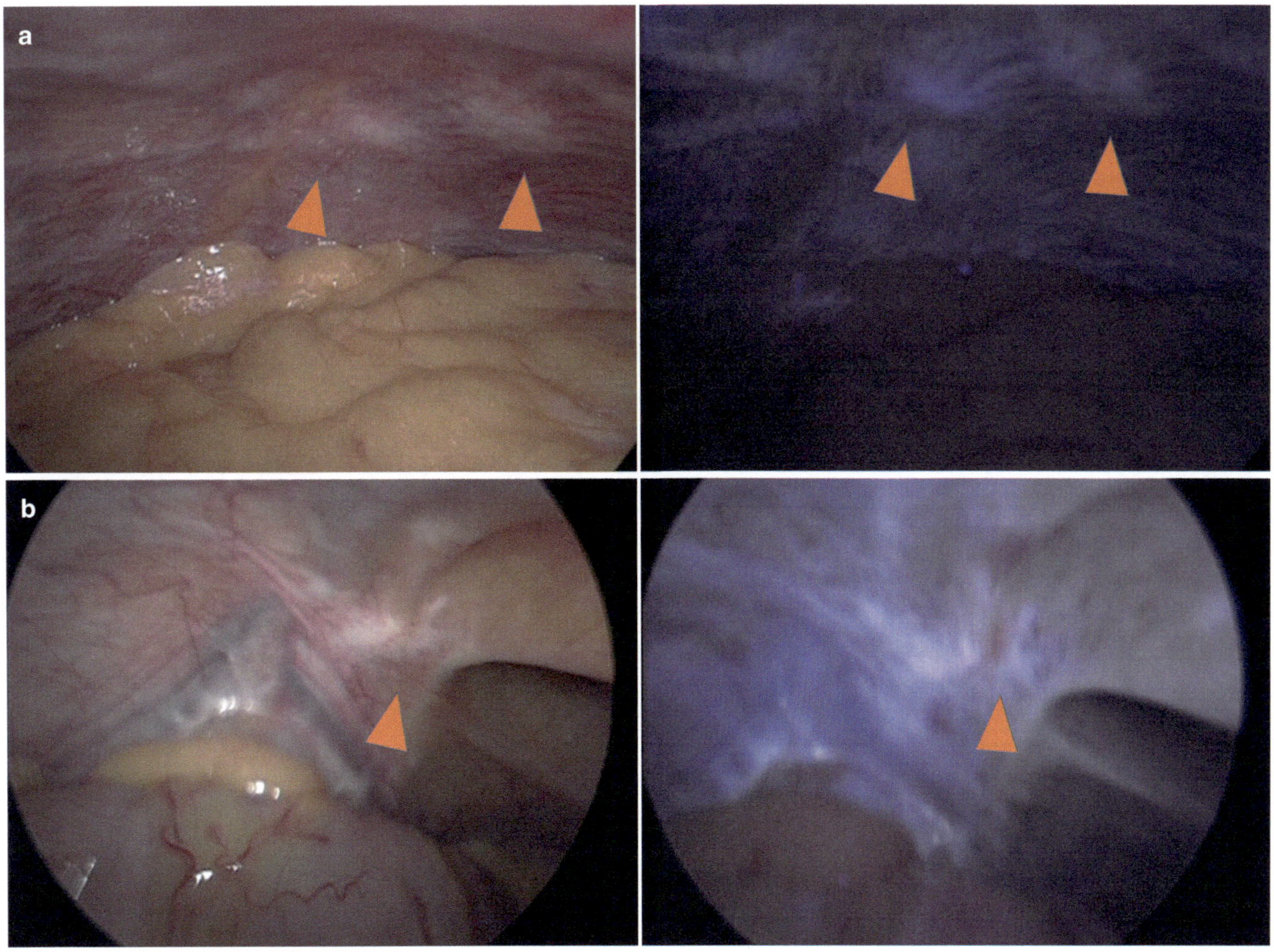

Fig. 16.4 Staging laparoscopy after chemotherapy. (**a**) White-light color imaging (*left*) and fluorescence imaging (*right*) of the left subdiaphragmatic field. White nodules are identified without red fluorescence signals in photodynamic diagnosis. They were all negative for viable cancer cells in postoperative pathological evaluations. (**b**) White-light color imaging (*left*) and fluorescence imaging (*right*) of the left subdiaphragmatic field. The visible white nodule has a partial red fluorescence signal of 5-ALA. Pathological examination of biopsied specimen was positive for viable cancer cells

7 Toward the Spread of Clinical Applications in Gastric Cancer Surgery

We have been conducting investigator-initiated clinical trials to expand the insurance coverage of 5-ALA in Japan. In the exploratory study, the safety, efficacy, and optimal dosage were demonstrated. Furthermore, in 2017, we conducted a multi-center phase III validation study consisting of 105 patients with gastric cancer, with "the number of subjects who are negative for peritoneal dissemination by normal light observation and can be diagnosed correctly by photodynamic diagnosis" as the primary endpoint, taking into account the clinical significance for patients. We are currently analyzing the outcomes.

8 Conclusions

The combined use of photodynamic diagnosis using 5-ALA may increase the sensitivity of detection of peritoneal dissemination of gastric cancer during staging laparoscopy and enable the planning of an appropriate treatment strategy.

References

1. Stell DA, Carter CR, Stewart I, et al. Prospective comparison of laparoscopy, ultrasonography and computed tomography in staging gastric cancer. Br J Surg. 1996;83:1260–2.
2. Tsuchida K, Yoshikawa T, Tsuburaya A, et al. Indications for staging laparoscopy in clinical T4M0 gastric cancer. World J Surg. 2011;35:2703–9.

3. Song KY, Kim JJ, Kim SN, et al. Staging laparoscopy for advanced gastric cancer: is it also useful for a group which has an aggressive surgical strategy? World J Surg. 2015;39:2742–7.
4. Miki Y, Tokunaga M, Tanizawa Y, et al. Staging laparoscopy for patients with cM0, type 4, and large type3 gastric cancer. World J Surg. 2015;39:2742–7.
5. Hayashi M, Fukuhara H, Inoue K, et al. The effect of iron ion on the specificity of photodynamic therapy with 5-aminolevulinic acid. PLoS One. 2015;10:e 0122351.
6. Hagiya Y, Endo Y, Yonemura Y, et al. Pivotal roles of peptide transporter PEPT1 and ATP-binding cassette (ABC) transporter ABCG2 in 5-aminolevulinic acid (ALA)-based photocyto-toxicity of gastric cancer cells in vitro. Photodiagnosis Hotodyn Ther. 2012;9:204–14.
7. Gibson SL, Cupriks DJ, Havens JJ, et al. A regulatory role for porphobilinogen deaminase (PBGD) in 5-aminolaevulinic acid (5-ALA)-induced photosensitization? Br J Cancer. 1998;77:235–42.
8. Ohgari Y, Miyata Y, Chau TT, et al. Mechanisms involved in delta-aminolevulinic acid (ALA)-induced photosensitivity of tumor cells: relation of ferrochelatase and uptake of ALA to the accumulation of protoporphyrin. Biochem Pharmacol. 2005;71:42–9.
9. Kishi K, Fujiwara Y, Yano M, et al. Diagnostic laparoscopy with 5-aminolevulinic- acid-mediated photodynamic diagnosis enhances the detection of peritoneal micrometastases in advanced gastric cancer. Oncology. 2014;87:257–65.
10. Inoue K, Anai S, Fujimoto K, et al. Oral 5-aminolevulinic acid mediated photodynamic diagnosis using fluorescence cystoscopy for non-muscle- invasive bladder cancer: a randomized, double-blind, multicenter phase II/III study. Photodiagnosis Photodyn Ther. 2015;12:193–200.
11. Ushimaru Y, Fujiwara Y, Kishi K, et al. Prognostic significance of basing treatment strategy on the result of photodynamic diagnosis in advanced gastric cancer. Ann Surg Oncol. 2017;24:983–9.
12. Kishi K, Fujiwara Y, Yano M, et al. Usefulness of diagnostic laparoscopy with 5-aminolevulinic acid (ALA)-mediated photodynamic diagnosis for the detection of peritoneal micrometastasis in advanced gastric cancer after chemotherapy. Surg Today. 2016;46:1427–34.

Brain Tumor

17

Toshihiko Kuroiwa

Summary
- The usefulness of fluorescence-guided surgery for malignant gliomas has been verified.
- Fluorescence imaging of brain tumors uses fluorophores leaking from the disrupted blood-brain barrier or those taken up and metabolized by tumor cells.
- Fluorescence-guided surgery has increased the resection rate and progression-free survival in patients with malignant glioma, but it has not yet enhanced overall survival.

1 Introduction

Overall survival in patients with malignant glioma, including glioblastoma, is still very poor; however, the higher the surgical resection rate, the longer the long-term survivals. During resection of these tumors, it is important to identify tumor regions with a high density of malignant cells. However, because of the highly invasive nature of these tumors, the border between the tumor and the surrounding brain is often unclear based on intraoperative findings under white light. For the identification of tumor tissues to be resected, fluorescence-guided surgery based on PDD (photodynamic diagnosis or photodynamic detection) using fluorophores has been clinically applied.

Fluorescent dyes used in fluorescence-guided surgery include fluorophores that utilize leakage from tumor blood vessels that lack the blood-brain barrier (BBB) and those that are selectively taken up and metabolized by brain tumor cells. Among these fluorophores, only 5-aminolevulinic acid (5-ALA) has been covered by the Japanese insurance system for the identification of malignant glioma since 2013. Recently, 5-ALA has been used in fluorescence-guided surgery not only for malignant gliomas but also for other brain tumors such as meningiomas, metastatic brain tumors, and malignant lymphomas. Fluorophores incorporated into tumor cells may also be used for treatments such as photodynamic therapy (PDT). Moreover the efficacy of PDT for brain tumors using various fluorophores has been investigated. Currently, only talaporfin sodium (Npe6) has been approved for PDT of brain tumors in Japan.

T. Kuroiwa (✉)
Tesseikai Neurosurgical Hospital, Shijonawate, Osaka, Japan
e-mail: toshihiko.kuroiwa@ompu.ac.jp

2 Conventional Techniques and Limitations

Frozen pathological examinations, ultrasound, intraoperative CT/MRI, and other diagnostic modalities have been used to identify the boundary between tumor and normal brain tissues during surgery for extremely-invasive malignant gliomas. However, these methods have problems such as the time required to obtain results, unclear images, the inability to obtain real-time information, and expensive equipment. PDD using fluorophores has advantages over conventional techniques in providing real-time information on the distribution of tumor tissues using inexpensive equipment.

3 History of Development and Clinical Applications of Photodynamic Diagnosis in Brain Tumor Surgery

Moore (1947) was the first to attempt intraoperative visualization of brain tumors using PDD, and reported its usefulness using fluorescein sodium (FS) in 12 cases [1]. PDD using FS did not attracted surgeons' attention until a surgical microscope with a built-in filter enabling clear visualization of FS signals was developed in the late 1990s [2, 3]. To overcome limitations in tissue permeability due to the short wavelength FS signals, a surgical microscope targeting indocyanine green (ICG) fluorescence was also developed for visualization of deeply-located lesions [4]. Instead of these techniques based on fluorophore leakage from the disrupted

T. Ishizawa (ed.), *Fluorescence-Guided Surgery*, https://doi.org/10.1007/978-981-19-7372-7_17

BBB, Stummer et al. reported the efficacy of intraoperative PDD for malignant gliomas based on the relatively tumor-specific uptake of 5-ALA in 1998 [5]. As 5-ALA, a precursor of heme and endogenous amino acids, can be administered safely and perorally, it has been mainly used in fluorescence-guided surgery for brain tumors.

Around the same time, the usefulness of 5-aminofluorescein-human serum albumin as an intraoperative PDD for malignant gliomas was evaluated in Germany [6]. This fluorescent substance is used intravenously, and the maximum excitation and fluorescence wavelengths are the same as those of FS. Although a report of clinical cases from a single institution is available [7], this technique has not gained widespread use. In addition, there have been reports of fluorescence-guided surgery for malignant gliomas using other hematoporphyrin derivatives, such as talaporfin sodium (mono-L-aspartyl chlorin e6; Npe6) [8], hypericin [9], IRDye800CW-BBN [10], and tozuleristide (BLZ-100) [11] and PDD for pituitary adenomas using OTL38 (a folic acid analog conjugated to an analog of ICG) [12]. Recently, a topical fluorescent probe has also been developed [13]. Further development is required for these PDD techniques to be used more widely.

At the same time, many surgical microscopes for fluorescence-guided surgery, allow observation of some or all of FS, 5-ALA, and ICG. Some devices enable the superimposition of ICG fluorescence images on white-light images during microscopic observation. Fluorescence imaging systems for exoscope are also commercially available.

4 Clinical Practice of Photodynamic Diagnosis for Brain Tumors

4.1 Fluorescein Sodium

This fluorophore has been used mainly in the field of ophthalmology, but it has also been used in neurosurgery and otolaryngology for the examination of cerebrospinal fluid leakage. As mentioned above, this fluorophore leaks from tumor blood vessels lacking the BBB, as well as from contrast media for CT and MRI. Therefore, fluorescence imaging following intravenous (8 mg/kg) [2, 3] or intra-arterial administration (mainly for intraoperative angiography; 0.06 mg/mL, 2 mL) [14] of this dye during surgery allows visualization of brain tumors identified by preoperative CT or MRI through leakage of FS from the BBB [14]. Although a weak green color can be observed by direct observation of the leaked dye without a filter [15], surgical microscopes equipped with the fluorescence imaging system optimized for FS fluorescence (a peak wavelength at 520 nm under illumination around 493 nm) enables clearer visualization of FS signals (Figs. 17.1 and 17.2a) [2, 3]. Since this technique is based on the leakage of FS from abnormal blood vessels in the tumor, it is important to note that fluorescence signals can diffuse into intercellular spaces over time and leak from blood vessels damaged by surgical manipulation. FS remains in the blood vessel for a relatively long time; therefore, it is necessary to allow sufficient time for repeated imaging.

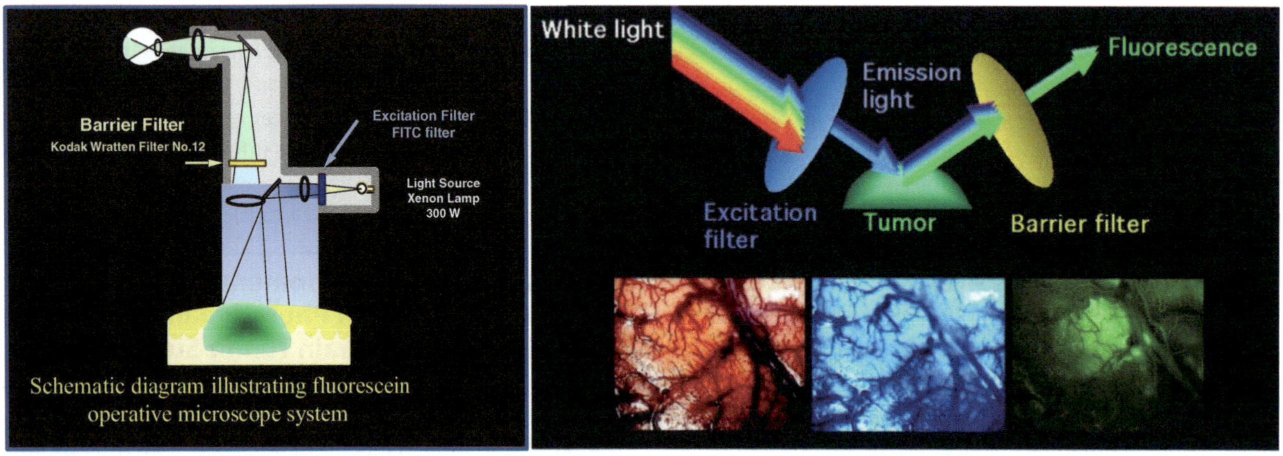

Fig. 17.1 Schema of the fluorescence microscope for fluorescein sodium. When a filter is incorporated into the light source and lens barrel, only the fluorescence of fluorescein sodium can be visualized

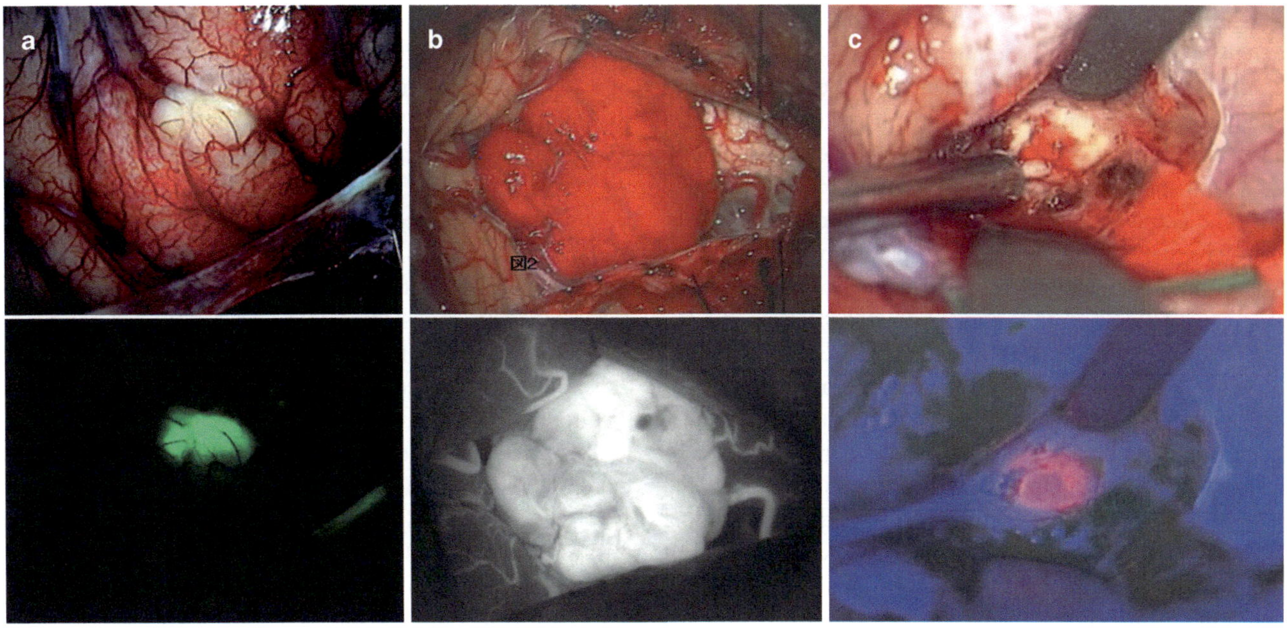

Fig. 17.2 Fluorescence navigation surgery for brain tumors (upper: white-light color imaging, lower: fluorescence imaging). (**a**) Observation of malignant glioma by FS. (**b**) Observation of hemangioblastoma by ICG; the relationship between the tumor and the surrounding blood vessels is clearly visible. (**c**) Observation of glioblastoma by 5-ALA

Compared with 5-ALA, FS is cheaper, can be administered during surgery without the need for preoperative administration, behaves in the same way as contrast media, and has no risk of photosensitivity. Recently, some facilities are using FS again for malignant gliomas for these reasons [16, 17]. Some papers have reported that FS is as useful as 5-ALA [18, 19]; however, the problem is that FS does not have a high level of evidence such as that of 5-ALA, as described by Stummer et al [20]

4.2 Indocyanine Green

Indocyanine green emits near-infrared light in the range of 820–920 nm, with excitation light in the range of 760–810 nm. An attempt to discriminate between brain tumors and normal surrounding tissues using ICG in clinical cases was first reported in 1996 [21]. In this technique, the differences in wavelengths between normal and tumor tissues after ICG administration were evaluated using analytical instruments. However, due to technical difficulties, PDD with ICG has not been widely used for the identification of brain tumors. In 2001, we developed a surgical microscope system capable of observing ICG (25 mg intravenous infusion) [4]. However, this technique is not suitable for tumor identification because ICG has a large molecular weight, and protein-bound ICG does not immediately leak out of blood vessels. Since the report on ICG videoangiography in 2003 [22], the near-infrared microscopic system has been widely used for fluorescence angiography in neurosurgery. ICG fluorescence imaging can be used to identify peri-tumor networks of blood vessels during surgery for tumors with abundant blood flow, such as hemangioblastomas (Fig. 17.2b) [23]. ICG can be used repeatedly during surgery because of its quick discharge from blood vessels.

Recently, a method called second window ICG has been reported in which ICG (5 mg/kg) is administered 24 h before surgery in order to allow it to accumulate in the tumor over time [24]. For malignant gliomas near the surface of the brain, this method allows 100% observation of tumors through the dura mater with near-infrared light visualizing even small residual tumors, and may increase removal rates. Near-infrared light is also observed in all cases of metastatic brain tumors and is useful for removal [25]. Another advantage of ICG is that it is not affected by surrounding autofluorescence.

4.3 5-Aminolevulinic Acid

5-aminolevulinic acid itself does not emit fluorescence but it is mainly taken up by tumor cells via PEPT1 and PEPT2 transporters, where it is metabolized to coproporphyrinogen III and enters mitochondria via ABCB6 to form protoporphyrin IX (PpIX). PpIX emits 635-nm red light when irradiated with 405 nm excitation light (Fig. 17.2c). However, the theoretical background of PpIX accumulation in tumor cells without being metabolized to heme remains unclear. There are many theories, including Fe^{2+} deficiency and ferrochelatase deficiency [26–28], but none have been proven completely.

Since intratumoral PpIX concentration peaks approximately 7 ~ 8 h after oral administration of 5-ALA, a dose of 20 mg/kg of 5-ALA should be taken orally before the patient is admitted to the operating room (2 h before induction of anesthesia) [29]. 5-ALA needs to be kept acidic; therefore, the use of antacids should be avoided. Steroids may interfere with the extravasation of 5-ALA from the collapsed BBB; therefore, they should not be used preoperatively. In addition, if 5-ALA is administered immediately before anesthesia, care should be taken to prevent drainage from the gastric tube after induction of anesthesia.

If the tumor is exposed on the surface of the brain, the operation is performed under fluorescence-guided imaging from the beginning. Otherwise, the tumor that emits fluorescence in the deep part of the brain is identified and removed by dividing the boundary with the surrounding brain under fluorescence-guided imaging. After removal of the tumor bulk, the fluorescing regions are checked for evidence of residual tumor; if any, it is removed from within the range minimizing the risk of neurological deficits. Therefore, monitoring of nerve function is required. If the illumination of the microscope is insufficient, the incidence of false negatives can increase. In such cases, it is useful to add an external light source, which also makes the brightness of the surgical fields sufficient for operations. Originally, laser light was used as an external light source; however, for safety reasons, we developed and use an LED light source. Regardless of the light source, it is important to note that the fluorescence intensity varies from case to case, even in the same type of tumor, and that the fluorescence intensity varies depending on the location, even within the same tumor. In the literature, 5-ALA PDD for malignant glioma has a sensitivity of 75–94% and a specificity of 71–100%. In other words, it reflects the histological findings (Figs. 17.3 and 17.4) and correlates with cell density, CD31 positivity, Ki-67 positivity, etc. In addition to the MIB-1 labeling index, genetic findings such as IDH-1 mutation and 1p19q status, contrast effects on MRI, and heterogeneity in T2-weighted

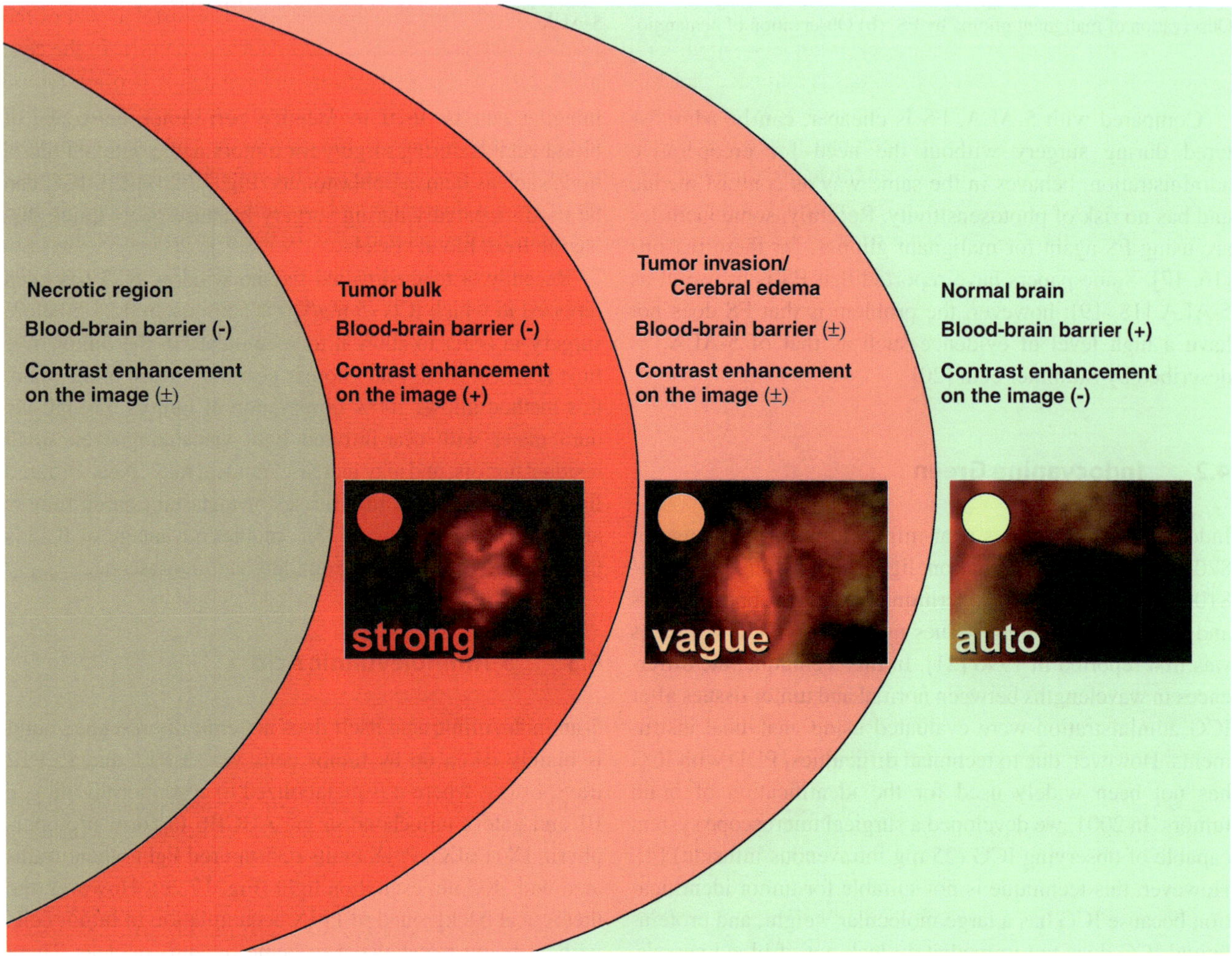

Fig. 17.3 Schema of the relationship between gross and fluorescence findings in glioblastoma. Regions with strong fluorescence (strong), weak fluorescence (vague), and autofluorescence (auto) are shown in contrast to each tumor region

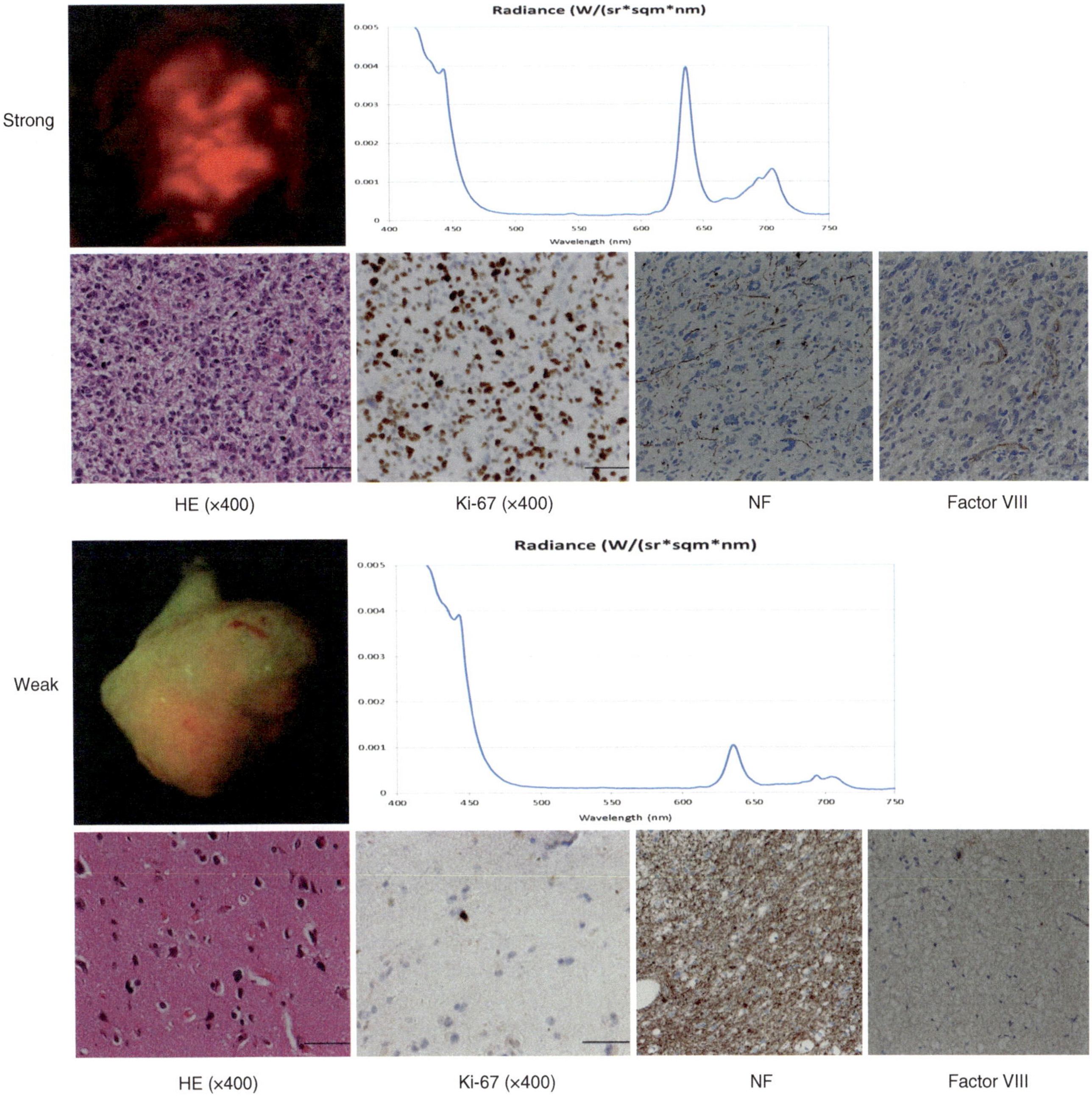

Fluorescence intensity	Cell density	Proliferation rate	Nervous tissue	Vascular density
Strong	(+++)	High	Destruction	(+++)
Weak	(+)	Low	Preservation	(+)

Fig. 17.4 PpIX fluorescence intensity and pathological examinations. In glioblastoma tissues, relationships between PpIX fluorescence, wavelengths, and pathological findings have been demonstrated

images of tumors were also reported to be significantly associated [30]. In the multivariate analysis, IDH-1 mutation was the only independent factor associated with tumor fluorescence [30]. This result suggests that intraoperative fluorescence findings may provide important information for the selection of postoperative adjuvant therapy.

We have reported that PpIX is released extracellularly from within cultured cells in vitro [31], and in clinical cases, discharged PpIX may spread to peritumoral brain tissues. Since several drugs, such as gefitinib, have been confirmed to increase intracellular PpIX concentration by suppressing efflux, it is necessary to devise a way to increase the intracellular PpIX concentration to improve PDD and PDT in the future. Alternatively, we need to know that there are factors that decrease intracellular PpIX concentration, such as phenytoin [32].

As mentioned above, 5-ALA fluorescence-guided surgery has been applied to malignant brain tumors other than gliomas, such as metastatic brain tumors [33–36] and malignant lymphoma [37], although the positive rate is not as high as that for malignant gliomas. In cases of metastatic brain tumors, the 5-ALA fluorescence rate was 62% [34] - 70% [35], but heterogeneous fluorescence was 84% [36] - 92% [34]. The fluorescence-positive rate of the surrounding brain ranged from 67% [36] - 76% [34]. Fluorescence in the surrounding brain was observed in 83% of cases if the tumor was positive, and in 64% if the tumor was negative [34]. Peritumor brain fluorescence did not correlate with cell infiltration but was significantly associated with angiogenesis, which was significantly associated with local progression/recurrence time and 1-year survival. Angiogenesis in the fluorescent region of the peritumoral brain affects prognosis [34]. Benign tumors, such as meningiomas, have also been reported [38–40]. According to previous reports on 5-ALA PDD for meningioma, outcomes of fluorescence-guided surgery were not associated with WHO grade or histological subtype, and the sensitivity of tumor identification was 92–98% with a specificity of 95% [39, 40].

To improve the visibility of 5-ALA PDD, a new illumination device combined with a suction tube was devised and used for the identification of residual tumors in deep lesions within the surgical field [41]. In addition to the surgical microscope, a fluorescent neuroendoscope with a built-in filter that enables PpIX observation has been developed [42]. These novel techniques would be useful for improving the safety and curability of fluorescence navigation surgery using 5-ALA for malignant gliomas, including glioblastoma.

4.4 Talaporfin Sodium (Mono-L-aspartyl chlorin e6; Npe6)

Talaporfin sodium was developed in Japan, and its efficacy as a PDT agent for lung cancer was reported for the first time in 2003. In 2013, it was approved as a PDT agent for malignant gliomas in Japan. Although PDD using talaporfin sodium for brain tumors is not yet covered by insurance in Japan, there have been reports on this technique in a limited number of institutions [8]. Recently, many institutions in Japan have introduced talaporfin sodium for PDT of brain tumors.

Point

- Fluorescein sodium leaks from tumor vessels that lack the BBB and are identified as green fluorescence signals (520 nm peak) under illumination with wavelength centered at 493 nm.
- The near-infrared light of 820–920 nm emitted by ICG has high tissue permeability and was expected to visualize deeply-located tumors, but does not leak immediately outside the blood vessels due to its large molecular weight. Therefore, it is more useful for intraoperative angiography than for tumor identification. However, a method of administering ICG 24 h before surgery and waiting for it to accumulate in the tumor over time has recently been reported [24].
- 5-aminolevulinic acid is converted to PpIX in tumor cells and emits red fluorescence with a 635 nm peak under 405 nm centered excitation light.
- Talaporfin sodium is selectively taken up by tumor cells and can be used for PDD and PDT of malignant brain tumors.

5 Expected Effects of Fluorescence Imaging

A randomized controlled multicenter phase III trial of 5-ALA in malignant gliomas was conducted by Stummer et al., as published in 2006 [20]. They concluded that the total resection and progression-free survival rates at six months were significantly higher in the 5-ALA group than in the control group (white light group), but there was no significant difference in overall survival. Based on this high level of evidence, 5-ALA was approved by the European Medical Agency in September 2007. In Japan, the Japanese Neurosurgical Society submitted a request for early approval of 5-ALA in 2009, and after a company trial, the Ministry of Health, Labour, and Welfare approved it as a PDD drug for malignant gliomas in March 2013. In June 2017, it was approved by the U.S. FDA for malignant gliomas.

According to a previous study comparing fluorescence-guided surgery with 5-ALA and intraoperative MRI in malignant gliomas, the ranges of fluorescence signals detected by 5-ALA PDD exceeded the enhanced areas identified by MRI [43]. This could be a reason for the higher resection rate in fluorescence-guided surgery using 5-ALA than in intraoperative MRI. In addition, the concomitant use of intraoperative MRI was not associated with the incidence of postoperative neurological deficits or a reduction in quality of life [44]. Therefore, fluorescence-guided surgery using 5-ALA in combination with intraoperative monitoring is considered more useful in terms of simplicity, improvement in resection rate, and avoidance of postoperative complications, except for in the case of low-grade malignant gliomas with little accumulation of PpIX. Intraoperative MRI also plays a role in the identification of deeply located residual tumors and tumor resection in a narrow surgical field.

Regarding the combined use of multiple fluorophores in fluorescence-guided surgery, there is a report on the use of FS and 5-ALA in malignant gliomas, suggesting that 5-ALA PDD was more useful than FS in identifying areas of tumor invasion around contrast-enhanced MRI lesions [45]. Concomitant use of FS with 5-ALA may also be useful in improving the tumor-to-background contrast of 5-ALA signals by staining the surrounding brain with green fluorescence [46].

6 Limitations and Challenges of 5-Aminolevulinic Acid Photodynamic Diagnosis for Brain Tumors

Although 5-ALA is currently the most widely used intraoperative PDD drug in neurosurgery, it is important to note that the evaluation of fluorescence intensity is based on a subjective visual inspection by the naked eye. To solve this problem, PpIX has recently been quantified and its spectrum measured and calculated by comparing it with the spectra of known PpIX concentrations [29, 47]. Consequently, the threshold value for a surgeon to judge that fluorescence is positive with the naked eye is 0.9 μg/ml [48]. Areas that cannot be assessed with the naked eye can be quantified and evaluated using this method. Therefore, quantitative PpIX measurements will provide useful information for surgical decision-making.

Additionally, there are several points to consider when 5-ALA is used for PDD. First, benign tumors such as meningiomas and other malignant brain tumors, such as metastatic brain tumors, and non-tumor tissues, such as degenerative diseases and brain edema, can emit strong red fluorescence [49]. Photobleaching is another issue that must be addressed. When excitation light is irradiated to a fluorescence tissue, the fluorescence intensity is halved for approximately 10 min. However, we have proved that a decrease in fluorescence intensity can be prevented by using an RGB laser [50]. Therefore, photobleaching can be avoided by using a microscope with an RGB laser for normal observation and surgical operations, and by using an external light source for fluorescence-guided surgery.

It should also be noted that the fluorescence intensity is significantly affected by various observation conditions, such as the distance and angle between the light source and the object, and the intensity of the light source. For example, if the angle is 0° when the light is irradiated perpendicular to the object and 90° when the object is parallel to the light, the light intensity per unit area is halved at an angle of 60°. Furthermore, the light intensity per unit area decreases as the distance from the microscope tip (excitation light source) increases.

7 Future Perspectives

Regarding the use of 5-ALA, which is currently the most frequently used drug for brain tumor imaging, it will be useful to elucidate the mechanism of selective increase of PpIX in tumor cells and to develop strategies for increasing its intracellular concentration to enhance the efficacy of PDD and PDT. To date, 5-ALA is the only approved intraoperative PDD drug for malignant gliomas not only in Japan but also in Europe and the United States. In the future, new photosensitive drugs with higher specificity and sensitivity for glioma cells, especially those that use long wavelengths with high tissue permeability, will be developed and installed in clinical settings.

References

1. Moore GE. Fluorescein as an Agent in the differentiation of normal and malignant tissues. Science. 1947;106:130–1.
2. Kuroiwa T, Kajimoto Y, Ohta T. A new fluorescein operative microscope for use in malignant glioma surgery. 11th International Congress of 1997, pp. 599–603.
3. Kuroiwa T, Kajimoto Y, Ohta T. Development of a fluorescein operative microscope for use during malignant glioma surgery. A technical note and A technical note and preliminary report. Surg Neurol. 1998;50:41–9.
4. Kuroiwa T, Kajimoto Y, Ohta T. Developmental and clinical application of near-infrared surgical microscope-preliminary report. Minim Invas Neurosurg. 2001;44:240–2.
5. Stummer W, Stocker S, Wagner S, et al. Intraoperative detection of malignant gliomas by 5-aminolevulinic acid-induced porphyrin fluorescence. Neurosurgery. 1998;42:518–26.
6. Kremer P, Wunder A, Sinn H, et al. Laser-induced fluorescence detection of malignant gliomas using fluorescein-labeled serum albumin: experimental and preliminary clinical results. Neurol Res. 2000;22:481–9.
7. Kremer P, Fardanesh M, Ding R, et al. Intraoperative fluorescence staining of malignant brain tumors using 5-aminofluorescein-labeled albumin. Neurosurgery. 2009;64:53–60.
8. Shimizu K, Nitta M, Komori T, et al. Using talaporfin sodium simultaneously applied for photodynamic therapy against malignant glioma: a prospective clinical study. Front Neurol. 2018;9 https://doi.org/10.3389/fneur.2018.00024.
9. Ritz R, Daniels R, Noell S, et al. Hypericin for visualization of high grade gliomas: first clinical experience. Eur J Surg Oncol. 2012;38:352–60.
10. Li D, Zhang J, Chi C, et al. First-in-human study of PET and optical dual-modality image-guided surgery in glioblastoma using 68Ga-IRDye800CW-BBN. Theranostics. 2018;8:2508–20.

11. Patil CG, Walker DG, Miller DM, et al. Phase 1 safety, pharmacokinetics, and fluorescence imaging study of Tozuleristide (BLZ-100) in adults with newly diagnosed or recurrent gliomas. Neurosurgery. 2019;85:E641–9.
12. Lee JYK, Cho SS, Zeh R. Folate receptor overexpression can be visualized in real time during pituitary adenoma endoscopic transsphenoidal surgery with near-infrared imaging. J Neurosurg. 2018;129:390–403.
13. Kitagawa Y, Tanaka S, Kamiya M, et al. A novel topical fluorescent probe for detection of glioblastoma. Clin Cancer Res. 2021;27:3936–47.
14. Kuroiwa T, Kajimoto Y, Ohta T. Usefulness of a new fluorescein operative microscope. 11th International Congress of Neurological Surgery. 1997, pp. 1641–6.
15. Shinoda J, Yano H, Yoshimura SI, et al. Fluorescein-guided resection of glioblastoma multiforme by using high-dose fluorescein sodium. note. J Neurosurg. 2003;99:597–603.
16. Katsevman GA, Turner RC, Urhie O, et al. Utility of fluorescein for achieving resection targets in glioblastoma: increased gross- or near-total resections and prolonged survival. J Neurosurg. 2019;132:914–20.
17. Smith EJ, Gohil K, et al. Fluorescein-guided resection of high grade gliomas: a meta-analysis. World Neurosurg. 2021;155:181–8.
18. Hansen RW, Pederson CB, Halle B, et al. Comparison of 5-aminolevulinic acid and sodium fluorescein for intraoperative tumor visualization in patients with high-grade gliomas: a single-center retrospective study. J Neurosurg. 2019;133:1324–31.
19. Naik A, Smith EJ, Barreau A, et al. Comparison of fluorescein sodium, 5-ALA and intraoperative MRI for resection of high-grade gliomas: a systematic review and network meta-analysis. J Clin Neurosci. 2022;98:240–7.
20. Stummer W, Pichlmeier U, Meinel T, et al; ALA-Glioma Study Group. Fluorescence-guided surgery with 5-aminolevulinic acid for resection of Malignant glioma: a randomised controlled multicentre phase III trial. Lancet Oncol. 2006;7:392–401.
21. Haglund MM, Berger MS, Hochman DW. Enhanced optical imaging of human gliomas and tumor margins. Neurosurgery. 1996;38:308–17.
22. Raabe A, Beck J, Gerlach R, et al. Near-infrared indocyanine green video angiography: a new method for intraoperative assessment of vascular flow. Neurosurgery. 2003;52:132–9.
23. Tamura Y, Hirota Y, Miyata S, et al. The use of intraoperative neat-infrared indocyanine green videoangiography in the microscopic resection of hemangioblastomas. Acta Neurochir (Wien). 2012;154:1407–12.
24. Cho SS, Salinas R, De Ravin E, Teng CW, et al. Near-infrared imaging with second-window indocyanine green in newly diagnosed high-grade gliomas predicts gadolinium enhancement on postoperative magnetic resonance imaging. Mol Imaging Biol. 2020;22:1427–37.
25. Teng CW, Cho SS, Singh Y, et al. Second window ICG predicts gross-total resection and progression-free survival during brain metastasis surgery. J Neurosurg. 2021;135:1026–35.
26. Hefti M, Holenstein F, Albert I, et al. Susceptibility to 5-aminolevulinic acid based photodynamic therapy in WHO I meningioma cells corresponds to ferrochelatase activity. Photochem Photobiol. 2011;87:235–41.
27. Suzuki T, Wada S, Eguchi H, et al. Cadherin 13 overexpression as an important factor related to the absence of tumor fluorescence in 5-aminolevulinic acid-guided resection of glioma. J Neurosurg. 2013;119:1331–9.
28. Kim S, Kim JE, Kim YH, et al. Glutaminase 2 expression is associated with regional heterogeneity of 5-aminolevulinic acid fluorescence in glioblastoma. Sci Rep. 2017;7:12221. https://doi.org/10.1038/s41598-017-12557-3.
29. Kaneko S, Molina ES, Ewelt C, et al. Fluorescence-based measurement of real-time kinetics of protoporphyrin IX after 5-aminolevulinic acid administration in human in situ malignant gliomas. Neurosurgery. 2019;85:E739–46.
30. Saito K, Hirai T, Takeshima H, et al. Genetic factors affecting intraoperative 5-aminolevulinic acid-induced fluorescence of diffuse gliomas. Radiol Oncol. 2017;51:142–50.
31. Sun W, Kajimoto Y, Inoue H, et al. Gefitinib enhances the efficacy of photodynamic therapy using 5-aminolevulinic acid in malignant brain tumor cells. Photodiagnosis Photodyn Ther. 2013;10:42–50.
32. Hefti M, Albert I, Luginbuehl V. Phenytoin reduces 5-aminolevulinic acid-induced protoporphyrin IX accumulation in malignant glioma cells. J Neuro-Oncol. 2012;108:443–50.
33. Yagi R, Kawabata S, Ikeda N, et al. Intraoperative 5-aminolevulinic acid-induced photodynamic diagnosis of metastatic brain tumors with histopathological analysis. World J Surg Oncol. 2017;15:175–87.
34. Mercea P, Mischkulnig M, Kiesel B, et al. Prognostic value of 5-ALA fluorescence, tumor cell infiltration and angiogenesis in the peritumoral brain tissue of brain metastases. Cancers. 2021;13:603. https://doi.org/10.3390/cancers13040603.
35. Hussein A, Rohde V, Wolfert C, et al. Survival after resection of brain metastases with white light microscopy versus fluorescence-guidence: a matched cohort analysis of the metastasys study data. Oncotarget. 2020;11:3026–34.
36. Marhold F, Mercea PA, Scheichel F, et al. Detailed analysis of 5-aminolevulinic acid induced fluorescence in different brain metastases at two specialized neurosurgical centers: experience in 157 cases. J Neurosurg. 2019;133:1032–43.
37. Widhalm G, Minchev G, Woehrer A, et al. Strong 5-aminolevulinic acid-induced fluorescence is a novel intraoperative marker for representative tissue samples in stereotactic brain tumor biopsies. Neurosurg Rev. 2012;35:381–91.
38. Kajimoto Y, Kuroiwa T, Miyatake S, et al. Use of 5-aminolevulinic acid in fluorescence-guided resection of meningioma with high risk of recurrence. Neurosurgery. 2007;106:1070–4.
39. Dijkstra BM, Jeltema HJR, Kruijff S. The application of fluorescence techniques in meningioma surgery-a review. Neurosurg Rev. 2019;42:799–809.
40. Foster N, Eljamel S. ALA-induced fluorescence image guided surgery of meningiomas: a meta-analyses. Photodiagn Photodyn Ther. 2016;15:73–8.
41. Morshed RA, Han SJ, Lau D. Wavelength-specific lighted suction instrument for 5-aminolevulinic acid fluorescence-guided resection of deep- seated malignant glioma: technical note. J Neurosurg. 2018;128:1448–53.
42. Tamura Y, Kuroiwa T, Kajimoto Y, et al. Endoscopic identification and biopsy sampling of an intraventricular malignant glioma using a 5 aminolevulinic acid- induced protoporphyrin IX fluorescence imaging system. J Neurosurg. 2007;106:507–10.
43. Suero Molina E, Schipmann S, Stummer W. Maximizing safe resections: the roles of 5-aminolevulinic acid and intraoperative MR imaging in glioma surgery-review of the literature. Neurosurg Rev. 2019;42:197–208.
44. Coburger J, Wirtz CR. Fluorescence guided surgery by 5-ALA and intraoperative MRI in high grade glioma: a systematic review. J Neuro-Oncol. 2019;141:533–46.

45. Yano H, Nakayama N, Ohe N, et al. Pathological analysis of the surgical margins of resected glioblastomas excised using photodynamic visualization with both 5-aminolevulinic acid and fluorescein sodium. J Neuro-Oncol. 2017;133:389–97.
46. Suero Molina E, Wölfer J, Ewelt C. Dual-labeling with 5-aminolevulinic acid and fluorescein for fluorescence-guided resection of high-grade gliomas: technical note. J Neurosurg. 2018;128:399–405.
47. Widhalm G, Olson J, Weller J, et al. The value of visible 5-ALA fluorescence and quantitative protoporphyrin IX analysis for improved surgery of suspected low-grade gliomas. J Neurosurg. 2020;133:79–88.
48. Suero Molina E, Kaneko S, Black D, et al. 5-aminolevulinic acid-induced porphyrin contents in various brain tumors: implications regarding imaging device design and their validation. Neurosurgery. 2021;89:1132–40.
49. Miyatake S, Kuroiwa T, Kajimoto Y, et al. Fluorescence of non-neoplastic, magnetic resonance imaging-enhancing tissue by 5-aminolevulinic acid: a case report. Neurosurgery. 2007;61:1101–4.
50. Matsuda F, Ikeda N, Kajimoto Y, et al. Neurosurgical microscopic solid laser-based light inhibits photobleaching during fluorescence-guided brain tumor removal with 5-aminolevulinic acid. Photodiagn Photodyn Ther. 2017;20:120–4.

Bladder Cancer

18

Keiji Inoue, Hideo Fukuhara, and Shinkuro Yamamoto

Summary

- After oral administration of 5-aminolevulinic acid (5-ALA), protoporphyrin IX (Pp IX) specifically accumulates in cancer cells and emits red fluorescence (600–740 nm) when excited with blue visible light (375–445 nm) (photodynamic diagnosis; PDD).
- Excitation of Pp IX in cancer cells with red (600–740 nm) or green (450–580 nm) visible light generates reactive oxygen species (ROS) that damage mitochondria and induce apoptosis, leading to cell death (photodynamic therapy, PDT).
- ALA-PDD (PDD using 5-ALA) is useful for the visualization of non-muscle-invasive bladder cancer during transurethral resection.
- Adverse reactions of 5-ALA include hepatic dysfunction, hypotension and hypotension, nausea and vomiting, and photosensitivity.
- False positives and false negatives are challenges in ALA-PDD for bladder cancer.

1 Introduction

Since more than 100 years ago, photodynamic diagnosis (PDD) and photodynamic therapy (PDT) using photosensitizers and special light sources have been focused on for the diagnosis and treatment of cancerous lesions. Specifically, porphyrin derivatives and lasers and light irradiation devices such as argon lasers have been developed in basic and clinical studies. In recent years, 5-aminolevulinic acid (5-ALA), a natural amino acid found in animals and plants, has been introduced as a safe and effective photosensitizer, evoking interest in PDD and PDT again. In Japan, 5-ALA was approved in 2013 as an intraoperative diagnostic agent for visualization of tumor tissue during resection for malignant glioma, followed by approval in 2017 for identification of non-muscle-invasive bladder tumor during transurethral resection. Currently, a physician-led clinical trial is underway to evaluate the efficacy and safety of PDD using 5-ALA during staging laparoscopy in patients with advanced gastric cancer.

In this chapter, we outline the current status and future development of 5-ALA-based PDD (ALA-PDD) in non-muscle invasive bladder cancer.

2 Conventional Surgical Treatment for Bladder Cancer

Non-muscle-invasive bladder cancer (Ta, T1) accounts for 75–85% of all bladder cancers [1]. Transurethral resection of bladder tumor (TURBT) is the first choice for the initial treatment of bladder cancer, which not only allows bladder preservation but also has a favorable prognosis. However, the recurrence rate of bladder cancer after TURBT is high, with Ta 15% and T1 61% within 1 year and Ta 31% and T1 78% within 5 years, respectively [2]. In addition, the probability of progression to invasive cancer is also high, at 5% and 50%, respectively, [3] and is considered as a major clinical problem.

There would be a lot of factors, such as multicentric development and intravesical metastasis, for the early intravesical recurrence of bladder cancer. Among them, residual lesions that are difficult to identify by conventional endoscopic examinations based on white-light imaging (non-visible lesions) at the time of surgery, including micro lesions and flat lesions such as dysplasia and intraepithelial carcinoma, are known to play a major role in causing early post-operative intravesical recurrence. In other words, the

K. Inoue (✉) · H. Fukuhara
Department of Urology, Kochi Medical School, Nankoku-shi, Kochi, Japan

Center for Photodynamic Medicine, Kochi Medical School, Nankoku-shi, Kochi, Japan
e-mail: keiji@kochi-u.ac.jp

S. Yamamoto
Department of Urology, Kochi Medical School, Nankoku-shi, Kochi, Japan

T. Ishizawa (ed.), *Fluorescence-Guided Surgery*, https://doi.org/10.1007/978-981-19-7372-7_18

identification of these endoscopic non-visible lesions is essential for improving the diagnostic accuracy and therapeutic outcome of non-muscle-invasive bladder cancer. That's why we introduced ALA-PDD into clinical practice as a remedy for this major clinical problem in the surgical treatment of bladder cancer.

3 Development History of Photodynamic Diagnosis and Photodynamic Therapy

The history of medical applications of photodynamic technology dates back to the beginning of the twentieth century. The first study on PDD is thought to be the experiment conducted by Raab [4] in 1900, in which acridine dye was used as a photosensitizer to show its killing effect on the weevil using sunlight as a light source. In 1960, Lipson et al. [5] developed a hematoporphyrin derivative (HpD), which is a photosensitizer with high tumor affinity without toxicity or mutagenicity. This can be regarded as the prototype of the current PDD. In 1978, Dougherty et al. [6] developed a red light source (630 nm) using an argon laser, and in 1984, Dougherty et al. [7] developed dihematoporphyrin ether, an active ingredient with higher tumor affinity. In Japan, Hayata et al. [8] developed the excimer dye laser, which is a high-power pulsed laser superior to the conventional light source for use in photodynamic techniques. In other words, the history of PDD and PDT can be regarded as the development history of photosensitizers and excitation sources.

Photodynamic techniques in the field of urology were reported by Rall et al. [9] in 1957 and started with the first clinical trial of PDD using tetracycline and UV light in 21 cases of bladder cancer, published by Whitmore et al. [10] in 1964. Kelly et al. [11] performed PDT using HpD and mercury arc lamps for bladder cancer. In Japan, Hisazumi et al. [12] first reported PDT using HpD and argon dye laser for 46 tumors in 9 cases of superficial bladder cancer in 1983. This treatment was vigorously tried and showed favorable antitumor efficacy. However, it did not become widely used because of the high incidence of adverse events, such as the appearance of phototoxic reactions, mainly photosensitivity, due to systemic administration of hematoporphyrin derivatives and atrophic bladder caused by laser irradiation of the entire bladder surface.

In 1987, Malik et al. [13] developed 5-ALA as a third-generation photosensitizer. Furthermore, a white xenon light source system with a bandpass filter system that transmits blue light for the excitation of red fluorescence and a long-pass filter that blocks blue light for observation of red fluorescence was developed. This enables PDD by instantly switching between conventional white light and blue light to excite red fluorescence through the same endoscope. The development of a dedicated PDD system using 5-ALA led to renewed interest in PDD and PDT. First, in 1990, Kennedy et al. [14] performed PDT using topical 5-ALA administration for skin cancer. In the urological field, Kriegmair et al. [15] first performed PDD and PDT by intravesical administration of 5-ALA for bladder cancer in 1992 and showed antitumor effects. Since then, many clinical trials of PDD and PDT have been conducted in Japan and overseas for many diseases in many fields. Now, 5-ALA and 5-ALA derivatives are approved by the pharmaceutical affairs bodies worldwide as photosensitizers for PDD and PDT (see below).

4 Principles of Photodynamic Diagnosis for Bladder Cancer

The natural amino acid 5-ALA, which has been endogenous to plants and animals since 3.6 billion years ago, is a common precursor of hemoglobin and chlorophyll. Both endogenous 5-ALA synthesized from succinyl CoA and glycine in mitochondria and 5-ALA administered from outside the body follow the same metabolic pathway, passing through a precursor in the cytoplasm to be biosynthesized again into protoporphyrin IX (Pp IX) in mitochondria. Subsequently, Pp IX is metabolized to heme and bilirubin by the insertion of divalent iron catalyzed by ferrochelatase. In cancer cells, abnormalities in the activity of various transporters and enzymes promote Pp IX production and inhibit its metabolism, leading to hyperaccumulation of Pp IX [13, 16]. In particular, Pp IX accumulation is 17-fold higher in urothelial carcinoma than in normal epithelium [15–18]. Pp IX is photoactive and emits red fluorescence (600–740 nm) when excited with blue visible light (375–445 nm). The 5-ALA-based photodynamic diagnosis (5-ALA-PDD) is a method of cancer diagnosis by fluorescence navigation using 5-ALA as a photosensitizer. In contrast, low-power excitation of Pp IX in cancer cells with red (600–740 nm) or green (450–580 nm) visible light induces the generation of reactive oxygen species (ROS), such as singlet oxygen, through the energy conversion of Pp IX from the excited state to the ground state. The mitochondria are damaged, and apoptosis is induced, leading to cell death. This is photodynamic therapy using 5-ALA (ALA-PDT) [19]. These ALA-PDD and ALA-PDT are considered to be photodynamic techniques based on the "Warburg effect," a fundamental biological characteristic common to all cancers, which states that cancers prefer anaerobic metabolism (Fig. 18.1).

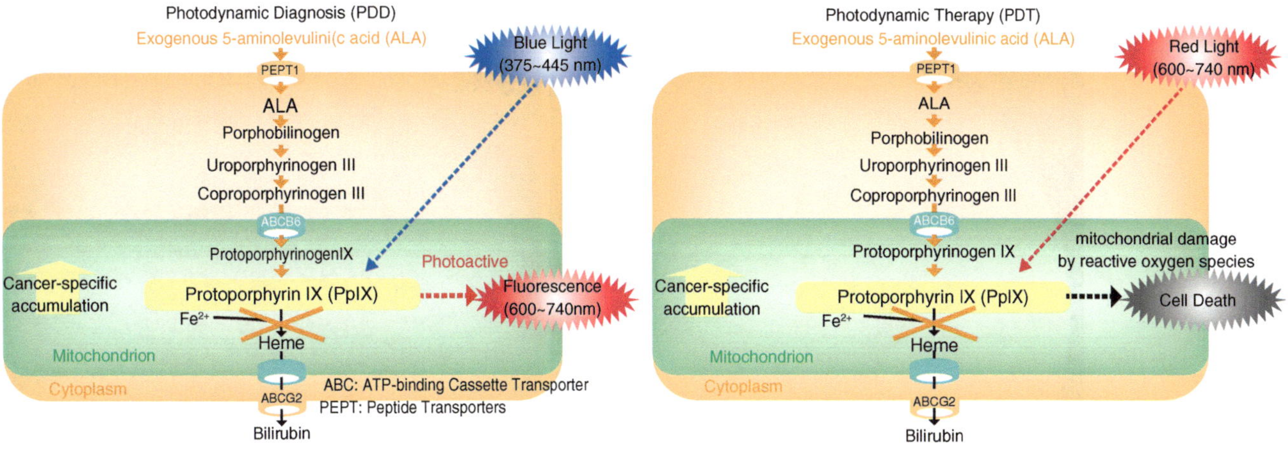

Fig. 18.1 Mechanistic background of PDD and PDT

5 Clinical Introduction of Photodynamic Diagnosis for Bladder Cancer

In 2004, the first physician-led clinical trial in Japan was conducted to introduce ALA-PDD into clinical practice as a remedy for a major clinical problem in bladder cancer treatment: endoscopic non-visible lesions, which are a cause of early intravesical recurrence after TURBT. As a result, it was shown that ALA-PDD improved the diagnostic accuracy of bladder cancer compared with conventional diagnosis using a white light source and, in particular, markedly improved the detection rate of endoscopically non-visible lesions such as microscopic cancers and flat intraepithelial cancers. Furthermore, it was demonstrated that TURBT with observation by ALA-PDD (PDD-TURBT) can reduce early postoperative intravesical recurrence [20]. In 2012, an investigator-initiated clinical trial (Phase II/III study) was conducted as a clinical trial promotion research project by the Japan Medical Association Clinical Trial Implementation Center [21, 22], and in 2015, a corporate trial (Phase III study) was conducted by SBI Pharmaceuticals Co. The results of these two clinical trials were combined into a single package and an application for a drug manufacturing license was submitted. Finally, on September 27, 2017, 5-ALA was approved in Japan for the first time in the world as an intraoperative diagnostic drug to be administered orally for PDD of bladder cancer. On December 19 of the same year, Alaglio® Granules 1.5 g (generic name: aminolevulinic acid hydrochloride) was launched as an orphan drug for the visualization of non-muscle-invasive bladder cancer during TURBT [22, 25] (Fig. 18.2).

In 2019, the Japanese "Guidelines for the Treatment of Bladder Cancer" were revised [24]. In the revised guideline, in response to CQ1, "Is tumor visualization technology recommended for the diagnosis of bladder cancer?", PDD and narrow-band imaging (NBI)are recommended with grade 1 strength (certainty of evidence is A for PDD and B for NBI). Furthermore, in response to CQ4, "Is PDD or NBI recommended for the treatment of non-muscle-invasive bladder cancer (NMIBC)?", PDD is recommended because it is associated with a lower bladder recurrence rate (strength of recommendation 1, certainty of evidence A). On the other hand, NBI improves the cancer detection rate, but whether it leads to a decrease in the bladder recurrence rate is undetermined (strength of recommendation 2, certainty of evidence B). The European and American guidelines [25, 26] also strongly recommend ALA-PDD as it markedly improves the diagnostic accuracy, especially the detection rate of bladder intraepithelial carcinoma, and also improves recurrence-free survival by PDD-TURBT.

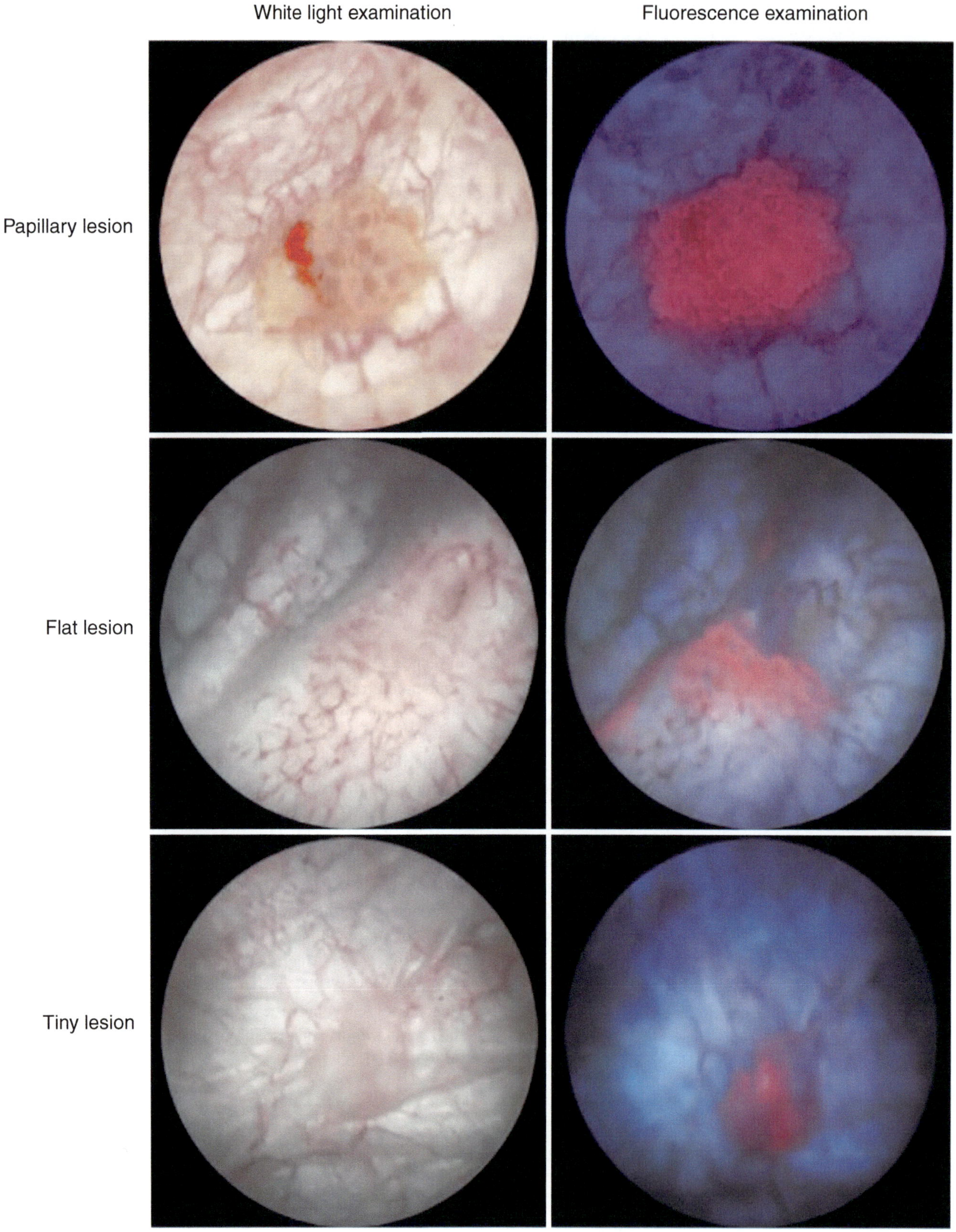

Fig. 18.2 Endoscopic image of PDD during TURBT for bladder cancer. (Cited from Ref. [24])

6 Preparation and Administration of 5-Aminolevulinic Acid in Photodynamic Diagnosis for Bladder Cancer

Patients with a history of hypersensitivity to porphyrins or porphyria should be excluded. If patients have undergone BCG, intravesical administration of anticancer drugs, bladder biopsy, or TURBT, ALA-PDD should be performed within 3 months after the procedure, considering the prolongation of bladder inflammation. First, 2–4 hours before the fluorescence observation, take one package (1.5 g) of Alaglio® granules in an appropriate container, add 50 mL of water, dissolve, and measure the volume of the administered solution calculated from 20 mg/kg of aminolevulinic acid hydrochloride, and administer the drug orally. The granules are not to be administered orally [22].

It is important to avoid exposure to strong light of 500 lux or more for at least 48 hours after taking Clozapine to avoid the adverse event of photosensitivity (photosensitive reaction). In addition, during the period from the time of taking the drug to the time of entering the operating room, and during the period between the time of moving from the operating room to the hospital ward, it is necessary to consider measures such as passing on the shaded side in the corridor exposed to direct sunlight [22].

7 Usefulness of Photodynamic Diagnosis for Bladder Cancer

We summarized the usefulness of ALA-PDD in bladder cancer based on meta-analysis and systematic review. The sensitivity of ALA-PDD is 90% or higher, but the specificity is low, ranging from 57% to 79%, and there are many false positives [27, 28]. In addition, the additional detection rate of cancer identified by ALA-PDD was high, ranging from 25% to 40.8% in per lesion, mainly in intraepithelial cancer, and 19% in per patient, supporting its effectiveness [29–32]. Furthermore, as a useful intraoperative diagnostic tool, it was shown to have a low postoperative residual tumor rate [27, 29, 30, 33] and a significantly high recurrence-free survival rate [27, 29–32, 34–36]. However, the efficacy of ALA-PDD on progression-free survival is controversial [27, 30, 34, 36, 37].

8 Side Effects and Limitations

The incidence of adverse drug reactions was reported in 2062 patients using Alaglio® Granules in 1.5 g packets registered in the results of use survey (all-case surveillance) from launch on December 19, 2017, to March 31, 2019 [23]. As a result, the total number of adverse reactions was 257 events in 197 cases, mainly including 112 events of hepatic dysfunction such as abnormal liver function in 101 cases, 38 events/cases of hypotension/low blood pressure, 28 events/cases of vomiting, and 7 events/cases of photosensitivity (photosensitive reaction). Of these, 26 events/26 cases were serious adverse reactions: 13 events/cases of hypotension, 4 events/3 cases of hepatic dysfunction such as hepatic function abnormality (tended to be relieved by symptomatic treatment within 2 to 3 days), 2 events/cases of myocardial infarction, and 1 event/case of photosensitivity (photosensitive reaction in an adult patient with atopic dermatitis). Although no drug-associated mortality was reported, there was serious hypotension prolonged after induction of anesthesia and required continuous administration of adrenaline, including a case of cardiac arrest that occurred temporarily in an elderly patient with cardiovascular disease. In patients with a risk of cardiovascular events, it is important to take measures in cooperation with an anesthesiologist in advance to the use of 5-ALA during surgery.

False positives and false negatives in ALA-PDD are also a challenge [22]. There are several reasons for false-positive results, including the accumulation of Pp IX in normal tissues and in inflamed mucosa with cell proliferation. In addition, when observing the bladder neck and prostatic urethra at the bladder outlet, the thickness of the normal mucosa increases due to the oblique angle of observation and may be visualized in red. The red fluorescence of PpIX gradually fades after exposure to blue light, a phenomenon known as photobleaching. Therefore, observation using a blue light source should be carried out as quickly and efficiently as possible. Some white light sources also contain wavelengths that cause fading, so it is advisable to avoid exposure to white light sources for longer than necessary. In addition, blue excitation light can penetrate only within 1 mm from the surface of the bladder wall. Therefore, in lesions where tumor tissue infiltrates the normal mucosa or where the surface of the tumor is necrotic, the excitation light cannot reach the lesion and red fluorescence does not emit, which may result in a false negative. Furthermore, thermal degeneration or hemorrhage of the resected surface during TURBT may cause false-negative results without emitting red fluorescence. PDD-TURBT should be performed keeping in mind that false-positive and false-negative results may occur in the above cases.

9 Future Perspectives

As mentioned above, the principle of photodynamic techniques is based on the "Warburg effect," a fundamental biological property common to malignant tumors, suggesting that PDD can be applied to various surgical procedures other than TURBT for bladder cancer. In fact, a physician-initiated clinical trial is being conducted for peritoneal dissemination of gastric cancer, following brain tumor and bladder cancer. In the field of urology, ALA-PDD has been reported for upper urinary tract cancers such as renal pelvis and ureter, which are the same type of urothelial carcinoma as bladder cancer, and has been shown to improve diagnostic accuracy compared with conventional pyelography using a white light source [38]. In addition, reports of ALA-PDT as well as ALA-PDD in bladder cancer are increasing, and a favorable antitumor effect has been demonstrated [39–41]. It is expected that these photodynamic techniques will be introduced into clinical practice in many cancers in the future.

Point

- Oral administration of 5-ALA induces tumor-specific hyperaccumulation of protoporphyrin IX (Pp IX).
- PDD and PDT have the potential to be used as effective therapeutic modalities in various cancers.

References

1. Babjuk M, Oosterlinck W, Sylvester R, et al.; European Association of Urology (EAU). EAU guidelines on non-muscle-invasive urothelial carcinoma of the bladder: European Association of Urology (EAU). Eur Urol. 2008;54:303–14.
2. Brausi M, Collette L, Kurth K, et al.; EORTC Genito-Urinary Tract Cancer Collaborative Group. Variability in the recurrence rate at first follow-up cystoscopy after TUR in stage Ta T1 transitional cell carcinoma of the bladder: a combined analysis of seven EORTC studies. Eur Urol. 2002;41:523–31.
3. Sylvester RJ, van der Meijden AP, Oosterlinck W, et al. Predicting recurrence and progression in individual patients with stage Ta T1 bladder cancer Using EORTC risk tables: a combined analysis of 2596 patients from seven EORTC trials. Eur Urol. 2006;49:466–75.
4. Raab O. Ueber die Wirkung fluorescierender Stoffe und Infusorien. ZBiol. 1900;39:524–6.
5. Lipson RL, Baldes EJ. The photodynamic properties of a particular hematoporphyrin derivative. Arch Dermatol. 1960;82:508–16.
6. Dougherty TJ, Kaufman JE, Goldfarb A, et al. Photoradiation therapy for the treatment of malignant tumors. Cancer Res. 1978;38:2628–35.
7. Dougherty TJ, Potter WR, Weishaupt KR. The structure of the active component of the hematoporphyrin derivative. In: Porphyrin localization and treatment of tumors. New York: Alan R. Liss Inc.; 1984. p. 301–14.
8. Hayata Y, Kato H, Noguchi M, et al. Photodiagnostic treatment of cancer using excimer-diode laser. BME. 1987;1:532–5.
9. Rall DP, Loo TL, Lane MG. Appearance and persistence of fluorescent material in tumor tissue after tetracycline administration. J National Cancer Institute. 1957;19:79–86.
10. Whitmore WF Jr, Bush IM, Esquivel E. Tetracycline ultraviolet fluorescence in bladder carcinoma. Cancer. 1964;17:1528–32.
11. Kelly JF, Snell ME. Hematoporphyrin derivative: a possible aid in the diagnosis and therapy of carcinoma of the bladder. J Urol. 1976;115:150–1.
12. Hisazumi H, Misaki T, Miyoshi N. Photoradiation therapy of bladder tumors. J Urol. 1983;130:685–7.
13. Malik Z, Lugaci H. Destruction of erythroleukaemic cells by photoactivation of endogenous porphyrins. Br J Cancer. 1987;56:589–95.
14. Kennedy JC, Pottier RH, Pross DC. Photodynamic therapy with endogenous protoporphyrin IX: basic principles and present clinical experience. J Photochem Photobiol B. 1990;6:143–8.
15. Kriegmair M, Waidelich R, Baumgartner R, et al. Photodynamic therapy of superficial bladder cancer: an alternative to radical cystectomy? Urologe A. 1990;33:276–80.
16. Inoue K, Takashi K, Kamada M, et al. Regulation of 5-aminolevulinic acid-mediated protoporphyrin IX-accumulation in human urothelial carcinomas. Pathobiology. 2009;76:303–14.
17. Steinbach P, Weingandt H, Baumgartner R, et al. Cellular fluorescence of the endogenous photosensitizer protoporphyrin IX following exposure to 5- aminolevulinic acid. Photochem Photobiol. 1995;62:887–95.
18. Steinbach P, Kriegmair M, Baumgartner R, et al. Fluorescence photodetection of neoplastic urothelial lesions following intravesical instillation of 5-aminolevulinic acid. Urology. 1994;44:836–41.
19. Ishizuka M, Abe F, Sano Y, et al. Novel development of 5-aminolevulinic acid (ALA) in cancer diagnoses and therapy. Int Immunopharmacol. 2011;11:358–65.
20. Inoue K, Fukuhara H, Shimamoto T, et al. Comparison between intravesical and oral administration of 5-aminolevulinic acid in the clinical benefit of photodynamic diagnosis for non-muscle invasive bladder cancer. Cancer. 2012;118:1062–74.
21. Inoue K, Anai S, Fujimoto K, et al. Oral 5-aminolevulinic acid mediated photodynamic diagnosis using fluorescence cystoscopy for non-muscle- invasive bladder cancer: a randomized, double-blind, multicentre phase II/III study. Photodiagn Photodyn Ther. 2015;12:193–200.
22. Alaglio® Granule Formulation Batch 1.5g Aminolevulinic Acid Hydrochloride Granules Appropriate Use Guide, November 2017.
23. Alaglio® Granule Formulation Batch 1.5g Aminolevulinic Acid Hydrochloride Granule Adverse Reactions, May 2019.
24. The Japanese Urological Association. Guidelines for the treatment of bladder cancer. Medical Library Publishing; 2019. edition.
25. EAU Oncology Guidelines. https://uroweb.org/individual-guidelines/oncology-guidelines/.
26. Diagnosis and Treatment of Non-Muscle Invasive Bladder Cancer. AUA/SUO Joint Guideline; 2016. https://www.auanet.org/guidelines/bladder-cancer-non-muscle-invasive-guideline.
27. Mowatt G, N'Dow J, Vale L, et al.; Aberdeen Technology Assessment Review (TAR) Group. Photodynamic diagnosis of bladder cancer compared with white light cystoscopy: systematic review and meta-analysis. Int J Technol Assess Health Care. 2011;27: 3–10.
28. Chen C, Huang H, Zhao Y, et al. Diagnostic performance of image technique based transurethral resection for non-muscle invasive bladder cancer: a systematic review and diagnostic meta-analysis. BMJ Open. 2019;9:e028173.
29. Kausch I, Sommerauer M, Montorsi F, et al. Photodynamic diagnosis in non-muscle- invasive bladder cancer: a systematic review and cumulative analysis of prospective studies. Eur Urol. 2010;57:595–606.
30. Rink M, Babjuk M, Catto JW, et al. Hexyl aminolevulinate-guided fluorescence cystoscopy in the diagnosis and follow-up of patients with non-muscle- invasive bladder cancer: a critical review of the current literature. Eur Urol. 2013;64:624–38.
31. Burger M, Grossman HB, Droller M, et al. Photodynamic diagnosis of non-muscle- invasive bladder cancer with hexaminolevulinate cystoscopy: a meta-analysis of detection and recurrence based on raw data. Eur Urol. 2013;64:846–54.
32. Di Stasi SM, De Carlo F, Pagliarulo V, et al. Hexaminolevulinate hydrochloride in the detection of nonmuscle invasive cancer of the bladder. Ther Adv Urol. 2015;7:339–50.

33. Shen P, Yang J, Wei W, et al. Effects of fluorescent light-guided transurethral resection on non-muscle-invasive bladder cancer: a systematic review and meta-analysis. BJU Int. 2012;110:E209–15.
34. Yuan H, Qiu J, Liu L, et al. Therapeutic outcome of fluorescence cystoscopy guided transurethral resection in patients with non-muscle invasive bladder cancer: a meta-analysis of randomized controlled trials. PLoS One. 2013;8:e74142.
35. Rolevich AI, Evmenenko AA. A systematic review and meta-analysis to assess the recurrence-free survival in non-muscle invasive bladder cancer after transurethral resection guided by 5-aminolevulinic acid-induced photodynamic diagnosis compared with white-light transurethral resection. Urologiia. 2016;4:137–46.
36. Chou R, Selph S, Buckley DI, et al. Comparative effectiveness of fluorescent versus white light cystoscopy for initial diagnosis or surveillance of bladder cancer on clinical outcomes: systematic review and meta-analysis. J Urol. 2017;197:548–58.
37. Gakis G, Fahmyb O. Systematic review and meta-analysis on the impact of hexaminolevulinate- versus white-light guided transurethral bladder tumor resection on progression in non-muscle invasive bladder cancer. Bladder Cancer. 2016;2:293–300.
38. Osman E, Alnaib Z, Kumar N. Photodynamic diagnosis in upper urinary tract urothelial carcinoma: a systematic review. Arab J Urol. 2017;15:100–1097.
39. Inoue K, Fukuhara H, Kurabayashi A, et al. Photodynamic therapy involves anti- angiogenic mechanism and is enhanced by ferrochelatase inhibitor in urothelial carcinoma. Cancer Sci. 2013;104:765–72.
40. Inoue K. 5-Aminolevulinic acid-mediated photodynamic therapy for bladder cancer. Int J Urol. 2017;24:97–101.
41. Lee JY, Diaz RR, Cho KS, et al. Efficacy and safety of photodynamic therapy for recurrent, high-grade nonmuscle invasive bladder cancer refractory or intolerant to bacille Calmette-Guérin immunotherapy. J Urol. 2013;190:1192–9.

Part IV

Intraoperative Fluorescence Imaging [Practice] – Imaging of Lymph Nodes and Lymph Vessels

Introduction

19

Masashi Yoshida

The sentinel node is the lymph node where the lymphatic flow from the primary tumor first enters. For the identification of the sentinel node, the dye method (injection of visible dye such as indigocarmine and indocyanine green (ICG) around the primary lesion) and/or radioisotope (RI) method (injection of ^{99m}Tc tin colloid, etc.) have been used. The dual tracer method (concomitant use of the dye method and the RI method) has recently been regarded as a standard because the dye method alone is associated with relatively high false-negative results due to fat tissues around the nodes. The use of fluorescence imaging following an injection of ICG may enhance the detectability of the sentinel node because near-infrared fluorescence signals can penetrate fat tissues up to about 10 mm. In addition, ICG fluorescence imaging has advantages over the RI technique in that it does not need the radiation exposure of the patient and medical staff, and the hurdles of handling radioactive isotopes (e.g., the need to maintain the endoscope set in a radiation-controlled area).

The advantage of the RI technique is that the particle size of 99mTc tin colloid is large and most of the particles stay in the sentinel node. In contrast, in the ICG fluorescence method, the particles are so fine that they pass through the sentinel node and secondary and tertiary lymph nodes. Therefore, it is important to observe the timing and distributions of fluorescence signals following an injection of ICG in order to identify the sentinel lymph node accurately [1]. Among various protocols of ICG fluorescence imaging for sentinel node navigation, the method introduced by Kinami et al. [2] [injection of 100-fold dilution (50 μg/mL ICG)] around the gastric cancer on a day before surgery, as demonstrated in Chap. 21, seems to be associated with favorable outcomes and may be useful in the surgical procedures for other organs. When the resected specimens are observed by fluorescence imaging according to Kinami's protocol, a lot of lymph nodes can be identified with different fluorescence intensities. Since ICG can be accumulated in the lymph nodes gradually over the course of a day, we can consider the node with the highest fluorescence intensity as the sentinel lymph node [3–5].

Chapter 24 also illustrates the use of ICG fluorescence imaging for lymphography and evaluation of lymphedema, which has been inspired by the sentinel node identification methods. Nowadays, ICG fluorescence imaging has widely been used as an essential technique for the evaluation and treatment of lymphoedema. This would be a good example indicating that cross-sectional exchange of information among specialties is indispensable for the development of new surgical techniques like intraoperative fluorescence imaging.

References

1. Takahashi N, Nimura H, Fujita T, et al. Laparoscopic sentinel node navigation surgery for early gastric cancer: a prospective multicenter trial. Langenbeck's Arch Surg. 2017;402:27–32.
2. Kinami S, Oonishi T, Fujita J, et al. Optimal settings and accuracy of indocyanine green fluorescence imaging for sentinel node biopsy in early gastric. Oncol Lett. 2016;11:4055–62.
3. Yoshida M, Kubota K, Kuroda J, et al. Indocyanine green injection for detecting sentinel nodes using color fluorescence camera in laparoscopy-assisted gastrectomy. J Gastroenterol Hepatol. 2012;27:29–33.
4. Ohdaira H, Yoshida M, Okada S, et al. New method of indocyanine green fluorescence sentinel node mapping for early gastric cancer. Ann Med Surg. 2017;20
5. Kamada T, Yoshida M, Takeuchi H, et al. A new method for sentinel node mapping for early gastric cancer using a fluorescent laparoscope that can adjust the intensity of excitation light and quantify the intensity of indocyanine green fluorescence: report of a case. Int J Surg Case Rep. 2020;73:248–52.

M. Yoshida (✉)
Department of Surgery, International University of Health and Welfare Hospital, Nasushiobara, Tochigi, Japan
e-mail: masashi@iuhw.ac.jp

T. Ishizawa (ed.), *Fluorescence-Guided Surgery*, https://doi.org/10.1007/978-981-19-7372-7_19

Identification of Sentinel Lymph Nodes in Breast Cancer Surgery

20

Manami Tada and Tomoharu Sugie

Summary

- Along with the RI method, ICG fluorescence imaging is a standard technique for sentinel node biopsy during breast cancer surgery.
- High feasibility and radiation-free nature are obvious advantages of ICG fluorescence imaging over the RI method.
- Applications of fluorescence imaging based on the optimal characteristics of ICG lead to highly sensitive identification of sentinel lymph nodes.

1 Introduction

The safety and usefulness of sentinel node biopsy have been established as a standard treatment for breast cancer. In Japan, the indocyanine green (ICG) fluorescence method has been developed as an alternative to the dye and radioisotope (RI) methods, which have been used as tracers in many institutions around the world. In this chapter, we outline the actual technique of sentinel lymph node biopsy using ICG fluorescence imaging in breast cancer surgery.

2 Indocyanine Green Fluorescence Method in the Breast Cancer Field

The Japanese Breast Cancer Treatment Guidelines [1], ASCO, and NCCN guidelines state that axillary lymph node dissection should be omitted when the sentinel node biopsy is negative in breast cancer patients without clinical findings suggesting lymph node metastases. For the sentinel node biopsy, concomitant use of the RI method and the dye method has recently been regarded as a standard technique.

In the RI method, a radioisotope (technetium phthenate [^{99m}Tc] in Japan) is administered and its uptake in lymph nodes is detected with lymphoscintigraphy before surgery. Intraoperatively, the nodes were picked up by identifying hot spots with the use of a RI detector. The RI method has a high identification rate (96%) and a low false-negative rate (7.3%) [2]. On the other hand, the use of radiopharmaceuticals is associated with major disadvantages, such as radiation exposure and limited access to hospitals with nuclear medicine facilities. The dye method is safe, feasible, and inexpensive, but its identification rate has been reported to be low (78%) [3].

In order to overcome these problems, sentinel node biopsy using ICG fluorescence imaging has been developed mainly in Japan. Recently, this technique has widely been used as a reliable technique for judging the need for lymph node dissection during breast cancer surgery.

3 Principle of Indocyanine Green Fluorescence Method

A sentinel lymph node is defined as one or a couple of lymph nodes where cancer cells first flow from the primary tumor. ICG fluorescence imaging is a method to visualize lymphatic flow by detecting fluorescence signals emitted from protein-bound ICG, with a near-infrared camera system. For example, a Photodynamic Eye (PDE, Hamamatsu Photonics) consists of a light-emitting diode (LED) light source emitting light at 760 nm and a charged-coupled device (CCD) camera with a filter that blocks light below 820 nm.

4 Indocyanine Green Fluorescence Method Procedure

First, 25 mg of ICG is dissolved in 5 mL of injection solution. The concentration of ICG has not been consistently determined through previous clinical studies, but recent

M. Tada · T. Sugie (✉)
Department of Breast Surgery, Kansai Medical University Hospital, Hirakata, Osaka, Japan
e-mail: tadam@hirakata.kmu.ac.jp; sugiet@hirakata.kmu.ac.jp

T. Ishizawa (ed.), *Fluorescence-Guided Surgery*, https://doi.org/10.1007/978-981-19-7372-7_20

meta-analyses have shown that ICG concentrations of less than 5 mg/mL lead to the highest sensitivity and the lowest false-negative rate [4]. ICG of 1 mL is injected into the nipple dermis using a syringe with a 27G intradermal needle (Fig. 20.1), massaged for 2–3 minutes, and the flow of ICG is immediately confirmed by PDE. ICG flow in the peripheral veins can be observed immediately after ICG injection, but it is quickly washed out. Several lymphatic streams may be observed, but they converge toward the axilla and flow to the sentinel lymph nodes (Fig. 20.2). If this lymphatic flow is traced on the skin with a dermal pen, it appears to be interrupted as it enters the axillary cavity and flows into the lymph nodes. Then, a 3–4 cm skin incision is made to follow the lymphatic flow (Fig. 20.3) by fluorescence imaging. The incision is extended bluntly as far as possible from the lateral border of the pectoralis major fascia to the contiguous superficial pectoral fascia to avoid injuring the lymphatic vessels. An incision is made through that membrane to reach the axillary cavity, which has a layer of fat that is softer than the subcutaneous layer. After bluntly widening the view of the axillary cavity with a muscle hook, the localization of the lymph nodes beyond the lymphatic vessels can be confirmed by adjusting the sensitivity of the PDE (Fig. 20.4). Once located, use forceps to pull the fluorescing fat layer forward. Grasping the lymph node with forceps allows for clearer visualization and differentiation of the node from the lymphatics. Finally, the sentinel node is removed along with the surrounding fat tissues. After the removal of the first node, fluorescence imaging as well as manual palpation should be used again for detecting residual sentinel nodes in the resected axillary space (Fig. 20.5).

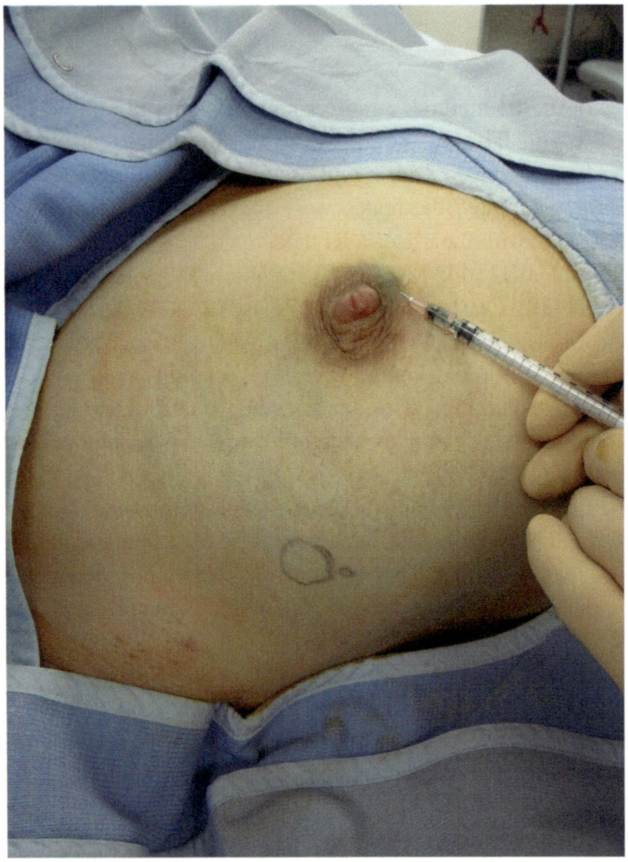

Fig. 20.1 Injection of ICG (原著)

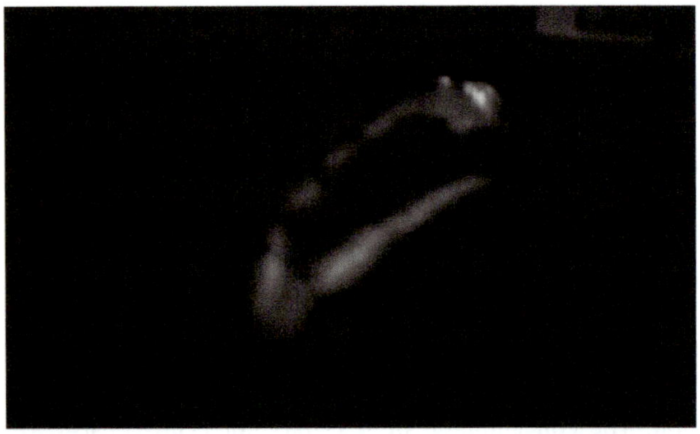

Fig. 20.2 Lymphatic flow confirmed by PDE (Hamamatsu Photonics K.K)

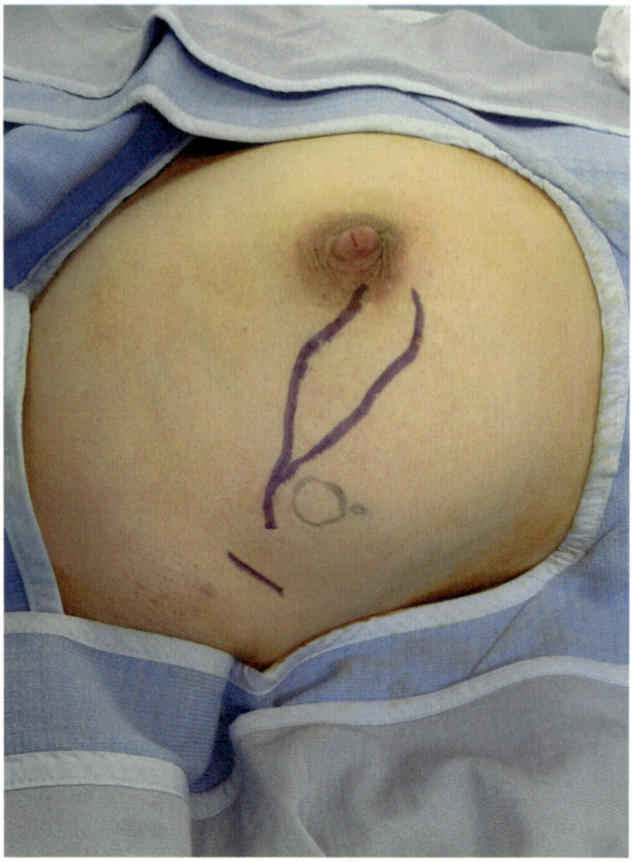

Fig. 20.3 Setting of the skin incision

Point

- Accurate intradermal injection of ICG solution in the areola of the nipple, followed by a sufficient massage

Fig. 20.4 Identification of the sentinel lymph node

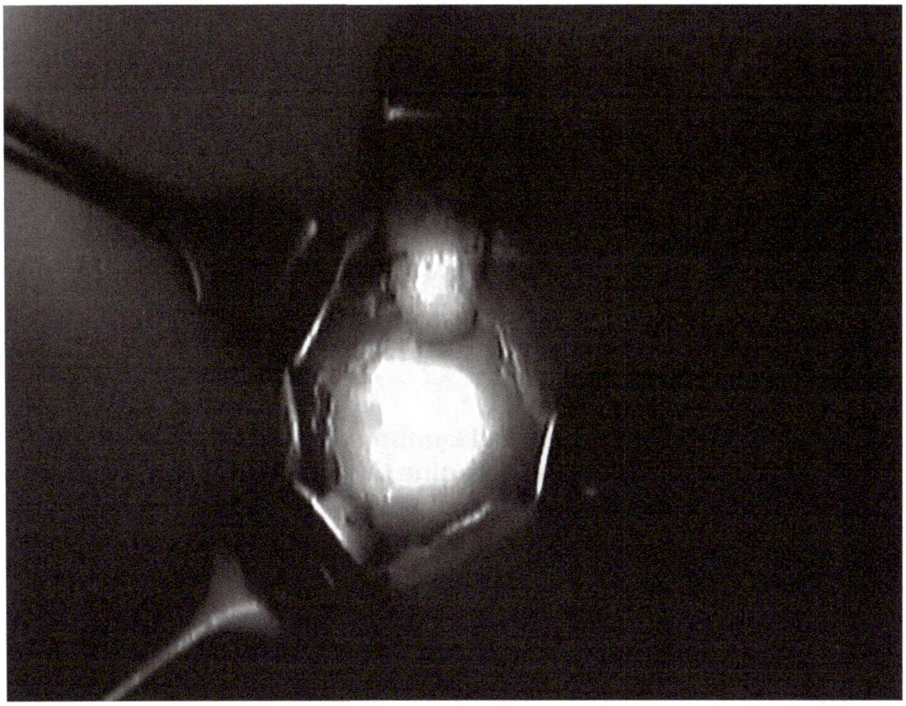

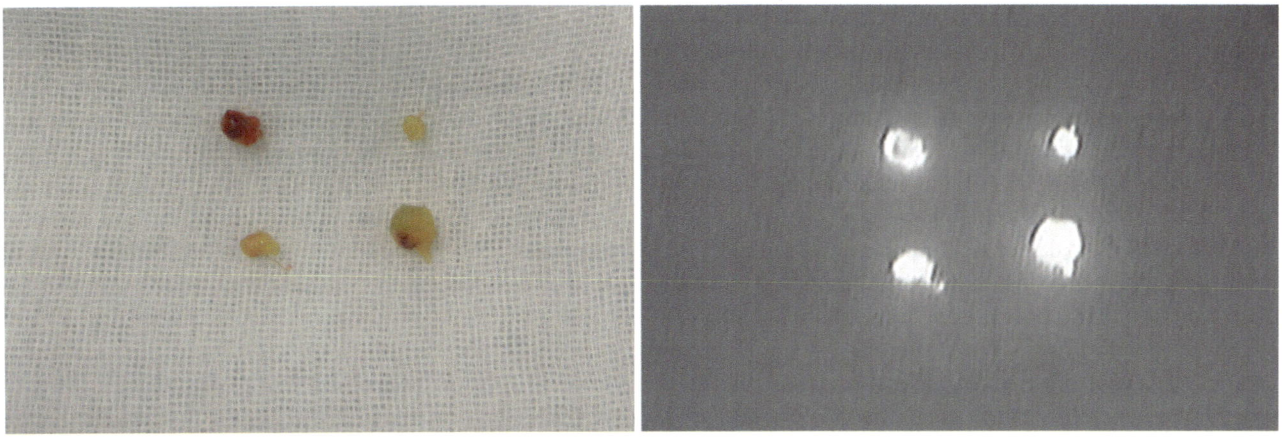

Fig. 20.5 Removed lymph nodes

- Follow lymphatic flows to the axilla using fluorescence imaging and determine the skin incision point
- Brunt dissection of subfascial spaces not to injure the lymphatic vessels
- Systemic removal of fluorescing lymph nodes

5 Knacks and Pitfalls of Indocyanine Green Fluorescence Imaging

For the effective use of this technique, we need to well understand the properties of ICG and fluorescence imaging. First, because ICG flows rapidly through the lymphatic systems over time, the procedure should be started immediately after ICG injection and completed quickly. If fluorescence images are obtained for a long time after ICG injection, lymph nodes distal to the true sentinel node may also fluoresce and be removed as sentinel lymph nodes. Therefore, it is important to share the procedure flow with the assistants and make all preparations prior to the injection of ICG.

Due to the limited tissue permeability and scattering of fluorescence signals, ICG fluorescence imaging usually allows the identification of the lymphatic systems in subcutaneous layers up to 1 cm. Therefore, in cases with thick subcutaneous fat or deep lymphatic vessels, it may be difficult to confirm the location of lymph nodes percutaneously. In such an occasion, skin compression may be effective for visualization of deeply located lymphatic vessels. If observation is still impossible, a skin incision line is determined by estimating the approximate location of the sentinel node based on its convergence with the extension of lymphatic flow and the fact that the sentinel node is often located anatomically between the line along the lateral border caudal to the pectoralis major muscle stop and the mid-axillary line. After reaching the layer of superficial pectoral fascia by blunt dissection, it is relatively easy to identify the node because the fat layer is usually soft and easy to extend. Care must be taken not to injure the lymphatic vessels during dissection, because it leads to leakage and attachment of ICG onto the surrounding fat tissues, making it difficult to detect lymph nodes by fluorescence imaging.

6 Evaluation of the Sentinel Lymph Node Removed by Indocyanine Green Fluorescence Imaging

Currently, the standard of care is to omit axillary lymph node dissection in patients with negative sentinel lymph nodes. In addition, based on the results of a randomized controlled trial (ACOSOG Z0011) [5], it is recommended that axillary lymph node dissection should be omitted for a partial mastectomy with two or fewer metastases, provided that postoperative radiation therapy and pharmacotherapy are appropriately administered. On the other hand, in the case of total mastectomy, axillary lymph node dissection is currently recommended in the presence of metastases. For these axillary treatment decisions, many centers use one-step nucleic acid amplification (OSNA), a rapid intraoperative diagnostic or molecular biological technique, as well as other tracer-based procedures, to determine the presence of sentinel node metastases. As mentioned above, there are cases in which dissection is not necessary even if the lymph nodes are positive. Therefore, in some institutions, sentinel lymph node metastasis is not evaluated intraoperatively and diagnosed by postoperative pathological examinations using permanent specimens.

Pathological assessment of sentinel lymph nodes is performed according to the College of American Pathologists (CAP) protocol based on conventional hematoxylin and eosin (H&E) staining. Metastases greater than 2 mm in largest diameter are considered macrometastasis, and therefore, slice specimens are prepared at least 2 mm apart and observed. Presence of isolated tumor cells (ITC; a single tumor cell or a small group of cells with a maximum diameter of less than 0.2 mm, or less than 200 cells in the largest histologic section of a lymph node, detectable by conventional H&E staining or immunohistochemistry) is not regarded as a lymph metastasis because they are not active in terms of penetration into the blood walls or lymphatic sinuses. Lymph nodes with a maximum diameter greater than 0.2 mm and/or a cell count greater than 200 but less than 2.0 mm are defined as micrometastasis. In the case of micrometastasis alone, axillary lymph node dissection should be avoided because it significantly increases complications without improving overall survival or axillary recurrence rate. The OSNA method using CK19mRNA as a marker has been shown to be highly specific for the detection of lymph node metastasis and can be used for rapid diagnosis without a pathologist's diagnosis.

The omission of axillary lymph node dissection by appropriate assessment of sentinel lymph nodes significantly reduces complications such as limited range of motion of the upper extremities, brachial edema, paresthesia, and pain, and avoids a decrease in the postoperative QOL of patients. Since lymph node metastasis is an important factor for surgical decision-making as well as for the prediction of long-term outcomes, an accurate diagnosis method with a higher identification rate and lower false-negative rate is required.

7 Clinical Outcomes of the Sentinel Node Biopsy

In 1992, Morton et al [6] identified sentinel lymph nodes using dye in malignant melanoma patients, and since then, sentinel lymph node biopsy has been applied to various types of carcinoma. In the field of breast cancer, Krag et al [7] and Giuliano et al [8] reported the results of RI and dye methods, respectively, in 1993 and 1994, and the combination of both methods has become the standard worldwide. Although the combination method shows a high identification rate, RI was not available for clinical use in Japan for a long time due to problems of radiation exposure and the insurance system. In 2005, Kitai et al [9] reported the world's first sentinel lymph node biopsy using ICG. Subsequently, results comparing ICG and conventional methods were reported in Japan and overseas. In a prospective clinical trial comparing dye and ICG methods, a significantly higher identification rate (99–100%) was demonstrated compared with dye. Between ICG and RI methods, a prospective, multicenter cohort study in 821 patients with early-stage breast cancer [10] identified no difference in identification rate (97.2% vs. 97.0%, $P = 0.88$). In general, the detection rate of metastasis-positive lymph nodes can decrease when the nodes are filled with cancer cells, leading to alteration of the lymphatic flows. However, a recent meta-analysis reported that the identification rate of metastasis-positive lymph nodes by ICG fluorescence imaging tended to be higher than that by the RI method (OR 1.87, 95% CI: 1.00–3.49) [11].

8 Limitations and Future Challenges

Based on the favorable outcomes of sentinel lymph node biopsy by ICG fluorescence imaging, this technique was covered by the national insurance system in Japan in 2018. Since then, it has been widely performed in clinical practice in our country. However, many of the past reports were from small institutions, and long-term studies of safety and other aspects are still insufficient.

In a technical aspect, the room needs to be darkened to obtain fluorescence images with conventional near-infrared camera systems like PDE. In addition, an operating surgeon has to change the eye directions from the operation field to a TV monitor to confirm fluorescence images. For the solution to these problems, a new imaging system enabling the projection of fluorescence signals directly onto the surgical fields has recently been launched in Japan [12]. Further development of near-infrared imaging systems would enhance the feasibility of sentinel node biopsy by ICG fluorescence imaging.

In breast cancer patients after preoperative chemotherapy, the omission of axillary lymph node dissection based on sentinel node biopsy has not been accepted as a standard of care in guidelines due to the lack of sufficient accuracy data. In contrast, the indication of preoperative chemotherapy has been extended, especially for triple-negative and HER2-positive breast cancer. Therefore, it is important to establish the role of sentinel node biopsy in patients after preoperative treatments. The identification and false-negative rates of sentinel lymph nodes in patients with cN0 at diagnosis have been shown to be similar to those without preoperative chemotherapy. However, the false-negative rate is still high, and the prognosis is unclear when a cN+ patient becomes cN0 after preoperative chemotherapy. In such cases, the false-negative rate may decrease as the number of sentinel nodes removed increases (more than three), and the ICG method, which can be expected to have a higher identification rate, can be used to properly evaluate metastasis-positive lymph nodes.

As mentioned above, in the case of breast-conserving therapy, axillary lymph node dissection is being omitted in an increasing number of cases even when sentinel lymph node metastases are detected. Therefore, in the future, it may become possible to preoperatively determine cases in which sentinel lymph node biopsy itself will no longer be necessary based on imaging diagnosis and other factors. On the other hand, the number of cases in which axillary lymph node dissection can be omitted is increasing due to improvements in preoperative diagnosis and systemic therapy. With an increase in the number of facilities that perform sentinel node biopsy by ICG fluorescence imaging, we will be able to elucidate the role of this technique according to the tumor stage and perioperative treatments of each patient.

References

1. Japanese Breast Cancer Society. Breast cancer treatment guidelines (1) treatment edition 2018. Kinbara Shuppan; 2018. p. 228–42.
2. Kim T, Guiliano AE, Lyman GH, et al. Lymphatic mapping and sentinel lymph node biopsy in early-stage breast carcinoma: a meta-analysis. Cancer. 2006;106:4–16.
3. Sugie T, Sawada T, Tagaya N, et al. Comparison of indocyanine green fluorescence and blue dye methods in the detection of sentinel lymph nodes in early-stage breast cancer. Ann Surg Oncol. 2013;20:2213–8.
4. Xiong L, Gazyakan E, Yang W, et al. Indcyanine green fluorescence-guided sentinel node biopsy: a meta-analysis on detection rate and diagnostic performance. Eur J Surg Oncol. 2014;40:843–9.
5. Giuliano AE, Hunt KK, Ballman KV, et al. Axillary dissection vs no axillary dissection in women with invasive breast cancer and sentinel node metastasis: a randomized clinical trial. JAMA. 2011;305:569–75.
6. Morton DL, Wen DR, Wong JH, et al. Technical details of intraoperative lymphatic mapping for early stage melanoma. Arch Surg. 1992;127:392–9.
7. Krag DN, Anderson SJ, Julian TB, et al. Sentinel-lymph-node resection compared with conventional axillary-lymph-node dissection in clinically node-negative patients with breast cancer: overall survival findings from the NSABP B-32 randomised phase 3 trial. Lancet Oncol. 2010;11:927–33.
8. Giuliano AE, Kirgan DM, Guenther JM, et al. Lymphatic mapping and sentinel lymphadenectomy for breast cancer. Ann Surg. 1994;220:391–8.
9. Kitai T, Kawashima M. Transcutaneous detection and direct approach to the sentinel node using axillary compression technique in ICG fluorescence- navigated sentinel node biopsy for breast cancer. Breast Cancer. 2012;19:343–8.
10. Sugie T, Kinoshita T, Masyda N, et al. Evaluation of the clinical utility of the ICG fluorescence method compared with the radioisotope method for sentinel lymph node biopsy in breast cancer. Ann Surg Oncol. 2016;23:44–50.
11. Sugie T, Ikeda T, Kawaguchi A, et al. Sentinel lymph node biopsy using indocyanine green fluorescence in early-stage breast cancer: a meta-analysis. Int J Clin Oncol. 2017;22:11–7.
12. Takada M, Takeuchi M, Suzuki E, et al. Real-time navigation system for sentinel lymph node biopsy in breast cancer patients using projection mapping with indocyanine green fluorescence. Breast Cancer. 2018;25:650–5.

Identification of Sentinel Lymph Nodes in Gastric Cancer Surgery

21

Shinichi Kinami

Summary

- ICG fluorescence imaging is a promising alternative to the standard dye RI combination method for sentinel lymph node biopsy during gastric cancer surgery.
- The ICG is diluted 100-fold and administered submucosally in 0.5 mL doses to four surrounding sites using an endoscope on the day before surgery.
- Intraoperatively, the lymphatic flow is observed by ICG fluorescence laparoscopic system to pick up the sentinel lymph nodes based on the lymphatic basin dissection method.
- When sentinel node biopsy is negative for metastases, function-preserving gastrectomy with lymphatic basin dissection can be indicated.

1 Introduction

While laparoscopic surgery started out as a surgical technique with a small wound and minimal invasion, advances in imaging devices have made laparoscopic surgery more delicate. Currently, further advances in image processing technology have enabled fluorescence navigation surgery, in which grossly undetectable biological structures are visualized by fluorescence imaging.

A variety of applications of fluorescence imaging have been reported to make surgical treatments for gastric cancer more sensitive and precise. Among them, sentinel node navigation surgery (SNNS) would be the most promising technique.

Supplementary Information The online version contains supplementary material available at https://doi.org/10.1007/978-981-19-7372-7_21.

S. Kinami (✉)
Department of Surgical Oncology, Kanazawa Medical University, Kahoku, Ishikawa, Japan
e-mail: kinami@kanazawa-med.ac.jp

2 What Is a Sentinel Lymph Node?

For many carcinomas, lymph node metastasis is an important indicator of prognosis. If the number of lymph node metastases increases or if the metastases extend to distant lymph nodes, the risk of cancer recurrence after resection increases. Recent advances in cancer chemotherapy and immunotherapy have demonstrated that adjuvant chemotherapy after resection can improve the prognosis of a variety of carcinomas. The degree of lymph node metastasis is important in determining whether or not adjuvant chemotherapy can be applied.

Metastatic lymph nodes are sometimes enlarged and can be diagnosed by CT or other imaging techniques. However, there are many cases in which metastatic cancer cells are microscopically detected even in non-enlarged lymph nodes. In order to accurately diagnose the presence and extent of lymph node metastasis, it is necessary to resect the primary tumor as well as the regional lymph nodes followed by thorough microscopic examinations.

Lymph node metastasis occurs when a cluster of cancer cells from the primary tumor site drains into the lymphatic system and reaches the lymph nodes, where they evade the immune system and settle and proliferate. The UICC TMN classification defines micrometastasis (MM) as metastases less than 2 mm in diameter and isolated tumor cells (ITC) as metastases less than 0.2 mm in diameter. MM is considered as metastasis, but ITC can also affect the prognosis depending on the type of cancer. However, in order to accurately diagnose the presence or absence of ITC, the entire lymph node must be sampled and multiple sections with a thickness of 0.2 mm or less must be prepared. In the case of gastric cancer, the number of dissected lymph nodes is often 40 to 50. Given the difficulty of making multiple sections of 0.2 mm intervals for all of these lymph nodes, which lymph nodes should be examined closely for an accurate diagnosis? The answer and solution to this question is the sentinel node concept.

The term sentinel node refers to the lymph node that first receives lymph flow from the primary cancer site [1]. Lymph node metastasis should first develop in a lymph node that

T. Ishizawa (ed.), *Fluorescence-Guided Surgery*, https://doi.org/10.1007/978-981-19-7372-7_21

receives lymph flow directly from cancer, and this is called the sentinel node concept. If the sentinel node concept is theoretically correct in a particular cancer condition, then detailed pathological examinations following the sentinel node biopsy (SNB) enable accurate diagnosis of lymph metastasis at the level of micrometastasis. In addition, if the sentinel node biopsy is negative for metastasis, there should be no need to screen metastasis in the surrounding lymph nodes. The sentinel node navigation surgery (SNNS) is a strategy that uses the sentinel node for surgical decision-making.

3 Sentinel Lymph Node Biopsy in Gastric Cancer Surgery

The location of the sentinel lymph node varies from case to case. There is not always a single sentinel lymph node. Therefore, for identification of the sentinel lymph node, lymphatic flows should be confirmed in each patient by injecting a tracer around the primary cancer. For SNNS to be used effectively, the following six factors are important: selection of patients, adequate tracer selection, proper tracer administration method, objective detection of tracer uptake by lymph nodes, reliable biopsy technique, and precise intraoperative diagnosis of nodal metastasis at the level of micrometastasis [2].

Gastric cancer is the most common type of gastrointestinal cancer for which the adequacy of the sentinel node concept has been evaluated. One of the most reliable reports is a multicenter prospective study conducted by the Japanese Society for SNNS. This study was designed to eliminate factors that may cause false negatives, such as learning phase problems, sampling errors, and the accuracy of intraoperative rapid pathology. Specifically, the number of participating centers was limited to experienced ones, standardized curative surgery was indicated regardless of the results of SNB, and only permanent specimens were used for diagnosis of lymph node metastasis. The results were excellent, with a sensitivity of 93% and an accuracy of 99%, providing scientific proof of the sentinel node concept in the treatment of early gastric cancer [3]. In this study, conditions of sentinel node biopsy were as follows: the indication was cT1N0 gastric cancer of 4 cm or less, the tracer was a combination mapping of a blue dye (isosulfan blue) and 99mTc tin colloid, and the tracer administration method was a transendoscopic injection of 0.5 mL each into the submucosal layer at four locations around the tumor. The sentinel lymph nodes were identified with a gamma probe following visual observation of dye-stained lymphatic flow. The sentinel lymph nodes were sampled by the lymphatic basin dissection method described below, and the pathological diagnosis was made based on H&E staining of permanent specimens of one section for each node.

An expected role of SNB would be ultrastaging and optimization of the extent of lymph node dissection [4]. In malignant melanoma, head and neck cancer, and colorectal cancer, SNB is mainly used for determining the indication of adjuvant therapy. In breast cancer, on the other hand, the SNB has been developed as an indicator for the omission of axillary dissection. In gastric cancer, SNB would be used as an index for omitting lymph node dissection rather than ultrastaging.

4 Application of ICG Fluorescence Method to Sentinel Node Biopsy During Gastric Cancer Surgery

Although several reports have demonstrated the validity of the sentinel node concept in surgical treatments for gastric cancer, it still remains in an investigational stage, because of the following two reasons [5].

One is the problem of tracers. In order to search for lymph nodes that receive lymph flow directly from a primary tumor, we need a tracer that has an affinity to lymph fluid and stains the lymphatic system clearly and can be captured by the first lymph node without overflowing into subsequent nodes. However, such an ideal drug is not yet commercially available. The blue dye has the disadvantage of disappearing over time due to washing out rapidly from the site of administration and lymph nodes. RI colloid is resistant to change over time. In gastric cancer surgery, however, it is technically difficult to identify the sentinel lymph node by the RI method because the primary tumor and the lymph node are close to each other and the radiation activity of the primary tumor hides the sentinel lymph node (shine through effect). Another reason is that the current mainstream of early gastric cancer surgery has changed to laparoscopic surgery. Laparoscopic surgery is somewhat more time-consuming than laparotomy, and the angle and movement of surgical devices, such as forceps and cameras, are more limited, which makes use of a blue dye or the RI method for SNB more difficult than open surgery.

ICG fluorescence imaging is a promising alternative method to overcome the problems of SNB in gastric cancer surgery. ICG can drain into the lymphatic system, and protein-bound ICG emits fluorescence signals with a peak wavelength of 835 nm, enabling highly sensitive identification of the lymphatic ducts and lymph nodes. Unlike blue dyes, ICG remains in the lymphatic system for a long time, and laparoscopic camera systems capable of detecting ICG fluorescence signals are now launched by various medical device companies. These characteristics suit the use of SNB during laparoscopic surgery.

5 Development History of the ICG Fluorescence Method for Sentinel Node Biopsy During Gastric Cancer Surgery

ICG has been used for SNB of gastric cancer since relatively early times because it is inexpensive and easy to use in clinical settings in Japan [6, 7]. Originally, staining of the lymphatic system was identified by naked-eye observation. At that time, ICG was not recommended for SNB because of its poor contrast with fat and was considered inferior to blue dye. On the other hand, ICG has the property of absorbing light in the infrared region of 800 nm, and at the same time, an infrared gastrointestinal endoscope was developed to delineate the capillary network of the stomach. Using this principle, Nimura et al. developed a laparoscopic infrared observation system and reported the world's first case series of SNB during gastric cancer surgery using ICG infrared imaging (ICG absorption method) [8]. Unfortunately, the ICG absorption method was not widely used because of the unavailability of the imaging system. Subsequently, a fluorescence imaging system for open surgery (PDE, Hamamatsu Photonics) became commercially available, and Kusano et al. reported the first experience of using ICG fluorescence imaging for SNB during gastric cancer surgery [9].

Our group started using ICG fluorescence imaging for gastric cancer SNB in 2008. At first, ICG (5 mg/mL) was used as an alternative to blue dye by submucosal injection around the tumor (0.5 mL for each point). As a result, fluorescence imaging enabled highly sensitive identification of lymph nodes as compared with naked-eye examinations of stained nodes. Subsequently, we standardized a method of ICG administration by diluting ICG solution and setting 24-h interval between its administration and surgery to avoid spillage of fluorescing lymphatic fluid during SNB [10].

6 Standardized Method of ICG Fluorescence Imaging for Sentinel Node Biopsy During Gastric Cancer Surgery

Herein, our standard method of ICG fluorescence imaging for gastric cancer SNB is demonstrated along with tips and tricks.

For the detection of ICG fluorescence signals, laparoscopic fluorescence camera systems or open imaging systems like PDE are needed. The standard method described here has been adjusted and developed through experience with PDE and its succession system, PDE-neo®. PDE is standard definition quality and the fluorescence image cannot be superimposed on a normal color image, but its sensitivity for detection of ICG fluorescence signals is high, and the brightness and contrast can be easily adjusted. HEMS (Mizuho Co., Ltd.) and SPY-PHI (Stryker) may be used for open surgery. Both have ICG fluorescence detection sensitivity equivalent to that of PDE. For laparoscopic surgery, ICG fluorescence laparoscopes have been launched in the market by many manufacturers. We have tried all of these systems and found that the lymphatic system and sentinel lymph nodes can be seen with any of them. However, there is a large variation in the sensitivity of fluorescence detection and imaging quality among the laparoscopic systems. This is a field that is still under development, and it is difficult to rank the best or worst at present. In our opinion, the current ICG fluorescence laparoscopy should be used only for intraoperative identification of the lymphatic system and should not be used for evaluation of the sentinel lymph node on the resected specimens.

Under endoscopic examinations, ICG solution (50 μg/mL) is injected into four submucosal sites around the tumor at 0.5 mL each on the day before surgery. According to this "100-fold dilution" technique, ICG can also flow into secondary lymph nodes after the sentinel node. Even in a such condition, the sentinel node can easily be identified based on the difference in fluorescence intensity among the lymph nodes (usually, the sentinel node shows the highest fluorescence signal) [10]. Although intraoperative administration of ICG also has advantages in visualizing real lymphatic flows for detection of the sentinel lymph node, it may lead to massive spillage of ICG from the stumps of lymphatic ducts during nodal dissection procedures, which may interfere with subsequent surgical procedures. In addition, accurate submucosal administration of ICG by intraoperative endoscopy is technically difficult.

In order to identify the sentinel lymph node, the lymphatic basin dissection method is used. Lymphatic basin refers to the regions of the lymphatic system identified by SNB using dye during gastric cancer surgery [11]. Also in the present technique using ICG, the lymphatic basin can be identified by observing lymphatic flow by fluorescence imaging. The sentinel lymph node is included in this lymphatic basin. In the case of gastric cancer surgery, *ex vivo* examination of the lymphatic basin rather than *in vivo* sampling is easier to pick up the sentinel node. This unique method applied in gastric cancer surgery also has a role as a backup lymphatic dissection in case of a false negative, which may support the oncological safety of gastrectomy using SNB.

In the lymphatic basin dissection, the regions of connective tissues with fluorescence signals of lymphatic ducts and nodes are dissected *en bloc* (Movie 21.1). The proximal lymph node is usually the one attached to the gastric wall with a fluorescing spot due to submucosal ICG injection, and the distal side of dissection is determined to remove the most distant fluorescing lymph node [11]. The distal aspect of the connective tissues is dissected in a usual manner. In the case of the left gastric artery basin, the distal side is often lymph node No. 7, and the dissection end can be set at the root of the left gastric artery. In the case of the right gastroepiploic artery basin, lymph node No. 6 is often the most distal one, but if the fluorescing lymph node is confined to No. 4d and not found at lymph node No. 6, the first branch of the right gastroepiploic artery can be preserved. The proximal side may be dissected off from the gastric wall, or it may be resected together with the stomach without dissection. The latter method is more difficult to perform, but it is oncologically safer because it does not divide the lymphatic pathway, which can cause spillage of cancer cells.

Point

- ICG is administered submucosally in four locations around the tumor using an endoscope the day before surgery, enabling easy and accurate identification of the lymph nodes during surgery.
- The lymphatic basin is the region that is stained by the dye. The distal side of nodal dissection can be determined to remove the most distant fluorescing lymph node. The lymphatic basin should be dissected *en bloc* because the sentinel node is included in that region.
- The fluorescence signals of ICG are sufficiently high and stable, but they should be observed from multiple directions for reducing the risk of false negatives.

After removal of the lymphatic basin, the sentinel node is picked up on a back table in the operating room using fluorescence imaging with an open imaging system like PDE and submitted to rapid pathological examinations (Movie 21.2). Since the sensitivity of the open fluorescence imaging systems is very high, not only the sentinel node but also the lymphatic ducts and secondary lymph nodes (nodes downstream of the sentinel node) can also fluoresce. The sentinel lymph node and the secondary nodes can be distinguished based on their fluorescence intensities, because the sentinel node would have the largest amount of ICG, resulting in the highest intensity of fluorescence signals [10]. This theory should be proved by subsequent studies for the lymphatic basin technique using ICG fluorescence imaging to be used more widely in gastric cancer surgery. Lymphatic ducts, on the other hand, can easily be differentiated from lymph nodes on fluorescence imaging because the ductal structures are deformable by compression.

Point

The lymphatic basin dissection method is a technique in which whole the region stained by dye is removed *en bloc* and the sentinel lymph node is biopsied from the resected specimen on a back table in the OR.

The lymphatic basin dissection method is easy and can reduce the risk of false negatives in SNB in gastric cancer surgery.

The use of 100-fold dilution of ICG (50 μg/mL) enables differentiation of the sentinel lymph node from the other nodes according to fluorescence intensities.

Lymph nodes that clearly emit strong fluorescence signals are picked up as the sentinel lymph node and submitted to rapid pathological examinations.

7 Clinical Practice of SNNS for Early Gastric Cancer

Most early-stage gastric cancer can be cured by gastrectomy with lymph node dissection. However, patients who underwent standard gastrectomy with lymph node dissection often suffer from various postoperative symptoms and disorders such as heartburn, early satiety, and abnormal bowel movements, as well as gallstones and bone metabolism disorders. According to the Japanese guidelines, the standard extent of lymph node dissection for early gastric cancer beyond the indication of endoscopic submucosal dissection is defined as D1+. In D1+ dissection, the major gastric arteries should be divided at their root, requiring the removal of a large portion of the stomach. Therefore, even if laparoscopic surgery is indicated for early gastric cancer, postoperative symptoms in terms of the loss of the stomach would be the same as that by open surgery, as far as conventional D1+ dissection is applied. In order to prevent the occurence of postoperative gastrointestinal dysfunction, it is necessary to indicate stomach function-preserving procedures by accurately identifying patients without lymph node metastasis. SNB by ICG fluorescence imaging may have a role in selecting patients with early gastric cancer in whom the standard lymph dissection can be omitted.

Then, what kind of function-preserving radical surgery is appropriate for a patient who is intraoperatively diagnosed as node-negative by SNB? The first idea that comes to mind is to perform just SNB and omit further dissection, preserving all the major gastric vessels and limiting the extent of gastrectomy to the localized area of the primary tumor (or endoscopic mucosal resection alone). However, this strategy is premature at this time. There are two reasons for this; the first reason is that the accurate method for intraoperative rapid diagnosis of lymph node metastasis has not yet been established, and secondly, the conventional lymph node biopsy (pickup method) cannot identify deeply located lymph nodes in the fatty tissue or attached to the gastric wall, resulting in a false negative by SNB. Since reoperation for additional nodal dissection after final pathological diagnosis is nonsence in the treatment for gastric cancer, it is safe to apply lymphatic basin dissection for SNB at present. The remaining limitation of the lymphatic basin dissection technique is that some of the main arteries feeding the gastric wall are lost, making it difficult to indicate local resection of the stomach even if SNB is negative for metastasis.

The lymphatic basin is a concept specific to gastric cancer, and its distribution varies from case to case. They run from the gastric wall to the root of the major arteries and can be summarized into five main lymphatic compartments (Fig. 21.1). In distal gastrectomy and total gastrectomy, three and five lymphatic compartments, respectively, are dissected.

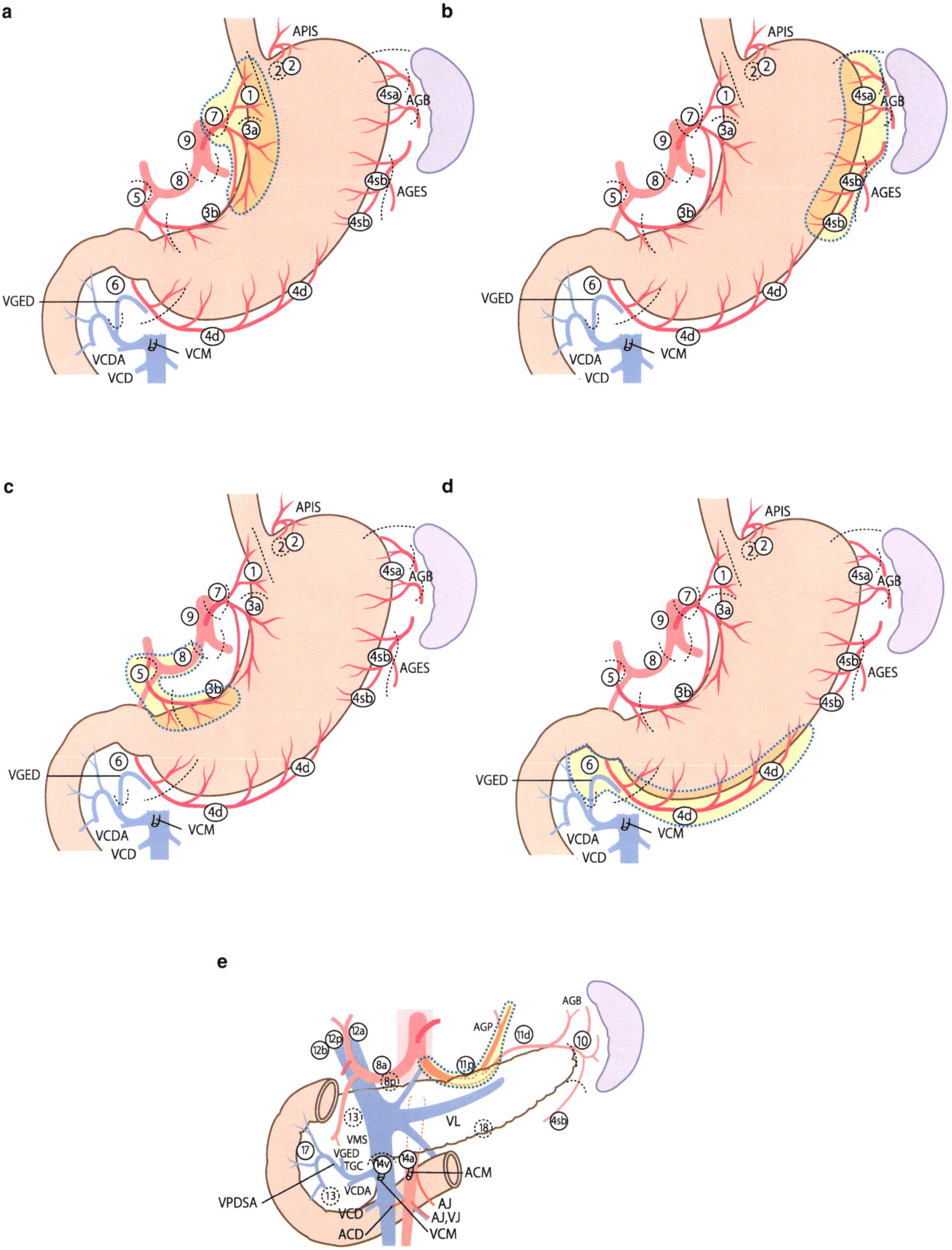

Fig. 21.1 Lymphatic compartments. The lymphatic basin is a concept of the lymphatic system specific to gastric cancer, and its distribution differs from case to case. These regions start from the gastric wall to the root of the dominant artery, which can be classified into five lymphatic basins, except for the lymphatic flow to the left cardiac lymph nodes (No. 2). (**a**) Left gastric artery basin (l-GA), (**b**) left gastroepiploic artery basin (l-GEA), (**c**) right gastric artery basin (r-GA), (**d**) right gastroepiploic artery basin (r-GEA), (**e**) posterior gastric artery basin (p-GA)

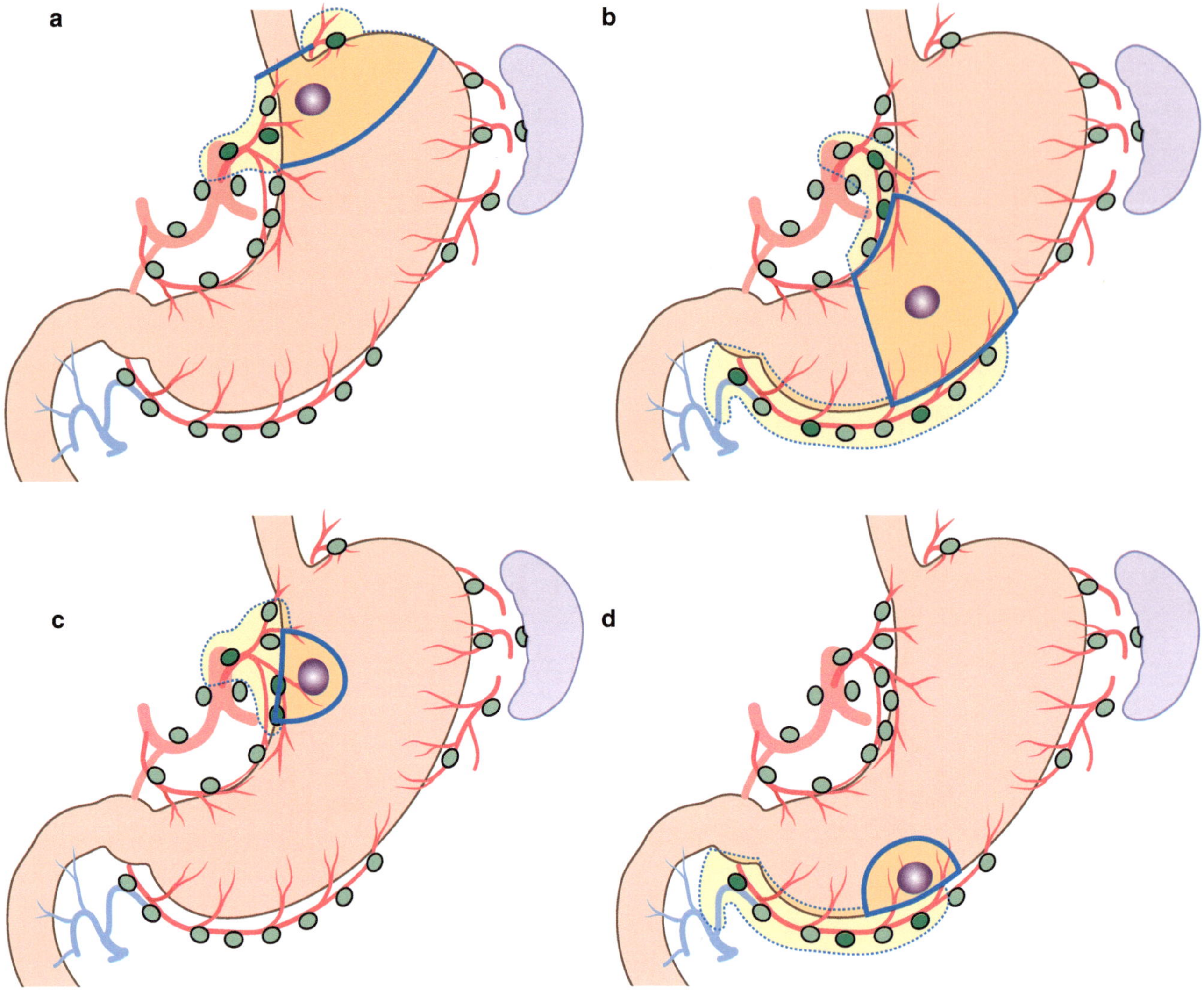

Fig. 21.2 Sentinel node navigation surgery for early gastric cancer. For patients with early gastric cancer diagnosed as node-negative by SNB, function-preserving gastrectomy only with the lymphatic basin dissection can be indicated. (**a**) Mini-proximal gastrectomy, (**b**) Segmental gastrectomy, (**c**) Local resection of stomach, (**d**) Mini-distal gastrectomy

On the other hand, if the cancer-associated lymphatic basin is less than two, lymphatic basin dissection can ensure gastric blood flow to some extent, enabling an indication of function-preserving radical surgery. Figure 21.2 shows examples of the extent of dissection and types of function-preserving radical surgery to be indicated when SNB is negative for metastasis. Previous reports demonstrated that no postoperative recurrence in lymph nodes developed when SNB according to the lymphatic basin method had been negative for metastasis, including false-negative cases.

8 Challenges and Future Perspectives

Our results of SNB using ICG fluorescence imaging alone were demonstrated [10]. In this clinical study (UMIN000010154), we were able to identify lymphatic basins and sentinel nodes in all 87 subjects, with a total of 456 sentinel nodes, a median of 6 nodes per patient, and 12 metastatic cases, of which 11 were diagnosed as positive by sentinel node biopsy. The sensitivity specificity, and accuracy of SNB were 91.7%, 100%, and 98.9%, respectively.

This series included a case of false negatives, in which postoperative pathological examinations on deep-cut specimens identified ITC in the sentinel node. These favorable outcomes suggest that ICG fluorescence imaging can be used as an alternative to the conventional RI-dye combination method for SNB during gastric cancer surgery.

The major challenge of the ICG fluorescence method for SNB is that it cannot prevent false-negative cases due to the limitations of intraoperative pathological examinations. That's why lymphatic basin dissection is recommended at this moment. Ideally, genetic approach like OSNA method (SYSMEX) should be used alone or in addition to the conventional H&E staining for intraoperative diagnosis of cancer cells in the sentinel nodes. Another problem is that the discrimination of sentinel lymph nodes from other nodes is somewhat subjective, which may be improved by the development of new tracers. In fact, RI-labeled ICG nanocolloids and ribosomal ICG have been developed and used in preclinical studies or in the field of animal medicine. We hope that these new tracers will become available for SNB in future.

References

1. Morton DL, Wen DR, Wong JH, et al. Technical details of intraoperative lymphatic mapping for early stage melanoma. Arch Surg. 1992;127:392–9.
2. Kinami S, Kosaka T. Laparoscopic sentinel node navigation surgery for early gastric cancer. Transl Gastroenterol Hepatol. 2017;2:42.
3. Kitagawa Y, Takeuchi H, Takagi Y, et al. Sentinel node mapping for gastric cancer: a prospective multicenter trial in Japan. J Clin Oncol. 2013;31:3704–10.
4. Miwa K. Sentinel node concept and its application for cancer surgery. Nihon Geka Gakkai Zasshi. 2000;101:307–10.
5. Kinami S, Nakamura N, Tomita Y, et al. Precision surgical approach with lymph-node dissection in early gastric cancer. World J Gastroenterol. 2019;25:1640–52.
6. Hiratsuka M, Miyashiro I, Ishikawa O, et al. Application of sentinel node biopsy to gastric cancer surgery. Surgery. 2001;129:335–40.
7. Ichikura T, Morita D, Uchida T, et al. Sentinel node concept in gastric carcinoma. World J Surg. 2002;26:318–22.
8. Nimura H, Narimiya N, Mitsumori N, et al. Infrared ray electronic endoscopy combined with indocyanine green injection for detection of sentinel nodes of patients with gastric cancer. Br J Surg. 2004;91:575–9.
9. Kusano M, Tajima Y, Yamazaki K, et al. Sentinel node mapping guided by indocyanine green fluorescence imaging: a new method for sentinel node navigation surgery in gastrointestinal cancer. Dig Surg. 2008;25:103–8.
10. Kinami S, Oonishi T, Fujita J, et al. Optimal settings and accuracy of indocyanine green fluorescence imaging for sentinel node biopsy in early gastric cancer. Oncol Lett. 2016;11:4055–62.
11. Kinami S, Fujimura T, Ojima E, et al. PTD classification: proposal for a new classification of gastric cancer location based on physiological lymphatic flow. Int J Clin Oncol. 2008;13:320–9.

Identification of Sentinel Lymph Nodes in Colorectal Cancer Surgery

22

Hironori Odaira, Masashi Yoshida, and Yutaka Suzuki

Summary

- Few reports have revealed the efficacy of sentinel node biopsy in colorectal cancer surgery.
- On the other hand, there are an increasing number of publications using fluorescence imaging to identify the lymphatic pathway in order to determine the appropriate extent of resection for colorectal cancer.
- Future development of colorectal cancer-specific fluorophores and highly sensitive imaging systems may make sentinel node biopsy more useful and effective.

1 Introduction

Reports of sentinel node biopsy in colorectal cancer surgery are quite fewer than those in the field of breast, skin, and gastric cancers. In this article, we describe the current status and future perspective of this technique based on literature reviews.

2 Significance of Sentinel Node Biopsy in Colorectal Cancer Surgery

As mentioned above, there are few reports on the identification of sentinel lymph nodes in colorectal cancer surgery [1]. The reason for this lies in the characteristics of surgical procedures in this field [2, 3].

First of all, resection of colorectal cancer usually requires wide mesenteric dissection and mobilization of digestive tracts for assuring anastomosis as well as resection of cancer, regardless of the nodal stage of cancer. Even if sentinel node biopsy is negative for metastasis, the extent of the dissection is not so much different from the standard lymphatic dissection. Second, surgical procedures for D2 to D3 lymph node dissection are not so complicated, especially in resection for colon cancer. Resection with D0 or D1 dissection is sometimes more demanding than standard lymph dissection in terms of the number of blood vessels to be divided. Third, the use of sentinel node biopsy in colorectal cancer surgery may not always lead to improvement of postoperative QOL. For example, in breast cancer, malignant melanoma, and gynecological cancer surgeries, identification of the sentinel lymph node is the basis for reducing the extent of resection and omitting lymph node dissection, leading to a lower incidence of postoperative lymphedema [4, 5]. In the case of gastric cancer, the use of sentinel node biopsy enables function-preserving gastrectomy, such as local resection and pylorus-preserving gastrectomy [6]. These surgical conditions seem to be different from the case of colorectal cancer surgery. Finally, a risk of false negatives due to obstruction of the lymphatic drainage in a case of advanced colorectal cancer [7] and more complicated lymphatic networks than in a case of breast cancer or malignant melanoma [8] should be considered. For these reasons, the significance of sentinel lymph node biopsy in colorectal cancer surgery seems to be low as compared with resections of other organs [9, 10].

On the other hand, in daily clinical practice, we often encounter cases in which a partial colon resection seems to be technically and oncologically more favorable than standard right or left hemicolectomy or cases in which lymph node dissection around the root of the middle colonic artery should be omitted for prevention of postoperative pancreatic leak, if possible. In such cases, the identification of sentinel lymph nodes and the direction of lymph flow by fluorescence imaging may be useful for surgical decision-making.

H. Odaira (✉) · M. Yoshida · Y. Suzuki
Department of Gastroenterological Surgery, International University of Health and Welfare Hospital, Nasushiobara-Shi, Tochigi-Ken, Japan
e-mail: ohdaira@iuhw.ac.jp

T. Ishizawa (ed.), *Fluorescence-Guided Surgery*, https://doi.org/10.1007/978-981-19-7372-7_22

3 Our Experience of Intraoperative Fluorescence Imaging of the Mesocolon Following Preoperative Submuscular Injection of ICG

In our center, intraoperative ICG fluorescence imaging is used for localizing small colorectal cancers. According to the protocol for sentinel node biopsy in gastric cancer surgery, ICG is injected in the submucosal layer around the colorectal cancer on the day before surgery (Fig. 22.1). Intraoperatively, fluorescence imaging is performed using PINPOINT (Stryker) (Figs. 22.2 and 22.3) [11]. In our previous series, however, fluorescing lymph nodes were undetectable in the patient's abdominal cavity or on the resected specimen (Fig. 22.4), while the main tumor was clearly identified as fluorescing spots on the colorectal wall.

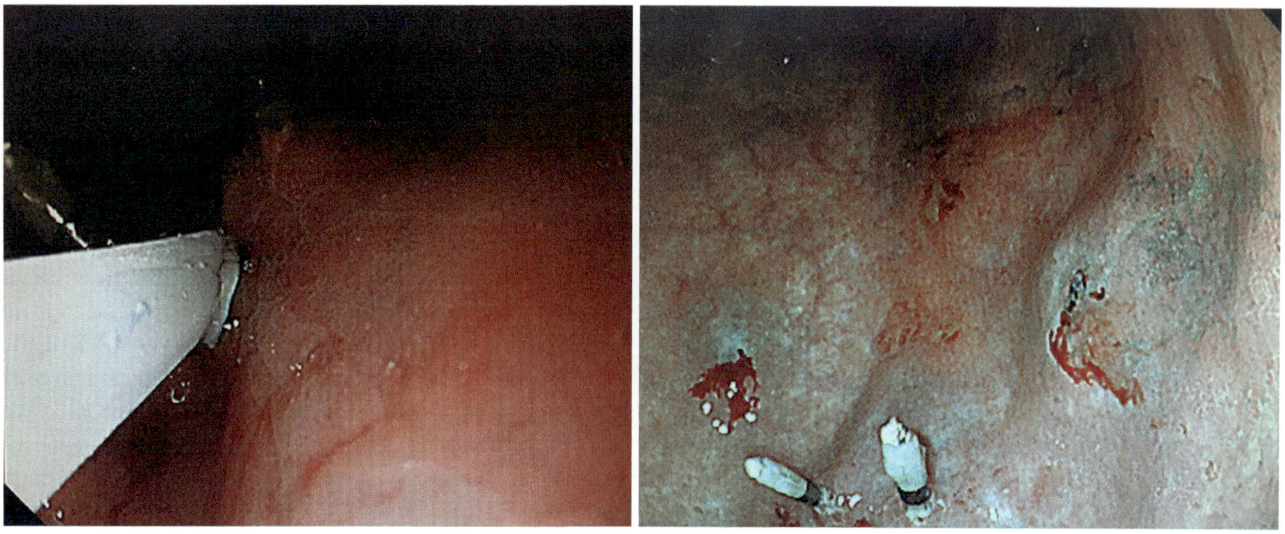

Fig. 22.1 Submucosal injection of ICG on the day before surgery in a case of early gastric cancer

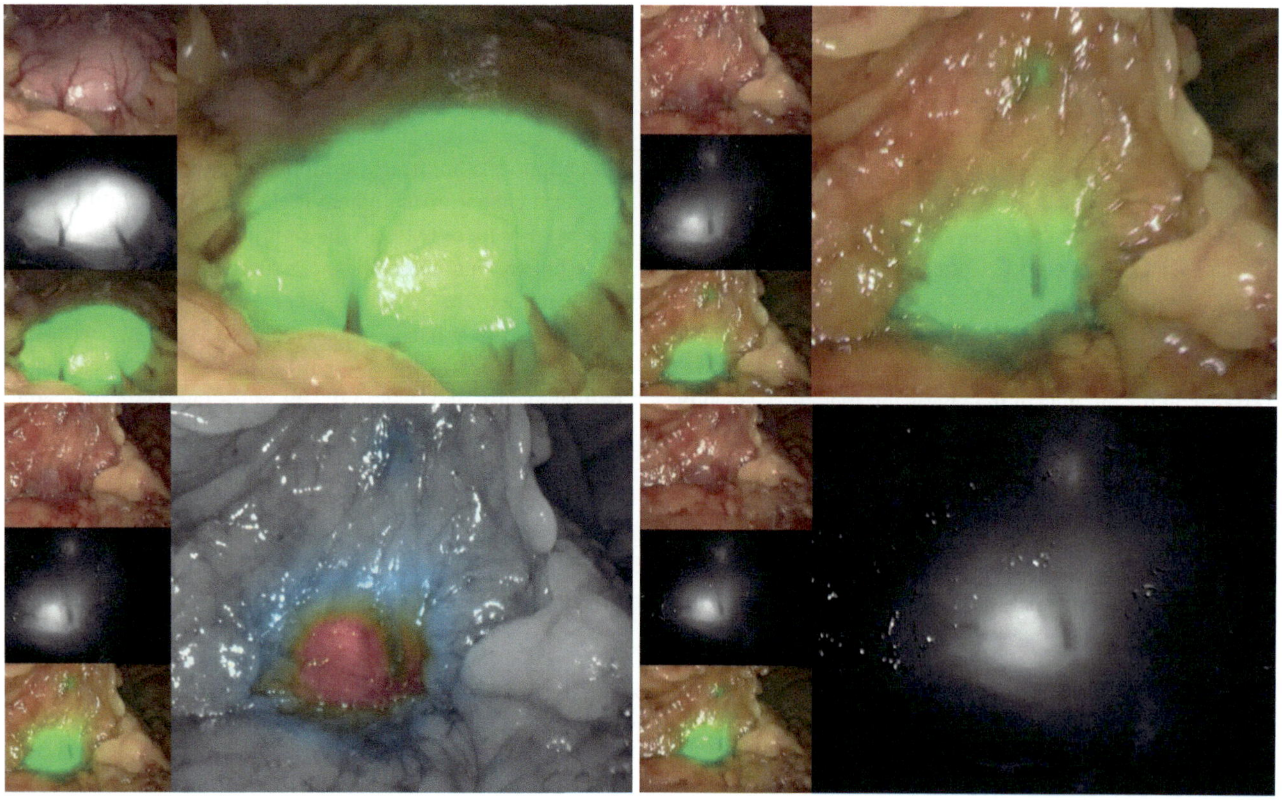

Fig. 22.2 Localization of the primary tumor and lymph nodes (No. 4d) in a case of early gastric cancer

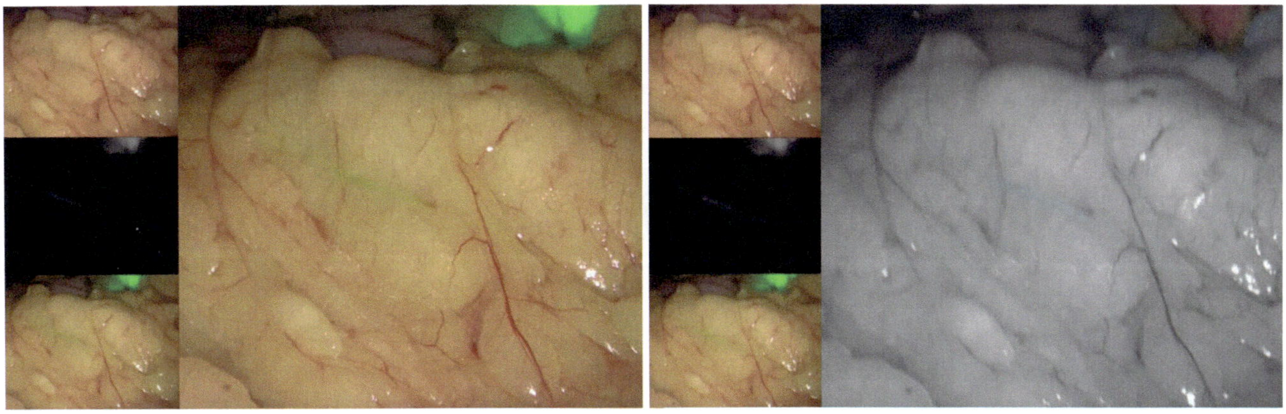

Fig. 22.3 Visualization of lymph flow toward No. 6 lymph node (same as the above case)

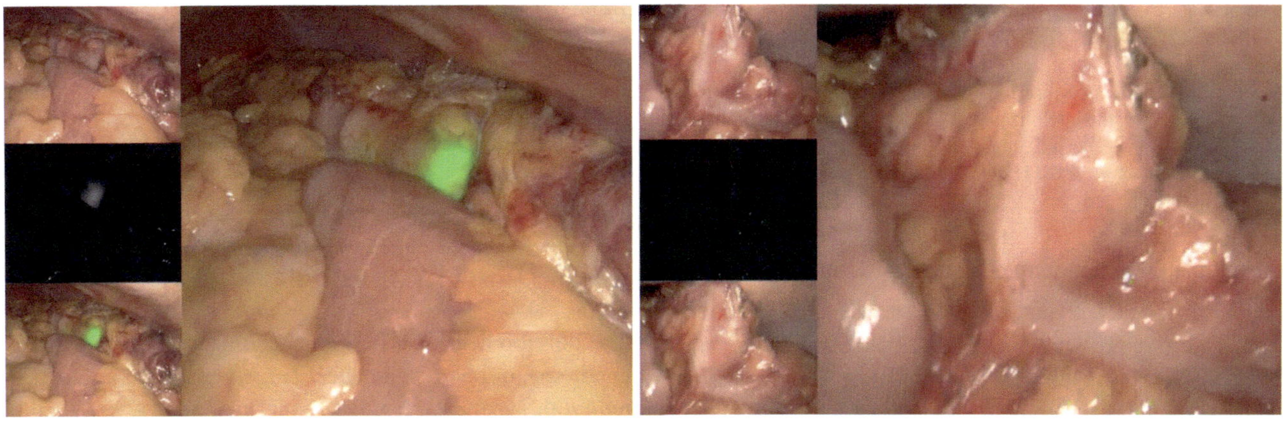

Fig. 22.4 A case of early-stage descending colon cancer. Although the primary tumor was identified by fluorescence imaging, lymph nodes or lymphatic ducts were not delineated

The above findings suggest that ICG fluorescence imaging according to our protocol, which was totally effective for the identification of sentinel lymph nodes of gastric cancer, is not applicable to the detection of lymphatic drainage from colorectal cancer. This may be due to the fact that the lymphatic systems in the colon and rectum tend to wash out ICG more quickly than those around the stomach, although no supporting data is available. If so, lymph node detectability by fluorescence imaging may be improved by shortening the interval between ICG administration and surgery. In this sense, a novel tracer retaining in the nodes for a long time and a highly sensitive fluorescence imaging system are awaited.

4 Clinical Outcomes Reported in Previous Studies

Some authors have reported outcomes of ex vivo identification of sentinel lymph nodes by fluorescence imaging on the resected specimens [12–15]. There were also reports of sentinel lymph node mapping after submucosal injection of ICG by intraoperative endoscopy using CO_2, like in the case of gastric and esophageal cancer surgeries [16]. In the technique reported by Sorares et al. [17], ICG (2.5 mg/mL) was injected into the submucosa near the tumor prior to colorectal mobilization, followed by fluorescence imaging of the mesentery with PINPOINT (Stryker) setting 30-minute interval from the ICG injection (Fig. 22.5). In the report by Ankersmit et al. [18], submucosal or subserosal injection of a tracer (25 mg of ICG diluted in 1.0 mL of 20% albumin +9.0 mL of 0.9% NaCl) just after induction of general anesthesia was applied. Following the mobilization, the mesentery was expanded, and the lymph nodes were searched with a 30° oblique fluorescent laparoscope (VESERA ELITE II, Olympus). In this series, the detection rate was 89.7%, and sensitivity was 44% (80% for submucosal injection and 0% for subserosal injection). According to the meta-analysis of eight articles reported by the same group, the lymph node detection rate of ICG fluorescence imaging was 94%, with a sensitivity of 63% (95% CI, 51–74%). The authors concluded that identification of the sentinel lymph node using ICG fluorescence imaging in colorectal cancer surgery has many challenges

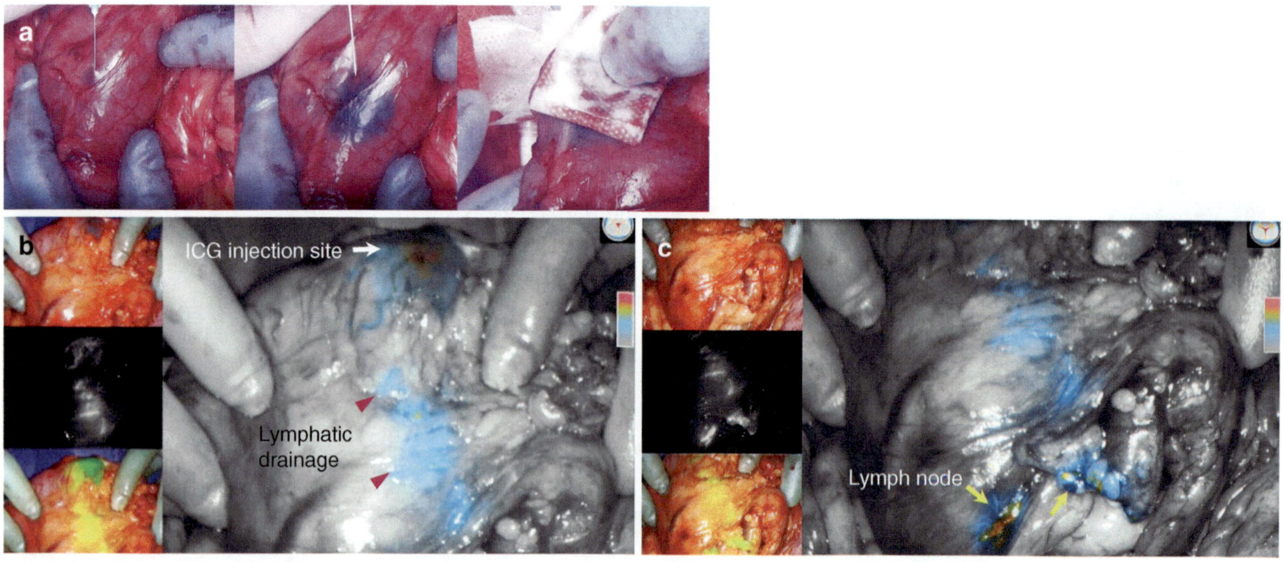

Fig. 22.5 Mapping of lymphatic flow pathways in colon cancer resection. (**a**) Prior to the mobilization of the mesocolon, ICG solution (2.5 mg/mL) was injected into the subserosa near the tumor. (**b**) Fluorescence imaging (PINPOINT) of mesocolon 30 min after the ICG administration, delineating lymphatic drainage from the injection site. (**c**) Tracing the lymphatic flow path identified lymph nodes in the mesocolon. (Cited from Ref. [17], courtesy of Dr. Manish Chand [University College London])

due to anatomical or technical difficulties and suggested to (1) focus on early-stage cancer, (2) change the tracer component, (3) reconsider the method of tracer injection, and (4) develop an optimal real-time guiding method for further improvement.

In upper gastrointestinal surgery, ICG fluorescence imaging is frequently used for the identification of lymph nodes. Although there are some variations, submucosal endoscopic injection of ICG on the day before or on the day of surgery is a common protocol. In contrast, there seems to be no such consensus yet for colorectal cancer. As mentioned above, intraoperative administration of a tracer under a colorectal endoscope is not always easy in both open and laparoscopic surgery because of the unusual environment for scope insertion. The operative field can also be interfered with due to intestinal dilatation by airflow. Therefore, the conditions of ICG administration and fluorescence imaging can largely be different among the cases especially in the situation of resection of colorectal cancer [19, 20].

If the metastatic lymph nodes can be identified just by intravenous administration of tracer (e.g., accumulation of ICG in cancerous tissues through EPR effect [21] in the future, fluorescence imaging can be used more easily and effectively for optimization of surgical procedures based on distributions of metastatic lymph nodes.

5 Future Perspectives

The development of a fluorescence tracer applicable to the conditions of colorectal cancer surgery and an imaging system with excellent sensitivity control would be key factors for intraoperative fluorescence imaging of the lymphatic system to develop into an essential technology that contributes to patients with colorectal cancer.

Point

- The concept and techniques of sentinel node biopsy in colorectal cancer surgery have not been established.
- Novel fluorescent tracers adjusted to colorectal surgery and highly sensitive imaging systems are key factors for the further development of fluorescence lymphatic mapping in colorectal cancer surgery.

References

1. Cahill RA. What's wrong with sentinel node mapping in colon cancer? World J Gastroenterol. 2007;13:6291–4.
2. Colorectal Cancer Study Group. Colorectal cancer treatment protocol. 9th ed. Kinbara Shuppan; 2018.
3. Colorectal Cancer Study Group. Colorectal cancer treatment guidelines for physicians., 2019 Edition. Kinbara Shuppan; 2019.

4. Giuliano AE, Ballman KV, McCall A, et al. Effect of axillary dissection vs no axillary dissection on 10-year overall survival among women with invasive breast cancer and sentinel node metastasis: the ACOSOG Z0011 (Alliance) randomized clinical trial. JAMA. 2017;318:918–26.
5. Leiter U, Stadler R, Mauch C, et al. Complete lymph node dissection versus no dissection in patients with sentinel lymph node biopsy positive melanoma (DeCOG- SLT): a multicentre, randomised, phase 3 trial. Lancet Oncol. 2016;17:757–67.
6. Kitagawa Y, Takeuchi H, Takagi Y, et al. Sentinel node mapping for gastric cancer: a prospective multicenter trial in Japan. J Clin Oncol. 2013;31:3704–10.
7. Liberale G, Lasser P, Sabourin JC, et al. Sentinel lymph nodes of colorectal carcinoma: reappraisal of 123 cases. Gastroenterol Clin Biol. 2007;31:281–25.
8. Shiozawa M, Akaike M, Yamada R, et al. Clinicopathological features of skip metastasis in colorectal cancer. Hepato-Gastroenterology. 2007;54:81–4.
9. Ronan AC. What's wrong with sentinel node mapping in colon cancer? World J Gastroenterol. 2007;13:6291–4.
10. van der Pas MH, Meijer S, Hoekstra OS, et al. Sentinel-lymph-node procedure in colon and rectal cancer: a systematic review and meta-analysis. Lancet Oncol. 2011;12:540–50.
11. Ohdaira H, Yoshida M, Okada S, et al. New method of indocyanine green fluorescence sentinel node mapping for early gastric cancer. Ann Med Surg. 2017;20:61–5.
12. Chand M, Keller DS, Joshi HM, et al. Feasibility of fluorescence lymph node imaging in colon cancer. Tech Coloproctol. 2018;22:271–7.
13. Hutteman M, Choi HS, Mieog JS, et al. Clinical translation of ex vivo sentinel lymph node mapping for colorectal cancer using invisible near-infrared fluorescence light. Ann Surg Oncol. 2011;18:1006–14.
14. Schaafsma BE, Verbeek FP, van der Vorst JR, et al. Ex vivo sentinel node mapping in colon cancer combining blue dye staining and fluorescence imaging. J Surg Res. 2013;183:253–7.
15. Liberale G, Vankerckhove S, Galdon MG, et al. Sentinel lymph node detection by blue dye versus indocyanine green fluorescence imaging in colon cancer. Anticancer Res. 2016;36:4853–8.
16. Currie AC, Brigic A, Thomas-Gibson S, et al. A pilot study to assess near infrared laparoscopy with indocyanine green (ICG) for intraoperative sentinel lymph node mapping in early colon cancer. Eur J Surg Oncol. 2017;43:2044–51.
17. Sorares AS, Lovat LB, Chand M. Intracorporeal lymph node mapping in colon cancer surgery. Eur J Surg Oncol. 2019;45:2316–8.
18. Ankersmit M, Bonjer HJ, Hannink G, et al. Near-infrared fluorescence imaging for sentinel lymph node identification in colon cancer: a prospective single-center study and systematic review with meta-analysis. Tech Coloproctol. 2019;23:1113–26.
19. Nimura H, Narimiya N, Mitsumori N, et al. Infrared ray electronic endoscopy combined with indocyanine green injection for detection of sentinel nodes of. Br J Surg. 2004;91:575–9.
20. Miyashiro I, Hiratsuka M, Sasako M, et al. High false-negative proportion of intraoperative histological examination as a serious problem for clinical application of sentinel node biopsy for early gastric cancer: final results of the Japan Clinical Oncology Group multicenter trial JCOG0302. Gastric Cancer. 2014;17:316–23.
21. Wei R, Jiang G, Lv M, et al. TMTP1-modified indocyanine green-loaded polymeric micelles for targeted imaging of cervical cancer and metastasis sentinel lymph node in vivo. Theranostics. 2019;9:7325–44.

Identification of Sentinel Lymph Nodes in Gynecologic Surgery

23

Kensuke Sakai, Wataru Yamagami, Nobuyuki Susumu, and Daisuke Aoki

Summary

- The use of sentinel lymph node biopsy in gynecologic malignancies such as cervical, endometrial, and vulvar cancers enables avoidance of unnecessary lymph node dissection, thus improving the QOL of cancer survivors.
- Evidence of sentinel node biopsy using ICG fluorescence has been accumulated sufficiently for clinical applications both in Japan and overseas, although there remain challenges such as insurance coverage and standardization of protocols.

1 Introduction

Gynecologic malignancies, especially cervical, endometrial, and vulvar cancers, mainly progress by lymphatic metastasis, and therefore, systematic lymph node dissection is performed in many patients. On the other hand, the rates of lymph node metastasis in pT1 and pT2 patients are about 15.8% in cervical cancer, 5.9% in endometrial cancer, and 25% in vulvar cancer [1, 2]. Therefore, the decrease in QOL due to postoperative lymphedema and lymphatic cysts caused by lymph node dissection is becoming a growing issue corresponding to the improvement of long-term outcomes.

The sentinel lymph node (SN) is the first lymph node to be reached by lymphatic flow from the primary tumor. Recently, the concept of SN navigation surgery, which omits systematic lymph node dissection based on the assumption that there are no other metastases to the involved lymph nodes, has been proposed and is already covered by social insurance system in Japan for breast cancer and malignant melanoma. In gynecologic cancers as well, the usefulness of SN biopsy has been recently reported both in Japan and overseas. In this article, we will discuss the actual methods of SN biopsy in gynecological malignancies, focusing on the use of fluorescence imaging (fluorescence method).

2 Conventional Methods and Problems

SNs are identified by visualizing the flow of tracer drugs into the primary organ. Tracers include radioisotopes (99mTc – phytic acid, 99mTc – tin colloid, etc.) and dyes (indigo carmine, patent blue [PB], indocyanine green [ICG], etc.), which have different characteristics (Table 23.1) [3]. The identification method using radioisotope (RI), dye, and ICG fluorescence is called the RI method, the dye method, and the fluorescence method, respectively. These methods are often used in combination to identify SNs.

The use of the RI method is limited to medical centers and areas which meet regulations for radiation safety control. On the other hand, this method has the advantage of identifying the SN before surgery with lymphoscintigraphy or SPECT-CT, since the tracer particles are large and remain in the lymph nodes for a long time. Simplicity and direct observation are the major advantages of the dye method, although the particles of dye are so small that they flow through the lymphatic system within minutes after injection, leading to missed SNs or misidentification of downstream lymph nodes as SNs. It has been reported that the fluorescence method has

Table 23.1 Characteristics of tracers for SN mapping. (Cited from Ref. [3] with revisions)

	RI	Dye	Fluorescence
Costs	+		–
Simplicity	–	+	+
Detection rate	++	+	+++
Physical Burden	–	+	+
Allergic reaction	–	++	–

K. Sakai (✉) · W. Yamagami · D. Aoki
Department of Obstetrics & Gynecology, Keio University School of Medicine, Shinjuku-ku, Tokyo, Japan
e-mail: k_sakai@a7.keio.jp

N. Susumu
Department of Obstetrics and Gynecology, International University of Health and Welfare, Narita, Chiba, Japan

T. Ishizawa (ed.), *Fluorescence-Guided Surgery*, https://doi.org/10.1007/978-981-19-7372-7_23

a higher SN identification rate than the dye method and is equivalent to the RI method, because even SN with a small amount of residual ICG can be visualized with a near-infrared fluorescence camera [4]. With the current development of near-infrared laparoscopic imaging systems and advances in laparoscopic surgery, the fluorescence method is expected to become the dominant technique for SN navigation in the field of gynecological surgery.

Point

- There are several techniques for the detection of SN biopsy, such as the RI, dye, and fluorescence methods, all of which have advantages and disadvantages.
- In addition to its high detection rate and safety, ICG has been widely used in recent years because of its simplicity and ease of use in laparoscopic surgery.

3 Evidence for SN Biopsy in the Treatment of Gynecologic Malignancies

3.1 Cervical Cancer

The SENTICOL study is known as a representative clinical trial of SN mapping in this field [5]. The SN identification rate was 97.8%, and the sensitivity and negative predictive values were 92.0% and 98.2%, respectively, for node-positive cases, indicating sufficient diagnostic accuracy. In a meta-analysis examining the usefulness of the ICG fluorescence method in six publications, the identification rate was 100%, which was equivalent to and sufficient compared with the RI method alone or the combination of the RI and dye methods [4]. In addition, several institutions have already reported the results of SN navigation surgery that omits lymph node dissection [6–8], and the omission of systematic lymph node dissection using SN biopsy is considered to be one of the useful options that can be considered.

The Japanese guidelines for the treatment of cervical cancer (2017 edition) state that "SN biopsy should be performed as an experimental procedure by a team skilled in the technique at a facility with a cooperative system of pathologists. Omission of lymph node dissection in SN-negative cases may be considered at facilities where the accuracy and safety of the procedure have been sufficiently confirmed (grade C1 recommendation)" [9]. The NCCN guidelines also state that SN mapping should be considered and introduces detailed techniques [10]. The European Society of Gynecologic Oncology (ESGO) guidelines are strongly recommended for SN mapping, which suggest that SN biopsy should be performed in stage IA cases with positive lymphovascular invasion and stage IB1 cases. Furthermore, the guidelines suggest that systematic lymph node dissection should be performed in stage IB1 cases with negative SN, while CCRT should be performed in stage IB1 cases with positive SN without performing a wide total hysterectomy. In particular, in stage IB1, systematic lymph node dissection is performed in SN-negative cases, while CCRT is performed in SN-positive cases without performing a radical hysterectomy as the standard treatment [11].

3.2 Endometrial Cancer

In the SENTI-ENDO study, a representative clinical trial of SN mapping was performed on 125 patients with stage I or II endometrial cancer using RI and dye methods. The SN detection rate was 89%, with a sensitivity of 84% for positive lymph nodes and a negative predictive value of 97% [12]. In the FIRES trial, SN mapping by ICG fluorescence was performed in 385 patients with stage I endometrial cancer. The SN detection rate was 86%, the sensitivity for node-positive patients was 97.2%, and the negative predictive value was 99.6%, all of which were satisfactory results [13]. In a meta-analysis of 44 papers, the detection rate by ICG was 93%. In a meta-analysis of 44 papers, the detection rate by ICG was 93%, which was better than the overall detection rate of 83% [14].

The Japanese guidelines for the treatment of uterine body neoplasms (2018 edition) state that "SN biopsy should be performed as an experimental procedure by a team skilled in the technique at a facility with a good cooperation system of pathologists and omission of lymph node dissection (biopsy) in SN-negative cases should be considered in the case of experimental procedure (grade C1 recommendation)" [15]. The NCCN guidelines also state that SN mapping should be considered and describes the specific procedures [16]. The ESGO guideline, on the other hand, states that SN mapping is still in the research stage, although it is considered feasible through large-scale clinical trials [17].

3.3 Vulvar Cancer

In a meta-analysis of 49 papers on SN mapping, all of which were performed by RI or dye methods, the SN detection rate was 94% (92–96%), sensitivity was 92% (90–95%), and negative predictive value was 97% (96–98%), all of which were excellent [18]. Although there have not been many reports on the use of the fluorescence method in vulvar cancer, in an SN mapping study of 27 vulvar cancers, the detection rate of the fluorescence method was 100%, the sensitivity was 100%, and the positive predictive value was 91.5%, which was better than that of the RI method [19]. In the GROINSS-V study, 403 patients with vulvar cancer were compared between SN biopsy alone and inguinal lymph node dissection, and although complications were significantly lower, there was no significant difference in local recurrence rate or survival rate, indicating the validity of SN navigation surgery [20].

Based on these considerations, the Japanese guidelines for the treatment of vulvar and vaginal cancer (2015 edition) classified this technique as a grade C1 recommendation, stating, "In cases where inguinal lymph node metastasis is not suspected, omission of inguinal lymph node dissection by SN biopsy may be considered, but should be performed on a trial basis in light of the current situation in Japan" [21]. In contrast, the NCCN guidelines clearly state that SN biopsy should be performed as a standard treatment, and it is strongly recommended compared with cervical cancer and endometrial cancer [22].

Point

- The evidence for SN mapping in gynecologic malignancy has been accumulating in recent years, with a sufficient detection rate by the ICG fluorescence method.
- In Europe and the United States, SN mapping is regarded as a standard treatment especially for vulvar cancer, while it is considered only in limited cases in the Japanese treatment guidelines for cervical cancer, endometrial cancer, and vulvar cancer.

4 Clinical Practice of SN Mapping by Fluorescence Imaging

The near-infrared fluorescence imaging system used in our department is the HyperEye Medical System (HEMS) (Fig. 23.1). It is known that ICG concentration and fluorescence intensity do not directly correlate and that excessive ICG concentrations reduce fluorescence intensity. Although the optimal concentration of ICG for this purpose has not been established, in our department, ICG is diluted to 2.5 mg/mL.

As shown in Fig. 23.1, for cervical cancer, ICG is administered intraoperatively in two to four locations in the cervix from 3 to 9 o'clock. Although there is some controversy in endometrial cancer, the NCCN guidelines adopt tracer administration in the cervix, and injections in two to four sites are commonly used [16]. In vulvar cancer, the tracer is administered at four points around the tumor at 2, 5, 7, and 10 o'clock.

In cervical and endometrial cancers, the lymphatic vessels can usually be detected within a few minutes, so the retroperitoneum space is carefully explored and visually confirmed (Fig. 23.2). The external iliac nodes are identified easily, but the obturator nodes, internal iliac nodes, and parametrial nodes are difficult to visualize unless the paravesical fossa and pararectal space are sufficiently explored.

The most proximal lymph node along the lymphatic vessel is considered as the SN for the biopsy, but over time the lymph nodes in the more distal region may fluoresce, so the SN is marked first and removed with the surrounding lymphatic tissues. After the lymph node is removed, HEMS is used to check for any remaining lymph nodes that emit fluorescence. The removed lymph nodes are checked for fluorescence and then submitted for pathological diagnosis.

Point

ICG is administered to the cervix for cervical or endometrial cancer and peri-tumorally for vulvar cancer.

In cervical cancer and endometrial cancer surgery, the rapid expansion of the retroperitoneal space is important for better identification.

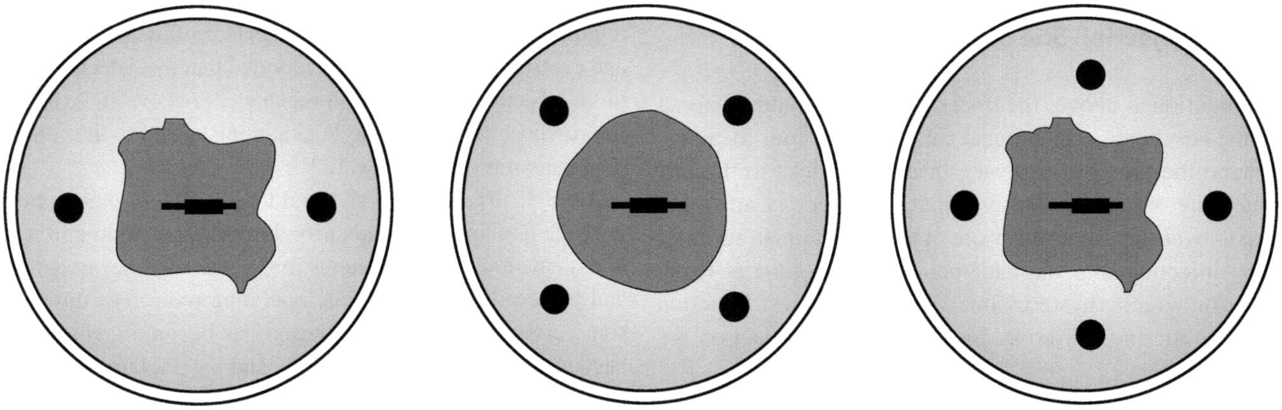

Fig. 23.1 Tracer injection sites for SN mapping in cervical and uterine body cancers. (Cited from Refs. [10, 16])

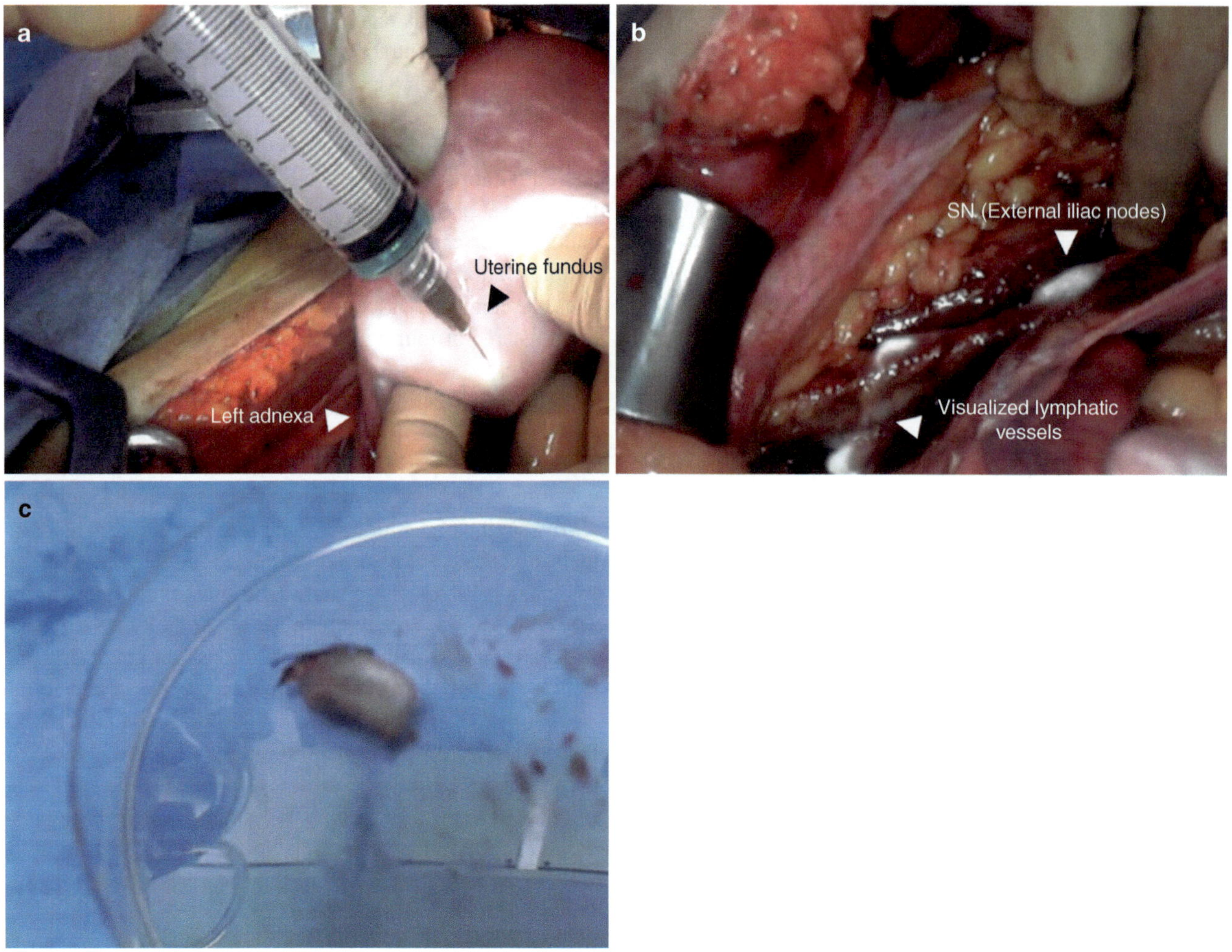

Fig. 23.2 Clinical practice of SN mapping using ICG fluorescence imaging. (**a**) Local administration of diluted ICG into the uterine subserosa. (**b**) Detection of SNs in the right external iliac lymph node region. (**c**) Confirmation of fluorescence signals in the excised SN

5 Pitfalls and Limitations

5.1 Injection Site of Tracers

As mentioned above, the tracer is generally administered to the cervix area for cervical cancer and in four locations around the tumor for vulvar cancer (i.e., close to the primary site of cancer). For endometrial cancer, controversy exists over the appropriate site of tracer administration. In fact, injection into the endometrium of the uterus using ultrasonography/hysteroscope, intraoperative injection into the uterine subserosa, and transcervical injection are available.

In general, SNs in the pelvic lymph node (PLN) as well as the para-aortic lymph node (PAN) region are identified in subendometrial and subserosal injections, but SNs in the PAN region are rarely identified in cervix injections. On the other hand, observation of the lymph nodes in the PAN region is more time-consuming than that of the PLN region, and even in our own study, we reported that the detection rate of the SNs in the PAN region was lower (74% vs 88%) in the fluorescence method using ICG administration to the subserosa than in the RI method [23].

Ideally, the RI method should be used in combination with the ICG method to identify the PAN area, although it is unrealistic to equip all systems for RI, fluorescence imaging, and hysteroscope in the same operating room. For this reason, cervical tracer administration has become mainstream because of its simplicity, stability, and patient burden, and it has been adopted in the NCCN guidelines [16]. Cervical tracer administration was also used in the latest multicenter prospective study (SENTI-ENDO study) [12].

It is also important to understand that cervical injection is not the best approach for detecting SN in the PAN region and a risk of false negatives is unignorable in a group of patients with a high probability of PAN metastasis.

5.2 Indications According to Cancer Types

Appropriate case selection is very important for SN biopsy and SN navigation surgery. In cervical cancer, SN biopsy should be limited to patients with a tumor less than 2 cm in diameter and no parametrial invasion, as described in the Japanese guidelines and the NCCN guidelines. The indications for SN biopsy in endometrial cancer are considered to be cases in which the disease is confined to the uterus with no extrauterine lesions. It is not necessary to apply SN biopsy in stage IA and endometrial carcinoma G1 cases without myometrial invasion where systematic lymph node dissection is generally omitted because of little risk of lymph node metastasis. It is important to note that the SENTI-ENDO study reported that 3 (16%) false-negative cases of type 2 endometrial cancer were observed in 125 patients with stage I or II endometrial cancer [12], suggesting that careful judgment of indication is necessary in high-risk cases such as type 2 endometrial cancer. In vulvar cancer, Japanese guideline states that SN biopsy is indicated only in resectable cases without obvious inguinal lymph node swelling and limited to only vulvar and perineum cases (T1 cases other than stage IA). The NCCN guidelines exclude locally advanced cancers and suggest that T1 or small T2 tumors should be included, as well as tumors less than 4 cm in diameter and no enlarged inguinal lymph nodes on imaging. The SN mapping should be applied with caution in patients with cancer in the midline region because lymphatic flow can drain into bilateral inguinal lymph nodes and increase the incidence of false-negative results.

5.3 Intraoperative Diagnosis of Metastasis in Sentinel Nodes

Because the number of SNs identified during SN biopsy is usually small, the presence of lymph node metastases can be examined in detail (ultrastaging) by serial step sections at narrow intervals and immunohistochemical staining. As a result, metastases smaller than 2 mm (micrometastasis) or 0.2 mm (isolated tumor cells [ITCs]) are occasionally detected. However, it is still inconclusive whether lymph dissection and adjuvant therapy for these small metastases, which are not often detected by conventional pathological examinations, improve the prognosis.

In order to compensate the risk of false negatives due to technical limitations of intraoperative pathological examinations, the role of molecular biological diagnosis (e.g., OSNA method) and imprint cytology has been discussed.

5.4 Learning Curve

When the fluorescence method is applied to cervical and endometrial cancers, the retroperitoneal spaces should be explored quickly without damaging the lymphatic flow for avoiding misidentification of the SNs, which requires a certain level of surgical skills. It has been reported that the false-negative rate can be reduced and a high detection rate can be maintained with experience of about 20 cases [3].

5.5 Safety of the Drug

The frequency of severe allergic reactions to ICG is very low (about 0.05% for intravenous administration), and it is considered to be a safe tracer [3, 24]. On the other hand, ICG should be used carefully for patients with a history of iodine allergy (ICG contains sodium iodide).

Point

Although the transcervical approach has recently been used for tracer injection during SN mapping for cervical cancer. And tracer injection is administered peri-tumorally for vulvar cancer. Especially for endometrial cancer, it should be noted that the detection rate in the paraaortic regions may decrease by the transcervical approach.

There remain some issues to be addressed, such as the selection criteria according to the cancer type and the management of micrometastases and ITCs which would not have been detected by conventional surgical procedures and pathological assessment.

6 Future Perspectives

In recent years, laparoscopic and robotic-assisted surgery has been widely applied to gynecological cancers. Since these minimally invasive approaches and fluorescence imaging are compatible with each other, the use of SN biopsy is expected to increase in the future. For further development of SN navigation surgery, we need to establish protocols such as tracer selection, administration/detection methods, and evaluation of intraoperative findings.

References

1. Treatment annual report of the Committee on Gynecologic Oncology, Japan Society of Obstetrics and Gynecology, 2012.
2. Patient annual report of Gynecologic Oncology Committee, Japan Society of Obstetrics and Gynecology, 2017.
3. Papadia A, Gasparri ML, Buda A, et al. Sentinel lymph node mapping in endometrial cancer: comparison of fluorescence dye with traditional radiocolloid and blue. J Cancer Res Clin Oncol. 2017;143:2039–48.

4. Ruscito I, Gasparri ML, Braicu EI, et al. Sentinel node mapping in cervical and endometrial cancer: indocyanine green versus other conventional dyes-a meta-analysis. Ann Surg Oncol. 2016;23:3749–56.
5. Lécuru F, Mathevet P, Querleu D, et al. Bilateral negative sentinel nodes accurately predict absence of lymph node metastasis in early cervical cancer: results of the SENTICOL study. J Clin Oncol. 2011;29:1686–91.
6. Niikura H, Okamoto S, Otsuki T, et al. Prospective study of sentinel lymph node biopsy without further pelvic lymphadenectomy in patients with sentinel. Int J Gynecol Cancer. 2012;22:1244–50.
7. Yahata H, Kobayashi H, Sonoda K, et al. Prognostic outcome and complications of sentinel lymph node navigation surgery for early-stage cervical cancer. Int J Clin Oncol. 2018;23:1167–72.
8. Gortzak-Uzan L, Jimenez W, Nofech-Mozes S, et al. Sentinel lymph node biopsy vs. pelvic lymphadenectomy in early stage cervical cancer: is it time to. Gynecol Oncol. 2010;116:28–32.
9. Ebina Y, Mikami M, Nagase S, et al. Japan Society of Gynecologic Oncology guidelines 2017 for the treatment of uterine cervical cancer. Int J Clin Oncol. 2019;24(1):1–19. https://doi.org/10.1007/s10147-018-1351-y.
10. National Comprehensive Cancer Network. nccn Clinical Practice Guidelines in Oncology Cervical cancer Ver. 4.2019 [Cited 7th Jul 2019]. Available from URL: https://www.nccn.org/professionals/physician_gls/pdf/cervical.pdf.
11. European Society of Gynaecological Oncology. POCKET GUIDELINES CERVICAL CANCER based on ESGO-ESTRO-ESP Guidelines for the Management of Patients with Cervical Cancer [cited 7th Jun 2019]. Available from URL: https://www.esgo.org/wp-content/uploads/2015/12/ESGO_Cervical-Cancer_A6.pdf.
12. Ballester M, Dubernard G, Lécuru F, et al. Detection rate and diagnostic accuracy of sentinel-node biopsy in early stage endometrial cancer: a prospective multicentre study (SENTI-ENDO). Lancet Oncol. 2011;12:469–76.
13. Rossi EC, Kowalski LD, Scalici J, et al. A comparison of sentinel lymph node biopsy to lymphadenectomy for endometrial cancer staging (FIRES trial): a multicentre, prospective, cohort study. Lancet Oncol. 2017;18:384–92.
14. Lin H, Ding Z, Kota VG, et al. Sentinel lymph node mapping in endometrial cancer: a systematic review and meta-analysis. Oncotarget. 2017;8:46601–10.
15. Yamagami W, Mikami M, Nagase S, et al. Japan Society of Gynecologic Oncology 2018 guidelines for treatment of uterine body neoplasms. J Gynecol Oncol. 2020;31(1):e18. https://doi.org/10.3802/jgo.2020.31.e18.
16. National Comprehensive Cancer Network: NCCN Clinical Practice Guidelines in Oncology Uterine Neoplasm Ver. 3.2019 [Cited 7th Jul 2019]. Available from URL: https://www.nccn.org/professionals/physician_gls/pdf/uterine.pdf
17. European Society of Gynaecological Oncology. POCKET GUIDELINES ENDOMETRIAL CANCER based on ESGO-ESTRO-ESP Guidelines for the Management of Patients with Endometrial Cancer [Cited 7th Jul 2019]. Available from URL: https://www.esgo.org/wpcontent/uploads/2015/12/Endometrial_broz_A6_b.pdf.
18. Hassanzade M, Attaran M, Treglia G, et al. Lymphatic mapping and sentinel node biopsy in squamous cell carcinoma of the vulva: systematic review and meta-analysis of the literature. Gynecol Oncol. 2013;130:237–45.
19. Soergel P, Hertel H, Nacke AK, et al. Sentinel lymphadenectomy in vulvar cancer using near-infrared fluorescence from indocyanine green compared with technetium 99m nanocolloid. Int J Gynecol Cancer. 2017;27:805–12.
20. Van der Zee AGJ, Oonk MH, De Hullu JA, et al. Sentinel node dissection is safe in the treatment of early-stage vulvar cancer. J Clin Oncol. 2008;26:884–9.
21. Saito T, Tabata T, Ikushima H, et al. Japan Society of Gynecologic Oncology guidelines 2015 for the treatment of vulvar cancer and vaginal cancer. Int J Clin Oncol. 2018;23(2):201–34. https://doi.org/10.1007/s10147-017-1193-z.
22. National Comprehensive Cancer Network. nccn Clinical Practice Guidelines in Oncology Vulva cancer Ver. 2.2019 [Cited 17th Jul 2019]. Available from URL: https://www.nccn.org/professionals/physician_gls/pdf/vulvar.pdf.
23. Yamagami W, Susumu N, Kataoka F, et al. A comparison of dye versus fluorescence methods for sentinel lymph node mapping in endometrial cancer. Int J Gynecol Cancer. 2017;27:1517–24.
24. Hope-Ross M, Yannuzzi LA, Gragoudas ES, et al. Adverse reactions due to indocyanine green. Ophthalmology. 1994;101:529–33.

Lymphography and Evaluation of Lymphedema

24

Takumi Yamamoto

Summary

- Fluorescence imaging following intradermal or subcutaneous injection of ICG visualizes the lymph vessels.
- Dynamic lymphography, dual-phase observation immediately after and 2–72 hours after ICG injection, is useful for the comprehensive evaluation of lymphedema.
- Pathophysiological severity staging system based on ICG fluorescence lymphography is useful for the selection of surgical indication and prediction of postoperative efficacy.
- ICG fluorescence lymphography is most useful for early diagnosis of lymphedema, allowing detection of subclinical lymphedema and early treatment for cure.

1 Introduction

This chapter includes an outline of the problems of conventional lymphedema evaluation methods and the applications of ICG fluorescence lymphography in terms of diagnosis and severity classification, prognostic evaluation, and intraoperative navigation techniques, as well as future prospects.

2 Conventional Lymphedema Evaluation Methods

Lymphedema is a progressive edematous disease, associated with regional immune deficiency due to abnormal lymph circulation. Management of lymphedema is challenging, because of the lack of detailed methods for lymph flow evaluation and the intractable nature of the disease. Lymphoscintigraphy is considered a gold standard method for lymph flow evaluation, but its images are obscure and have a risk of radiation exposure [1, 2]. Early diagnosis is critical in the appropriate management of lymphedema, but lymphoscintigraphy is not suitable for this purpose because of its low sensitivity to detect abnormal lymph circulation. It is difficult to apply lymphoscintigraphy for lymph flow evaluation of small regions such as the genitalia, the face, and the breast [2–5].

Other imaging methods include lipiodol lymphography and MR lymphography, but it is often difficult to apply these methods to patients with lymphedema because of a risk of lymph vessels' obstruction due to lipiodol and skin necrosis caused by subcutaneous gadolinium injection [2, 6, 7]. Bioimpedance measurement and ultrasound are also used to evaluate edema in subcutaneous tissues, but these methods do not evaluate lymph circulation, and cannot differentiate between lymphedema and other edematous conditions [2, 8, 9]. The International Society of Lymphology (ISL) classification, which is based on circadian changes and gross examinations of edema, is widely used to classify the severity of lymphedema. However, it is not suitable for early diagnosis or prediction of therapeutic efficacy [10]. Since lymphedema can be cured only at an early stage, it is warranted to develop a better lymph flow imaging useful for early diagnosis with clear lymph flow visualization.

3 Application of Fluorescent Lymphography

ICG (ICG) fluorescent lymphography has been applied in sentinel lymph node biopsy for various cancers [11–13]. After intradermal or subcutaneous injection of ICG, lymph flows up to approximately 2 cm from the body surface and can be visualized using a near-infrared camera. ICG lymphography is becoming popular among medical staff for the diagnosis and treatment of lymphedema [12, 13]. Unlike sentinel lymph node biopsy, which identifies lymph nodes, the main focus of ICG lymphography for lymphedema evaluation is the visualization of lymph flows from the lym-

T. Yamamoto (✉)
Department of Plastic and Reconstructive Surgery, National Center for Global Health and Medicine, Shinjuku, Tokyo, Japan
e-mail: tyamamoto-tky@umin.ac.jp

T. Ishizawa (ed.), *Fluorescence-Guided Surgery*, https://doi.org/10.1007/978-981-19-7372-7_24

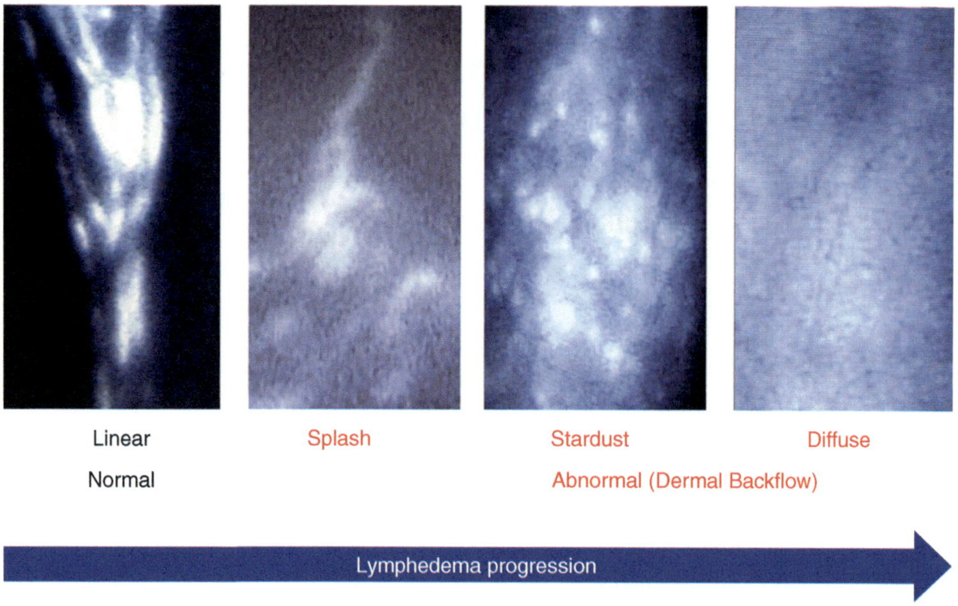

Fig. 24.1 ICG lymphography findings. With the progression of lymphedema, ICG lymphography findings change from a normal linear pattern to an abnormal dermal backflow (DB) pattern; DB pattern includes mild DB "splash," moderate DB "stardust," and severe DB "diffuse" pattern

phatic capillaries to the precollecting/collecting lymph vessels. Clinical studies elucidated characteristic findings of ICG fluorescence lymphography in lymphedema patients [12, 14, 15].

Lymphedema can be classified into primary lymphedema and secondary lymphedema. Approximately 90% of lymphedema seen in developed countries is secondary lymphedema caused by cancer treatment [10, 13, 16]. Secondary lymphedema is caused by obstruction of lymph flows due to lymph node dissection or irradiation. Lymphedema develops and progresses due to stagnation of lymph flow distal to the obstruction site. With the progression of lymphedema, ICG lymphography findings change from a normal linear pattern to abnormal dermal backflow (DB) pattern; DB pattern is subdivided into splash pattern, stardust pattern, and diffuse pattern (Fig. 24.1) [3, 4, 14, 15, 17].

Based on observational studies of patients with secondary lymphedema, pathophysiological severity staging systems using ICG lymphography, ICG stages, have been developed [3, 4, 14, 15, 17–21]. The stage is classified into six stages, from ICG stage 0 to stage V (Table 24.1). In ICG stage 0, only a linear pattern is seen: no lymphedema. At ICG stage I, a splash pattern is observed in addition to a linear pattern: subclinical lymphedema. Patients with ICG stage I have a progression risk of approximately 30% within 2 years. From ICG stage II, stardust/diffuse pattern is observed and then extends with lymphedema progression. Most patients with ICG stage II have mild edema but suffer from progressive lymphedema: early lymphedema. In ICG stages III to V, all patients have obvious edema and have a 100% risk of progression: advanced lymphedema. In ICG stage V, a linear pattern is absent and only a dermal backflow pattern is observed (Fig. 24.2).

Table 24.1 ICG stage: pathophysiological severity staging system based on ICG lymphography findings

Stage	ICG lymphography findings	Conditions
Stage 0	Linear pattern only (no DB pattern)	No lymphedema
Stage I	Linear pattern + splash pattern	Subclinical lymphedema
Stage II	Linear pattern + SD pattern (1 region)[a]	Early lymphedema
Stage III	Linear pattern + SD pattern (2 regions)[a]	Progressed lymphedema
Stage IV	Linear pattern + SD pattern (3 regions)[a]	
Stage V	SD pattern (no linear pattern)	

[a]Upper extremity, lower extremity, genitalia, and face are divided into three regions: upper arm/forearm/hand, thigh/lower leg/foot, lower abdomen/labia major (scrotum)/labia minora (penis), and neck/lower hemiface/upper hemiface, respectively

Primary lymphedema is a general term for lymphedema of unknown cause with no apparent triggers. As primary lymphedema includes all types of lymphedemas other than secondary lymphedema, its etiology includes a wide variety of conditions. With the accumulation of ICG lymphography experience in primary lymphedema, ICG lymphography classification has been developed for primary lymphedema; the ICG classification includes proximal DB (PDB) type, distal DB (DDB) type, less enhancement (LE) type, and no enhancement (NE) type (Fig. 24.3) [16]. The PDB type shows a DB pattern mainly in the proximal region (near the trunk), while the DDB type shows a DB pattern in the distal region. Both types of lymphedema are considered as obstructive lymphedema, and the therapeutic strategy is the same as that for secondary lymphedema. NE type is the most severe type, and is suspected to have localized lymphatic aplasia or

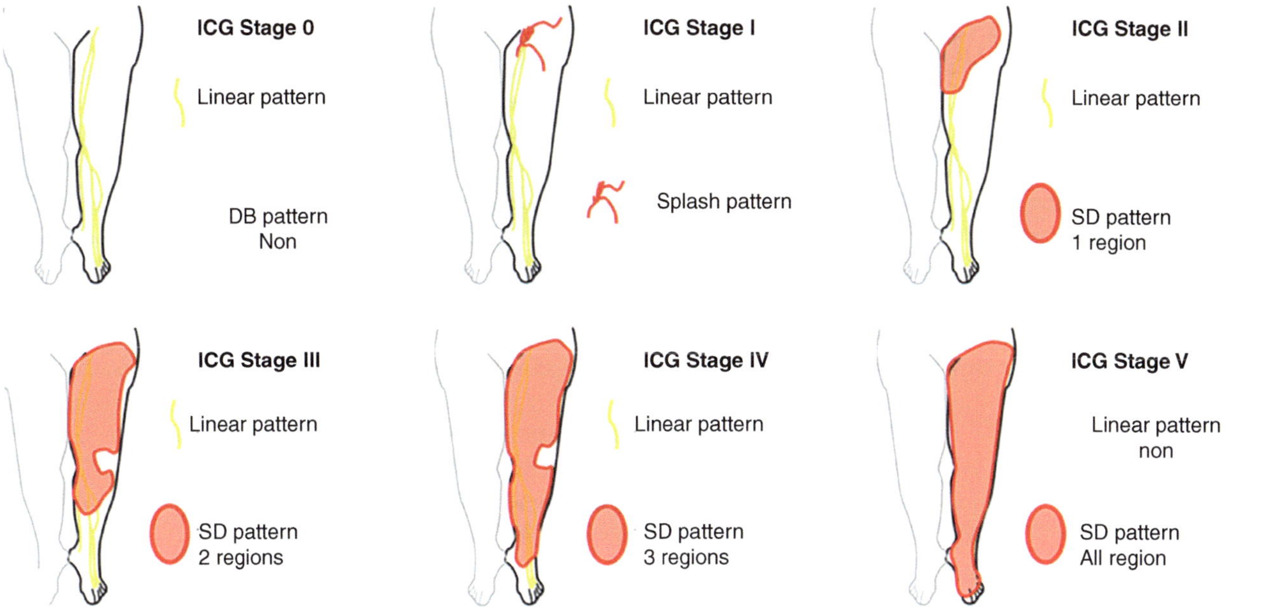

Fig. 24.2 ICG stage for lower extremity lymphedema. In stage 0, only linear pattern; in stage I, linear pattern and splash pattern; in stage II, linear pattern and stardust/diffuse pattern; in stage V, only stardust/diffuse pattern

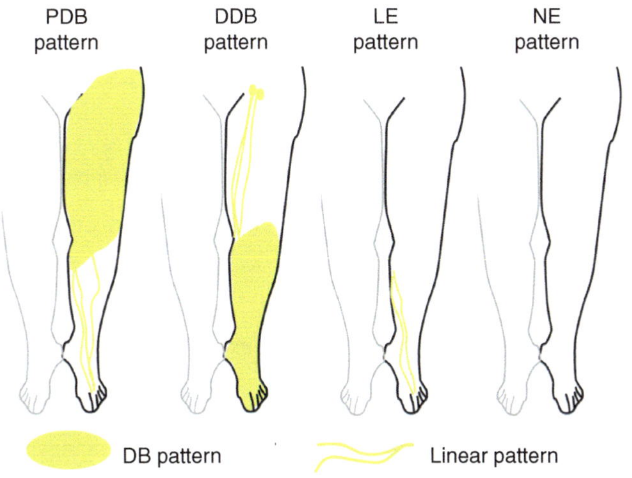

Fig. 24.3 ICG classification for primary lymphedema based on ICG lymphography findings

lymphatic malabsorption; lympho-venous shunt operation is hardly effective, and more invasive surgical treatment, such as vascularized lymph node transfer, is required (Table 24.2).

Point

- Characteristic ICG lymphography findings include a normal linear pattern and an abnormal DB pattern; DB pattern is subdivided into mild DB "splash," moderate DB "stardust," and severe DB "diffuse" patterns.
- Secondary lymphedema is classified by ICG stage into six stages, from stage 0 to V.

Table 24.2 Treatment strategy for primary lymphedema according to ICG classification

ICG classification	Possible pathophysiology	Treatment
PDB type	proximal obstruction (near the trunk) (≒secondary lymphedema)	Compression + LVA (±LNT ± LS)
DDB type	Distal obstruction Valvular insufficiency	Compression + LVA (±LNT ± LS)
LE type	Hypoplasia of the superficial lymphatics Lymphatic pump failure	Strict compression
NE type	Localized lymphatic aplasia/hypoplasia or malabsorption, etc.	LNT (±LS)

- Primary lymphedema is classified by ICG classification into four types: proximal DB (PDB) type, distal DB (DDB) type, less enhancement (LE) type, and no enhancement (NE) type.

4 Clinical Practice of ICG Lymphography

ICG lymphography requires ICG to be injected and a near-infrared camera for observation. There are two types of near-infrared cameras: one is a handheld type that allows the operator to easily change the observation area (e.g., PDE-neo®), and the other is a fixed type that allows the operator to fix the observation area on a stand (e.g., SPY® system) [3, 4, 11–16, 25]. The former is preferable for lymphedema evaluation in terms of observing lymph flows in various regions and from various angles. In addition to the above systems, surgical microscopic systems (e.g., Pentero® IR800) are useful for lymphatic navigation surgery [26–28].

The basic concentration of ICG is 2.5 mg/mL, but since this concentration often causes green coloration at the site of ICG injection, 0.25 mg/mL can be used for facial lymphedema evaluation [3, 4, 14–17]. Since the origin of lymph flow is the dermal capillary lymphatics, intradermal injection of ICG is ideal, but subcutaneous injection is also possible to reduce the pain of intradermal injection. It takes a little longer for the contrast agent to reach the collecting lymph vessels, but the same fluorescence images can be obtained.

In the upper extremities, 0.1 mL of ICG solution is injected at the second web space of the hand and at the wrist (just ulnar to the palmaris longus tendon); in the lower extremities, 0.2 mL of ICG is injected at the second web space of the foot and at the medial malleolus; in the face, 0.05 mL of ICG is injected at the midline points of the frontal area (around the hairline), at the glabella, and at the philtrum; in the male genitalia, injections are given at the tip of the foreskin on the dorsal side of the penis and at two sites in the middle of the scrotum [3, 4, 14–21].

The most important aspect of ICG lymphography is to perform two observations after each injection [29, 30] (Fig. 24.4). The first observation is immediately after the injection (transient phase) and is mainly to localize linear patterns. The second observation is performed 2 hours after injection when the ICG migration reaches equilibrium or plateau (plateau phase: up to 72 hours after injection). During the plateau phase, the type and extent of the DB pattern are determined, and the severity of the disease is assessed by the ICG stage. Observations in the transient phase are necessary to determine the presence or absence of Linear pattern in severe lymphedema (ICG stage IV or stage V), and observations in the plateau phase are essential for differentiating early stages of lymphedema (ICG stage 0 to II) as well as for severity classification. Since the plateau phase is essential for the diagnosis and severity classification, the latter phase observation should not be omitted. It is practical to inject ICG in the morning and observe the findings in the afternoon or the day after injection, when ICG lymphography is performed during a busy outpatient clinic.

It is recommended to mark findings of fluorescence imaging directly onto the body surface with a dermal pen and take a photo in addition to video recording for comparison of the edema status after the treatment. It is easy to understand the findings by marking the normal linear pattern with a dotted line and the abnormal findings of the splash/stardust/diffuse pattern with a solid line (Fig. 24.5).

Point
- Dynamic ICG lymphography with two observations allows comprehensive lymphedema evaluation.
- In the transient phase immediately after injection, a linear pattern is marked with dotted lines.
- Marking the extent of DB pattern with solid lines in the plateau phase 2 to 72 hours after injection.

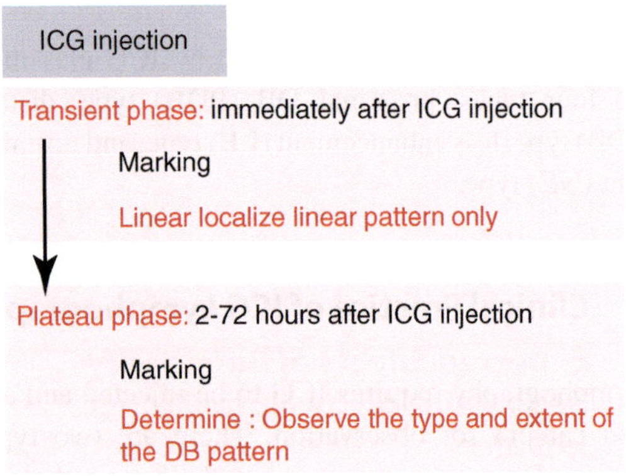

Fig. 24.4 Dynamic ICG lymphography. Observation in the transient phase immediately after ICG injection and in the plateau phase a few hours later

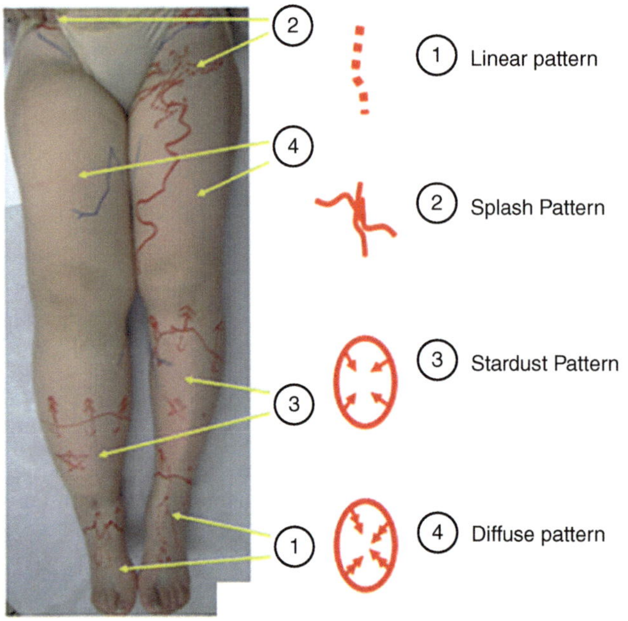

Fig. 24.5 Dermal marking of ICG lymphography findings. Linear pattern (dotted lines), range of abnormal findings (DB pattern, solid lines), splash pattern (tortuous lines), stardust pattern region (solid lines with arrows), diffuse pattern region (solid lines with double arrows). (i) Linear pattern; (ii) Splash Pattern; (iii) Stardust Pattern; (iv) Diffuse pattern

5 Applications of ICG Lymphography

The strength of ICG lymphography is that it has extremely high sensitivity and specificity in detecting abnormalities of lymphatic circulation [3, 22–24, 31]. It is the most useful test for the definitive diagnosis of lymphedema and early diagnosis as a screening test for lymphedema after cancer treatment. ICG lymphography allows early detection of abnormal DB patterns even before a patient or medical staff notices edema [15, 17–21]. ICG lymphography can subdivide the conditions before edema awareness (ISL stage 0) into three stages (ICG stage 0 to stage II). As mentioned above, the risk of lymphedema progression is completely different between ICG stage 0/ I/II, which is extremely useful for the prediction of prognosis.

The most important treatment for lymphedema is compression therapy. Although compression therapy can slow down the progression of lymphedema, it cannot prevent progression itself because it does not improve lymph circulation. At present, the only curative treatment for improving lymph circulation is physiologic or reconstructive surgery, such as lymphovenous bypass and lymph node transfer (LNT). The former is further classified into classic lymphovenous anastomosis (cLVA) and supermicrosurgical lymphaticovenular anastomosis (LVA). In cLVA, the lymphatic tissue is inserted into a vein (not a real anastomosis), whereas a lymph vessel is literally anastomosed to a venule or small vein in LVA. Classical anastomosis or cLVA is associated with a high incidence of thrombus occlusion, unstable efficacy, and risk of deep venous thrombosis and pulmonary embolism, although this method had been mainly used until the 1990s [32]. LVA, on the other hand, is a method that has become widespread since around 2000, in which lymph vessels of approximately 0.5 mm in diameter and small veins are literally anastomosed in an intima-to-intima coaptation manner, resulting in a minimally invasive and sustained bypass effect with a low risk of thrombotic occlusion [33–38].

ICG lymphography is also useful for predicting the effect of LVA treatment, and it was found that early diagnosis and early LVA can completely cure lymphedema [2, 23, 33, 36]. Lymphedema is progressive and refractory, and it is thought to never be completely cured once it develops. In these years, however, there are many cases reported where edema disappears without maintenance compression therapy. The cure rate after LVA differs according to the ICG stage; earlier is better [2, 23, 36, 39]. At present, there is a possibility of a complete cure by LVA up to ICG stage III, but there are no reports of a complete cure in ICG stage IV to V. Even if a complete cure is not possible, LVA can reduce edema and the burden of compression therapy, preventing cellulitis. Therefore, LVA is indicated even in advanced cases where a complete cure is not expected.

The most important point in LVA surgery is to find and anastomose high-flow lymph vessels. ICG lymphography is useful for the prediction of lymph vessels' conditions (Fig. 24.6). In regions with linear, splash, and stardust patterns on ICG lymphography, lymph vessels suitable for anastomosis can easily be found, whereas it is difficult to find such vessels in a diffuse pattern [18–21]. Therefore, in patients with a wide range of diffuse patterns, LNT rather than LVA is recommended. In LNT, normal lymph nodes are transplanted with feeding vessels and surrounding tissues by vascular anastomoses, and therapeutic effects can be expected even in severe cases where lymphosclerosis is too severe to perform effective LVA (Table 24.3).

ICG lymphography is also useful for intraoperative navigation, because it visualizes lymph vessels and nodes in real time during surgical procedures [26–28, 36]. In LVA, a near-

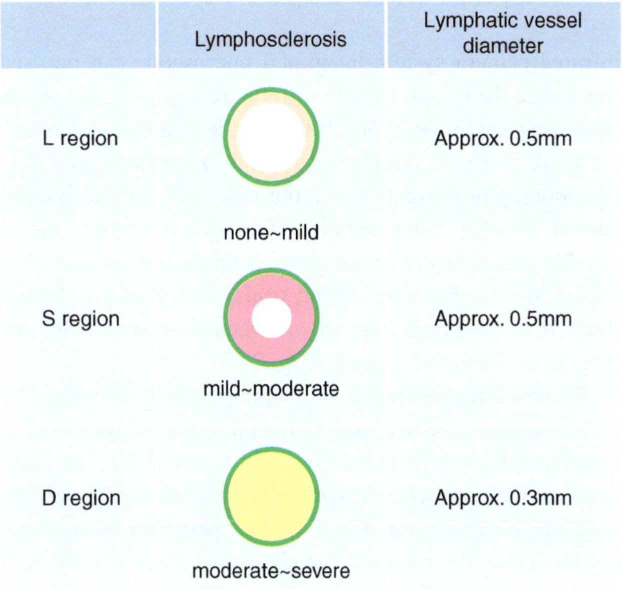

Fig. 24.6 ICG lymphography findings and lymph vessel conditions: L region (linear pattern), S region (splash/stardust pattern), and D region (diffuse pattern)

Table 24.3 ICG stage and treatment

ICG stage	Pathophysiology (risk of progression)	Treatment plan
Stage 0	No lymphedema (almost 0%)	No treatment required
Stage I	Subclinical lymphedema (approx. 30%)	Follow-up or LVA
Stage II	Early lymphedema (approx. 90%)	LVA
Stage III Stage IV Stage V	Progressed lymphedema (almost 100%)	Combined surgical treatment LVA LVA LS

ELVA efferent LVA, *LVA* lympahticovenular anastomosis, *LNT* lymph node transfer, *LS* liposuction

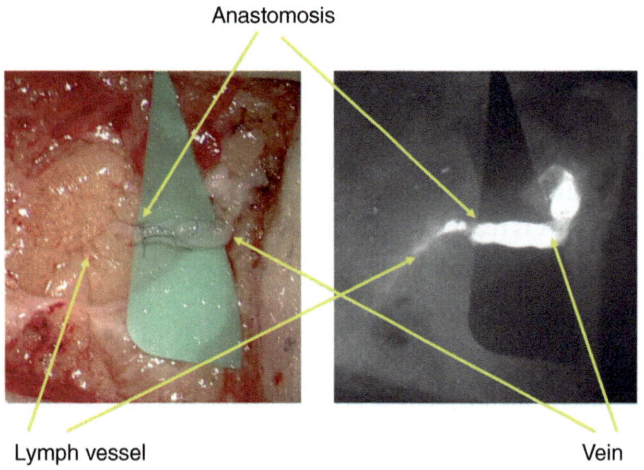

Fig. 24.7 Intraoperative ICG lymphography navigation. Evaluation of anastomotic patency in lymphatic microvascular anastomosis

infrared camera system-integrated microscope can be used to localize the lymph vessels with fluorescence signals while dissecting the fat layer. In LNT, only the lymph nodes which do not drain major lymph flows can be harvested under ICG lymphography navigation without risk of donor site lymphedema [39–45]. Intraoperative ICG lymphography allows reliable evaluation of anastomotic patency and leakage after LVA [26–28]. Because of the transparency of the lymph fluid, ICG lymphography enhancement is essential for the detection of a minor leakage (Fig. 24.7).

Besides lymphedema, ICG lymphography is also effective in the diagnosis and treatment of lymphorrhea and lymphocyst, lymphatic diseases with lymph vessel rupture [28]. The diagnosis can be made when leakage of fluorescing fluids is visualized following injection of ICG into the periphery of the target region. Then, the ruptured lymph vessels identified by ICG lymphography are ligated or anastomosed to treat the leakage.

Point

- ICG lymphography is the most effective tool for early diagnosis of lymphedema.
- ICG stage is useful in surgical decision-making based on the prediction of prognosis.
- Early diagnosis using ICG lymphography and treatment with LVA enables a complete cure of lymphedema.
- ICG lymphography can be used as a surgical navigation in both LVA and LNT procedures.
- ICG lymphography is also effective in the diagnosis and treatment of lymphorrhea and lymphocyst with lymph leakage.

6 Pitfalls and Limitations

While ICG lymphography has become an indispensable tool in the diagnosis and treatment of lymphedema, there are some contraindications: as ICG contains iodine, ICG cannot be used in patients with a history of iodine allergy or bronchial asthma [12–15]. The major advantage of ICG lymphography is that it can visualize lymph flows in detail, but the observation depth is limited to up to 2 cm from the body surface [12–21]. In order to visualize deeper lymph flows, other imaging tests such as lymphoscintigraphy, SPECT/CT, and MR lymphography are required [1, 2, 7–9]. Especially in primary lymphedema, it is important to evaluate deep lymph flows. It is essential to combine ICG lymphography with other imaging tests for comprehensive evaluation. Although ICG lymphography cannot directly visualize deep lymph flows, it indirectly reflects abnormalities in deep lymphatic circulation, enabling evaluation of the pathophysiology of secondary lymphedema when the site of lymphatic obstruction is clear. In fact, the conditions of deep lymph vessels can be predicted by the difference of three types of DB patterns, such as splash, stardust, and diffuse patterns, as shown in Fig. 24.6 [19–21].

Although not discussed in many papers, intradermal and subcutaneous injections of ICG can enhance not only the lymph vessels but also the veins [46]. The fluorescence signals in veins are characterized by the following features: veins appear as black lines on near-infrared camera observation before enhancement, fluorescence in veins tends to be clearer than lymph vessels because veins are usually located in more superficial layers, and the ICG fluorescence washes out and returns to black when squeezed along the enhanced structures. Therefore, we can differentiate the lymph vessels from the veins, based on the information that the lymph vessels are located in a slightly deeper layer and show a slightly blurred linear pattern, and the enhancement is not completely disappeared and continues to be fluorescent even after squeezing.

Point

- ICG lymphography is contraindicated in patients with iodine allergy or asthma.
- When evaluation of deep lymph flows is required, such as in primary lymphedema, ICG lymphography should be combined with other imaging studies.
- Intracutaneous and subcutaneous injections of ICG can visualize not only the lymph vessels but also the veins.

7 Future Perspectives

It is expected that ICG lymphography visualized deeply located lymph vessels in 3D, enabling comprehensive evaluation of the lymphatic system without the use of other imaging modalities.

In the treatment of lymphedema, various therapies such as new drug therapy, cell therapy, lymphangiogenesis therapy, and gene therapy are being developed. Even in these therapeutic approaches, one of the most important factors would be the accurate evaluation of lymph circulation before and after the treatment, which could be achieved by ICG fluorescence lymphography. While a complete cure of lymphedema by early diagnosis and treatment has been reported, it is necessary to establish an optimal protocol of ICG lymphography for lymphedema screening and appropriate early intervention in the future.

References

1. Yoshida RY, Kariya S, Ha-Kawa S, et al. Lymphoscintigraphy for imaging of lymphatic flow disorders. Tech Vasc Interv Radiol. 2016;19:273–6.
2. Greene AK, Goss JA. Diagnosis and staging of lymphedema. Semin Plast Surg. 2018;32:12–6.
3. Yamamoto T, Yamamoto N, Yoshimatsu H, et al. Indocyanine green lymphography for evaluation of genital lymphedema in secondary lower extremity. J Vasc Surg. 2013;1:400–5.
4. Yamamoto T, Iida T, Matsuda N, et al. Indocyanine green (ICG)-enhanced lymphography for evaluation of facial lymphoedema. J Plast Reconstr Aesthet Surg. 2011;64:1541–4.
5. Yamamoto T, Yamamoto N, Giacalone G. Supermicrosurgical lymphaticovenular anastomosis for a breast lymphedema secondary to vascularized axillary lymph node flap transfer. Lymphology. 2016;49:128–32.
6. O'Brien BM, Das SK, Franklin JD, et al. Effect of lymphangiography on lymphedema. Plast Reconstr Surg. 1981;68:922–6.
7. Pons G, Clavero JA, Alomar X, et al. Preoperative planning of lymphaticovenous anastomosis: the use of magnetic resonance lymphangiography as a complement to indocyanine green lymphography. J Plast Reconstr Aesthet Surg. 2019;72:884–91.
8. Svensson BJ, Dylke ES, Ward LC, et al. Electrode equivalence for use in bioimpedance spectroscopy assessment of lymphedema. Lymphat Res Biol. 2019;17:51–9.
9. Giray E, Yagci I. Diagnostic accuracy of interlimb differences of ultrasonographic subcutaneous tissue thickness measurements in breast carcinoma related arm lymphedema. Lymphology. 2019;52:1–10.
10. Executive Committee: the Diagnosis and Treatment of Peripheral Lymphedema: 2016 Consensus Document of the International Society of Lymphology. Lymphology. 2016;49:170–184.
11. Ahmed M, Purushotham AD, Douek M. Novel techniques for sentinel lymph node biopsy in breast cancer: a systematic review. Lancet Oncol. 2014;15:e351–62.
12. Unno N, Inuzuka K, Suzuki M, et al. Preliminary experience with a novel fluorescence lymphography using indocyanine green in patients with secondary. J Vasc Surg. 2007;45:1016–21.
13. Narushima M, Yamamoto T, Ogata F, et al. Indocyanine green lymphography findings in limb lymphedema. J Reconstr Microsurg. 2016;32:72–9.
14. Yamamoto T, Narushima M, Doi K, et al. Characteristic indocyanine green lymphography findings in lower extremity lymphedema: the generation of a novel lymphedema severity staging system using dermal backflow patterns. Plast Reconstr Surg. 2011;127:1979–86.
15. Yamamoto T, Yamamoto N, Doi K, et al. Indocyanine green (ICG)-enhanced lymphography for upper extremity lymphedema: a novel severity staging system. Plast Reconstr Surg. 2011;128:941–7.
16. Yamamoto T, Yoshimatsu H, Narushima M, et al. Indocyanine green lymphography findings in primary leg lymphedema. Eur J Vasc Endovasc Surg. 2015;49:95–102.
17. Yamamoto T, Matsuda N, Doi K, et al. The earliest finding of indocyanine green (ICG) lymphography in asymptomatic limbs of lower extremity lymphedema: the modified dermal backflow (DB) stage and concept of subclinical lymphedema. Plast Reconstr Surg. 2011;128:314e–21e.
18. Yamamoto T, Yamamoto N, Yoshimatsu H, et al. Factors associated with lower extremity dysmorphia caused by lower extremity lymphedema. Eur J Vasc Endovasc Surg. 2017;54:126.
19. Yamamoto T, Narushima M, Koshima I. Lymphatic vessel diameter in female pelvic cancer-related lower extremity lymphedematous limbs. J Surg Oncol. 2018;117:1157–63.
20. Yamamoto T, Yamamoto N, Yoshimatsu H, et al. Factors associated with lymphosclerosis: an analysis on 962 lymphatic vessels. Plast Reconstr Surg. 2017;140:734–41.
21. Yamamoto T, Yamamoto N, Fuse Y, et al. Optimal sites for supermicrosurgical lymphatic venular anastomosis: an analysis of lymphatic vessel detection rates on 840 surgical fields in lower extremity lymphedema. Plast Reconstr Surg. 2018;142:924e–30e.
22. Akita S, Mitsukawa N, Rikihisa N, et al. Early diagnosis and risk factors for lymphedema following lymph node dissection for gynecologic cancer. Plast Reconstr Surg. 2013;131:283–90.
23. Yamamoto T, Yamamoto N, Yamashita M, et al. Efferent lymphatic vessel anastomosis (ELVA): supermicrosurgical efferent lymphatic vessel -to- venous anastomosis for the prophylactic treatment of subclinical lymphedema. Ann Plast Surg. 2016;76:424–7.
24. Akita S, Nakamura R, Yamamoto N, et al. Early detection of lymphatic disorder and treatment for lymphedema following breast cancer. Plast Reconstr Surg. 2016;138:192e–202e.
25. Tsukuura R, Sakai H, Fuse Y, et al. Novel hands-free near-infrared fluorescence navigation and simultaneous combined imaging for elevation of vascularized lymph node flap. J Surg Oncol. 2018;118:588–9.
26. Yamamoto T, Yamamoto N, Azuma S, et al. Near-infrared illumination system- integrated microscope for supermicrosurgical lymphatic venular. Microsurgery. 2014;34:23–7.
27. Yamamoto T, Yamamoto N, Numahata T, et al. Navigation lymphatic supermicrosurgery for the treatment of cancer-related peripheral lymphedema. Vasc Endovasc Surg. 2014;48:139–43.
28. Yamamoto T, Yoshimatsu H, Koshima I. Navigation lymphatic supermicrosurgery for iatrogenic lymphorrhea: supermicrosurgical lymphaticolymphatic anastomosis and lymphaticovenular anastomosis under indocyanine green lymphography navigation. J Plast Reconstr Aesthet Surg. 2014;67:1573–9.
29. Yamamoto T, Narushima M, Yoshimatsu H, et al. Dynamic indocyanine green lymphography for breast cancer-related arm lymphedema. Ann Plast Surg. 2014;73:706–9.
30. Yamamoto T, Narushima M, Yoshimatsu H, et al. Indocyanine green velocity: Lymph transport capacity deterioration with progression of lymphedema. Ann Plast Surg. 2013;71:591–4.

31. Akita S, Mitsukawa N, Kazama T, et al. Comparison of lymphoscintigraphy and indocyanine green lymphography for the diagnosis of extremity. J Plast Reconstr Aesthet Surg. 2013;66:792–8.
32. Campisi C, Boccardo F. Lymphedema and microsurgery. Microsurgery. 2002;22:74–80.
33. Koshima I, Narushima M, Mihara M, et al. Lymphadiposal flaps and lymphaticovenular anastomoses for severe leg edema: Functional reconstruction for lymph drainage system. J Reconstr Microsurg. 2016;32:50–5.
34. Yamamoto T, Narushima M, Kikuchi K, et al. Lambda-shaped anastomosis with intravascular stenting method for safe and effective lymphaticovenular. Plast Reconstr Surg. 2011;127:1987–92.
35. Yamamoto T, Yoshimatsu H, Narushima M, et al. A modified side-to-end lymphaticovenular anastomosis. Microsurgery. 2013;33:130–3.
36. Yamamoto T, Narushima M, Yoshimatsu H, et al. Minimally invasive lymphatic supermicrosurgery (MILS): indocyanine green lymphography-guided simultaneous multi-site lymphaticovenular anastomoses via millimeter skin incisions.
37. Yamamoto T, Yoshimatsu H, Narushima M, et al. Sequential anastomosis for lymphatic supermicrosurgery: multiple lymphaticovenular anastomoses on one venule. Ann Plast Surg. 2014;73:46–9.
38. Yamamoto T, Yoshimatsu H, Yamamoto N, et al. Side-to-end lymphatic venular anastomosis through temporary lymphatic expansion. PLoS ONE. 2013;8:e59523.
39. Yamamoto T, Yoshimatsu H, Yamamoto N. Complete lymph flow reconstruction: a free vascularized lymph node true perforator flap transfer with efferent lymphaticolymphatic anastomosis. J Plast Reconstr Aesthet Surg. 2016;69:1227.
40. Yamamoto T, Iida T, Yoshimatsu H, et al. Lymph flow restoration after tissue replantation and transfer: importance of lymph axiality and possibility of lymph flow reconstruction using free flap transfer without lymph node or supermicrosurgical lymphatic anastomosis. Plast Reconstr Surg. 2018;142:796–804.
41. Yamamoto T, Saito T, Ishiura R, et al. Quadruple-component superficial circumflex iliac artery perforator (SCIP) flap: a chimeric SCIP flap for complex ankle reconstruction of an exposed artificial joint after total ankle arthroplasty. J Plast Reconstr Aesthet Surg. 2016;69:1260–5.
42. Yamamoto T. Onco-reconstructive supermicrosurgery. Eur J Surg Oncol. 2019;45:1146–51.
43. Brahma B, Yamamoto T. Breast cancer treatment-related lymphedema (BCRL): an overview of the literature and updates in microsurgery reconstruction. Eur J Surg Oncol. 2019;45:1138–45.
44. Yamamoto T, Yamamoto N, Kageyama T, et al. Supermicrosurgery for oncologic reconstructions. Glob Health Med. 2020;2:18–23.
45. Yamamoto T, Yamamoto N, Kageyama T, et al. Technical pearls in lymphatic supermicrosurgery. Glob Health Med. 2020;2:29–32.
46. Sumiya R, Fuse Y, Yamamoto T. Distinction between the lymph vessel and the vein on ICG lymphography: Intradermal or subcutaneous ICG injection also enhances the vein [published online ahead of print, 2020 May 5]. J Plast Reconstr Aesthet Surg. 2020;S1748-6815(20):30162–5.

Part V

Intraoperative Fluorescence Imaging [Practice] – Imaging of Anatomical Structures

Takeaki Ishizawa

Although anatomy textbooks neatly depict the "most typical" structures of the human body (vascular and organ morphologies), in reality there are many variations in each patient. In addition, because these structures are covered with fatty tissues and attached to neighboring organs, the "surgeon's eye" alone may not be able to sufficiently recognize the location information necessary for safe surgery.

Misrecognition of biological structures directly leads to complications such as hemorrhage and organ damage, and accurate resection "along the organ compartment" in liver and lung surgery is also important to reduce tumor recurrence. Therefore, fluorescent imaging has been applied to depict anatomical structures that are "difficult or impossible to see with the naked eye." Some of these techniques, such as fluorescence cholangiography, have become widely used in clinical settings for making surgery safer and more reliable.

Imaging of the Bile Ducts (Fluorescence Cholangiography) 25

Kazuhiro Matsuda, Takeshi Aoki, and Tomokazu Kusano

Summary

- Fluorescence cholangiography using ICG is a simple and real-time method to observe the bile ducts without X-ray exposure.
- Fluorescence cholangiography can be used by intravenous or intrabiliary (gallbladder or common bile duct) injection of ICG.
- The intravenous method has the advantage of not requiring cannulation of the bile duct.
- Fluorescence cholangiography is useful not only for identifying bile duct anatomy but also for confirming the presence or absence of bile leak after resection and biliary anastomosis.

1 Introduction

Fluorescence imaging using ICG is an intraoperative imaging technique that has been applied clinically in recent years due to the widespread use of near-infrared imaging systems. In this article, we review the history of bile duct imaging using ICG fluorescence and demonstrate the practice of intraoperative imaging with caution and future prospects.

2 Previous Methods and Problems of Intraoperative Bile Duct Imaging

Imaging of bile ducts is essential for the diagnosis, selection of the surgical procedures, and execution of the surgery with safety and curability for hepatobiliary diseases. Currently, CT with drip-infusion cholangiography (DIC-CT), magnetic resonance cholangiopancreatography (MRCP), and endoscopic cholangiography (ERCP) are widely used for preoperative imaging of the bile ducts. However, these cholangiography modalities are hard to be used during surgery. To perform intraoperative imaging of the bile ducts, a radiographic cholangiography has been conventionally performed in which a portable radiography system (C-arm) is brought into the operating room and an iodine contrast medium is administered into the bile ducts. This imaging method involves risks of X-ray exposure to patients and medical staff, technical difficulty in cannulating the bile duct for injecting the contrast medium into the biliary system, and time and manpower to interrupt the operation and place the imaging system in the operating field. In addition, the risk of bile duct injury associated with the cannulation procedure itself [1] is a major reason why intraoperative cholangiography is not routinely performed in these years. Furthermore, since only the monochromatic and two-dimensional X-ray image of the bile ducts is projected on the monitor, it is necessary to compare the monitor image with the field of view of the surgical field in order to understand the three-dimensional positional relationship between the contrasted bile ducts and the surrounding organs.

3 History of Fluorescence Imaging Using ICG

ICG has been widely applied clinically since the 1950s as a test drug for liver function. ICG binds in vivo to lipoproteins [2, 3] present in blood and bile and emits fluorescence signals with a peak wavelength at around 845 nm when illuminated with near-infrared light (750 to 810 nm, peaking at 805 nm [4]. Since the excitation light and fluorescence signals of ICG are in the near-infrared region, avoiding absorption by hemoglobin or water, near-infrared fluorescence imaging can visualize biological structures containing ICG penetrating covering tissues up to around 10 mm from tissue

Supplementary Information The online version contains supplementary material available at https://doi.org/10.1007/978-981-19-7372-7_25.

K. Matsuda · T. Aoki (✉) · T. Kusano
Division of Gastroenterological and General Surgery, Department of Surgery, School of Medicine, Showa University, Tokyo, Japan
e-mail: takejp@med.showa-u.ac.jp

T. Ishizawa (ed.), *Fluorescence-Guided Surgery*, https://doi.org/10.1007/978-981-19-7372-7_25

surfaces. The imaging system using the fluorescence property of ICG has been applied to many fields of clinical practice with the development and improvement of infrared observation systems, and now, many surgical device manufacturers have started to sell dedicated imaging systems.

In 2006, Kubota et al. reported for the first time that intraoperative administration of ICG during a living donor liver transplantation led to the production of bile with fluorescence signals, which could be used for rapid evaluation of graft function [5]. In 2008, Mitsuhashi et al. reported that the gallbladder, bile duct, and common bile duct could be visualized as fluorescing areas by ICG administration during open cholecystectomy [6]. In 2009, Ishizawa et al. first reported the application of intravenous ICG fluorescence cholangiography to laparoscopic cholecystectomy using a prototype of a near-infrared laparoscopic imaging system [7]. Also in 2009, Yasuda et al. reported the application of fluorescence imaging to laparoscopic cholecystectomy by direct injection of ICG into the gallbladder as well as intravenous injection [8]. In 2010, Aoki et al. reported a series of laparoscopic cholecystectomies using fluorescence cholangiography by intravenous injection of ICG (12.5 mg), resulting in a 71.4% identification rate of the common bile duct and cystic duct [9]. In the same year, Ishizawa et al. reported that bifurcation of bile ducts was identified before dissection of the Calot's triangle in 96.2% of patients undergoing laparoscopic cholecystectomy by intravenous injection of ICG (2.5 mg) [10], followed by a report by Tagaya et al. [11]. The usefulness of intraoperative ICG fluorescence cholangiography to confirm bile duct anatomy has become known worldwide, and finally in 2019, a multicenter randomized controlled trial demonstrated the superiority of ICG fluorescence imaging to conventional color imaging in visualizing the extrahepatic bile duct anatomy during laparoscopic cholecystectomy [12].

ICG fluorescence cholangiography can also be used during hepatectomy as first reported by Ishizawa et al. in 2009 [13]. In 2015, Kawaguchi et al. applied ICG fluorescence cholangiography to laparoscopic hepatectomy and reported that it was possible to confirm the confluence of the right and left hepatic ducts during hemi-hepatectomy [14]. In 2016, Tanaka et al. reported that ICG cholangiography was useful for identifying hepatic ducts running on the wall of hepatic cysts during laparoscopic fenestration procedures [15]. Not only fluorescence properties but also bile excretion properties of ICG can also be applied to the prevention of surgical complications: in 2011, Kaibori et al. reported that ICG fluorescence imaging was effective in identifying bile leak in hepatectomy [16], followed by reports by Mizuno et al. in 2014 [17] and Hong et al. in 2017 [18] demonstrating the efficacy of using fluorescence imaging in preventing bile leak and biliary stricture in living donor liver transplantation.

4 Clinical Practice

The ICG administration route for intraoperative fluorescence cholangiography includes the systemic injection (intravenous method) and the direct injection into the biliary system (intrabiliary methods) [6, 9, 10, 13] (Table 25.1). In the intravenous method, ICG is injected intravenously before the surgery, and the fluorescence signal of ICG excreted into the biliary tract is observed during the surgery (Figs. 25.1 and 25.2; Movies 25.1 and 25.2). In addition, 2.5 mg of ICG is often reported to be administered 1 hour before the surgery (Table 25.2) [19], and the signal-to-background ratio of the bile duct is maximal at 1 hour after intravenous administration [20]. The advantage of the intravenous method is that there is no need for cannulation in the biliary tract, and by administering ICG intravenously and selecting an appropriate imaging mode, bile duct anatomy can be easily observed as a superimposed image of pseudo-color fluorescence signals on full-color background images. The disadvantage is that it is sometimes difficult to recognize the fluorescence of the bile ducts when the background fluorescence signals are high. The contrast between the bile ducts and the surrounding organs may be improved by selecting a grayscale mode.

In the intrabiliary method, ICG diluted in saline (0.025 mg/mL) is injected into the gallbladder by puncturing the gallbladder during the operation (Fig. 25.3, Movie 25.3). When the ICG solution is mixed with the bile aspirated from the biliary drainage tube, injection of the mixture enables better and faster identification of the biliary tracts by fluorescence imaging. The direct injection of ICG into the gallblad-

Table 25.1 ICG administration methods for fluorescence cholangiography

	Intravenous method	Gallbladder puncture method	Intra-biliary injection method
ICG dosage	2.5 mg	0.025 mg	0.025 mg
Route	Intravenous	Gallbladder puncture	External biliary fistula tube (PTBD/ENBD)
Injection timing	1 hour before the operation	Perioperative period	Perioperative period

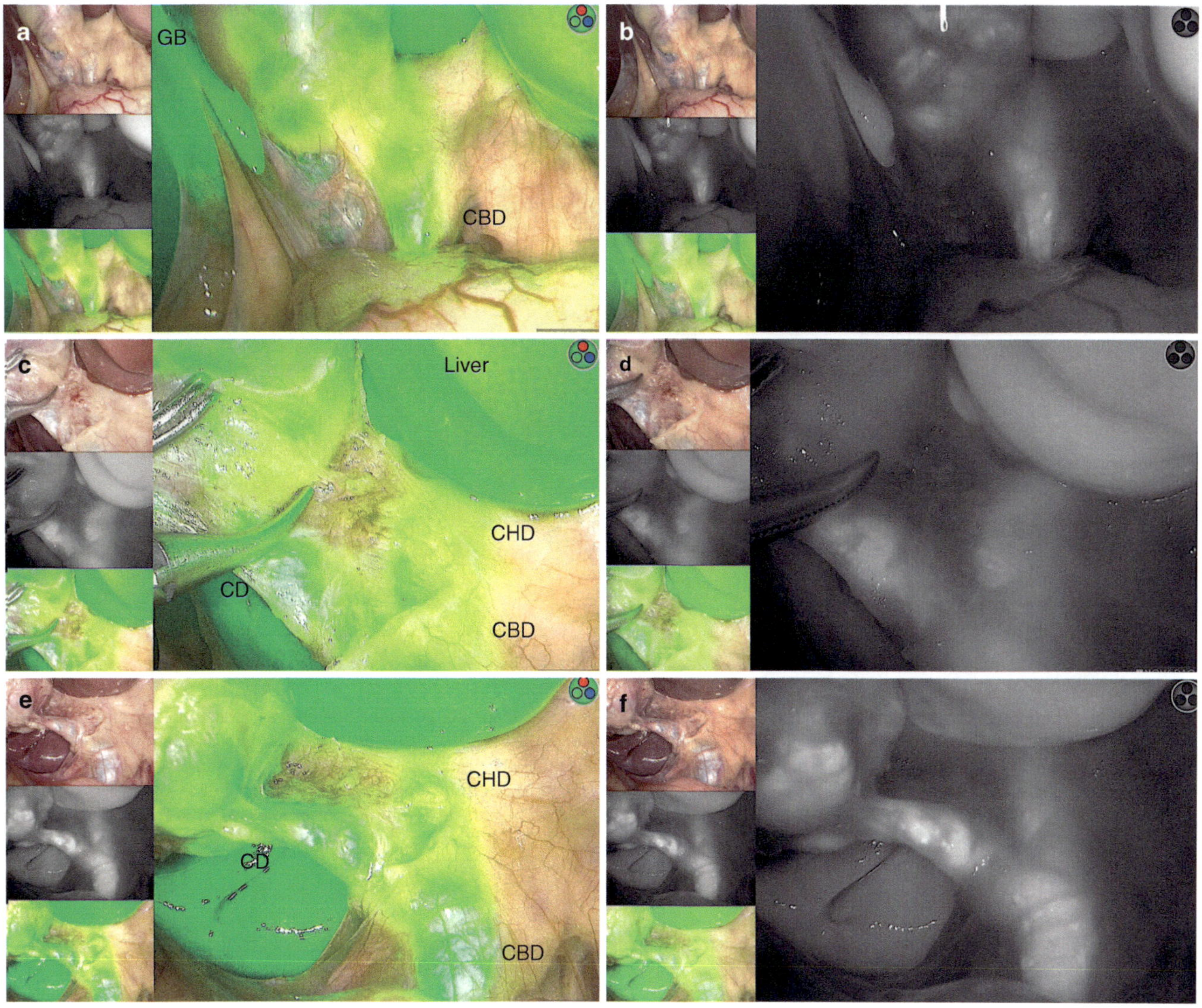

Fig. 25.1 ICG fluorescence cholangiography (intravenous method). A case of acute cholecystitis without anatomical variations in the biliary system. ICG (2.5 mg, 1 mL) was intravenously injected immediately after induction of anesthesia. Laparoscopic imaging system: PINPOINT (Stryker). (**a**, **b**) ICG fluorescence imaging clearly delineated the course of the common bile duct (**a**, overlay mode; **b**, spy mode). (**c**, **d**) Fluorescence cholangiography visualized the cystic duct, the common bile duct, and the common hepatic duct during the dissection of Calot's triangle (**c**, overlay mode; **d**, spy mode). (**e**, **f**) Calot's triangle was dissected with the use of fluorescence cholangiography as a guide to obtain a critical view of safety (**e**, overlay mode; **f**, spy mode). GB gallbladder; CBD common bile duct; CHD common hepatic duct; CD cystic duct

der or the bile ducts has the advantage that the ICG fluorescence of only the bile ducts is depicted without background fluorescence signals, resulting in clearer cholangiography images. On the other hand, if the ICG is injected by puncturing the gallbladder during the operation, spillage of bile containing ICG associated with puncture and injection procedures should be avoided. Otherwise, an increase in background fluorescence signals due to the attachment of ICG would make it difficult to identify biliary structures in Calot's triangle.

As a method for detection of bile leak during hepatectomy, it has been reported that after the common bile duct is temporarily clamped, 10 mL of 2.5 mg/mL ICG solution is injected into the bile duct, and the fluorescing spots on hepatic raw surfaces are confirmed with a near-infrared imaging device [16]. When the fluorescing spot is wiped off with gauze, the fluorescent signals adhering to the gauze are clearly observed, and the presence of a bile leak can be determined.

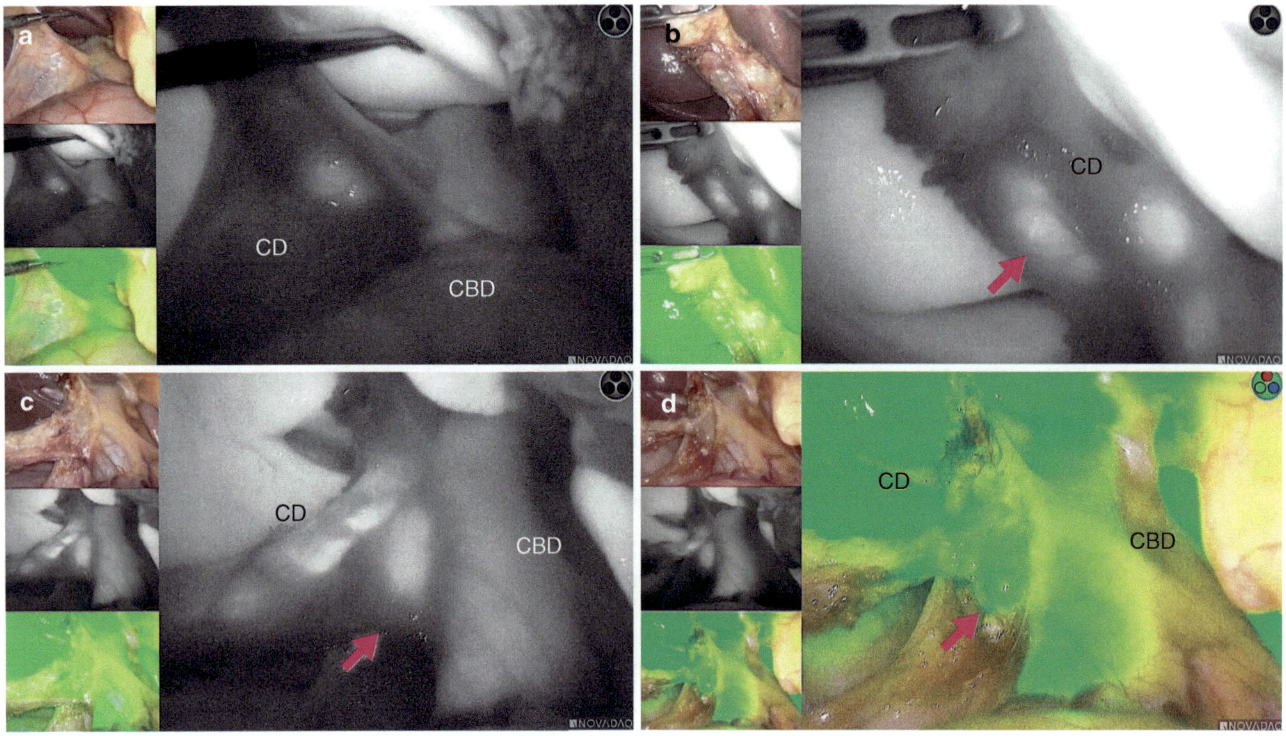

Fig. 25.2 ICG fluorescence cholangiography (intravenous method). Patients with acute cholecystitis with a cystic duct draining into the right posterior branch. ICG (2.5 mg, 1 mL) was injected intravenously immediately after induction of anesthesia. Laparoscopic system: PINPOINT (Stryker). (**a**) ICG fluorescence cholangiography clearly delineated the anatomy of the common bile duct (spy mode). (**b**) The independent right posterior branch was observed after dissection (spy mode). (**c**, **d**) The cystic duct draining into the right posterior branch was clearly visualized by ICG fluorescence cholangiography. (**c**, spy mode; **d**, overlay mode). CBD common bile duct; CD cystic duct. Arrow: hepatic duct draining posterior segment of the liver

Table 25.2 Previous reports on intraoperative fluorescence cholangiography using intravenous ICG (intravenous method)

						Identification accuracy (%)		
	Number of cases	Operation	Fluorescence imaging system	ICG dosage	Injection timing	Bile ducts	Common bile duct	Common hepatic duct
Ishizawa, 2008	10	Cholecystectomy	PDE	2.5 mg	1 hour before surgery or at the time of transition to laparotomy	90	–	100
Aoki, 2010	14	Laparoscopic cholecystectomy	Prototype	12.5 mg	30 minutes before surgery	71.4	71.4	–
Spinoglio, 2013	45	Cholecystectomy (robotic surgery)	da Vinci	2.5 mg	30–40 minutes before surgery	97	97	97
Daskalaki, 2014	184	Cholecystectomy (robotic surgery)	da Vinci	2.5 mg	45 minutes before surgery	97.8	96.1	94
Larsen, 2014	35	Laparoscopic cholecystectomy	Olympus	0.05 mg/kg (body weight)	After induction of anesthesia	100	100	100
Boni, 2015	52	Laparoscopic cholecystectomy	KARL STORZ	0.04 mg/kg (body weight)	At least 15 minutes before surgery	100	100	100
Osayi, 2015	82	Laparoscopic cholecystectomy	Stryker	2.5 mg	1 hour before surgery	95.1	76.8	69.5
Van dam, 2015	30	Laparoscopic cholecystectomy	Olympus	0.05 mg/kg (body weight)	After induction of anesthesia	96.7	86.7	Not reported

Table 25.2 (continued)

	Number of cases	Operation	Fluorescence imaging system	ICG dosage	Injection timing	Identification accuracy (%)		
						Bile ducts	Common bile duct	Common hepatic duct
Dip, 2016	71	Laparoscopic cholecystectomy	KARL STORZ	0.05 mg/kg (body weight)	1 hour before surgery	100	87.3	70.4
Diana, 2017	54	Cholecystectomy (robotic surgery)	Firefly	0.1–0.4 mg/kg (body weight)	45–60 minutes before surgery	98.2	98.2	Not reported
Liu, 2018	46	Laparoscopic cholecystectomy	KARL STORZ	1.25 mg	Perioperative period	84.7	78.2	73.9
Dip, 2019	321	Laparoscopic cholecystectomy	KARL STORZ	0.05 mg/kg (body weight)	At least 45 minutes before surgery	96.9	75.7	52.3

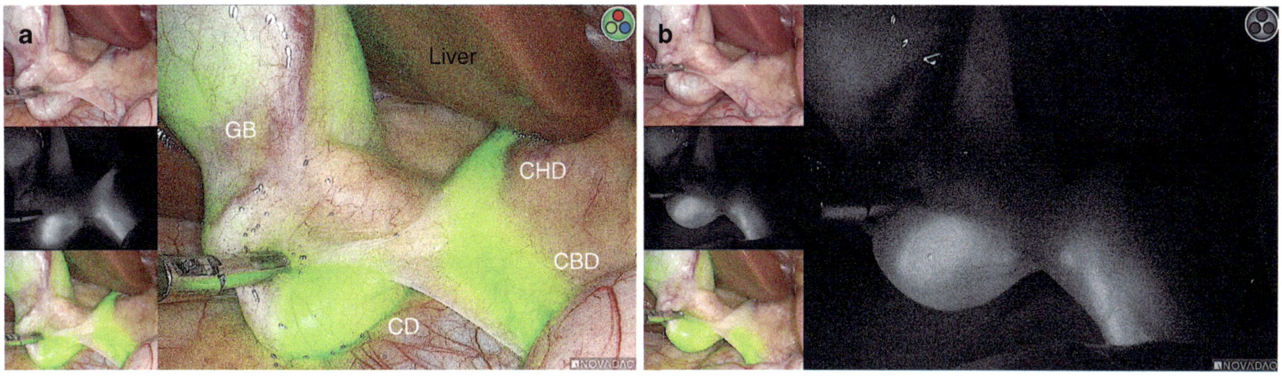

Fig. 25.3 ICG fluorescence cholangiography (intrabiliary method using an ENBD tube). A case of acute cholecystitis without anatomical variations. ICG solution (0.025 mg, 1 mL) was injected into the bile ducts through an ENBD tube after admixture with bile. Laparoscopic system; PINPOINT (Stryker). (**a**, **b**) ICG fluorescence imaging clearly delineated the cystic duct, common bile duct, and common hepatic duct (**a**, overlay mode; **b**, monochrome mode). GB gallbladder; CBD common bile duct; CHD common hepatic duct; CD cystic duct

Point

- The routes of ICG administration for intraoperative fluorescence cholangiography include intravenous injection and intrabiliary injection.
- In the intravenous method, 2.5 mg of ICG is administered 1 hour before surgery.
- In the intrabiliary method, the gallbladder is punctured intraoperatively, and ICG diluted in saline (0.025 mg/mL) is injected into the gallbladder or the common bile duct through a cystic tube.
- ICG fluorescence cholangiography can be applied to the detection of bile leak during hepatectomy.

5 Expected Effects of Fluorescence Cholangiography

Intraoperative ICG fluorescence cholangiography allows the bile ducts to be observed in real time in the operative field as superimposed images in pseudo color, and the bile ducts can be clearly identified in the same field of view during the operation. Fluorescence cholangiography is expected to be widely used for assuring the safety of laparoscopic cholecystectomy, in which bile duct injury still occurs with an incidence of <1% [21]. For the prevention of bile duct injury during cholecystectomy, intraoperative radiographic cholecystectomy has been recommended. However, this conventional procedure is usually applied selectively because of the complexity of the procedure, the time and manpower required for the C-arm manipulation, and the risk of bile duct injury due to the cannulation of the bile ducts for injection of contrast material [1]. Intraoperative fluorescence cholangiography, on the other hand, has great advantages in that the bile ducts can be observed in real time during the operation by administering ICG intravenously or through an external biliary drainage tube, the bile ducts can be observed in the same field of view as the surgical field, the observation can be easily performed by switching to near-infrared observation conditions, and the X-ray exposure of medical personnel and patients can be avoided.

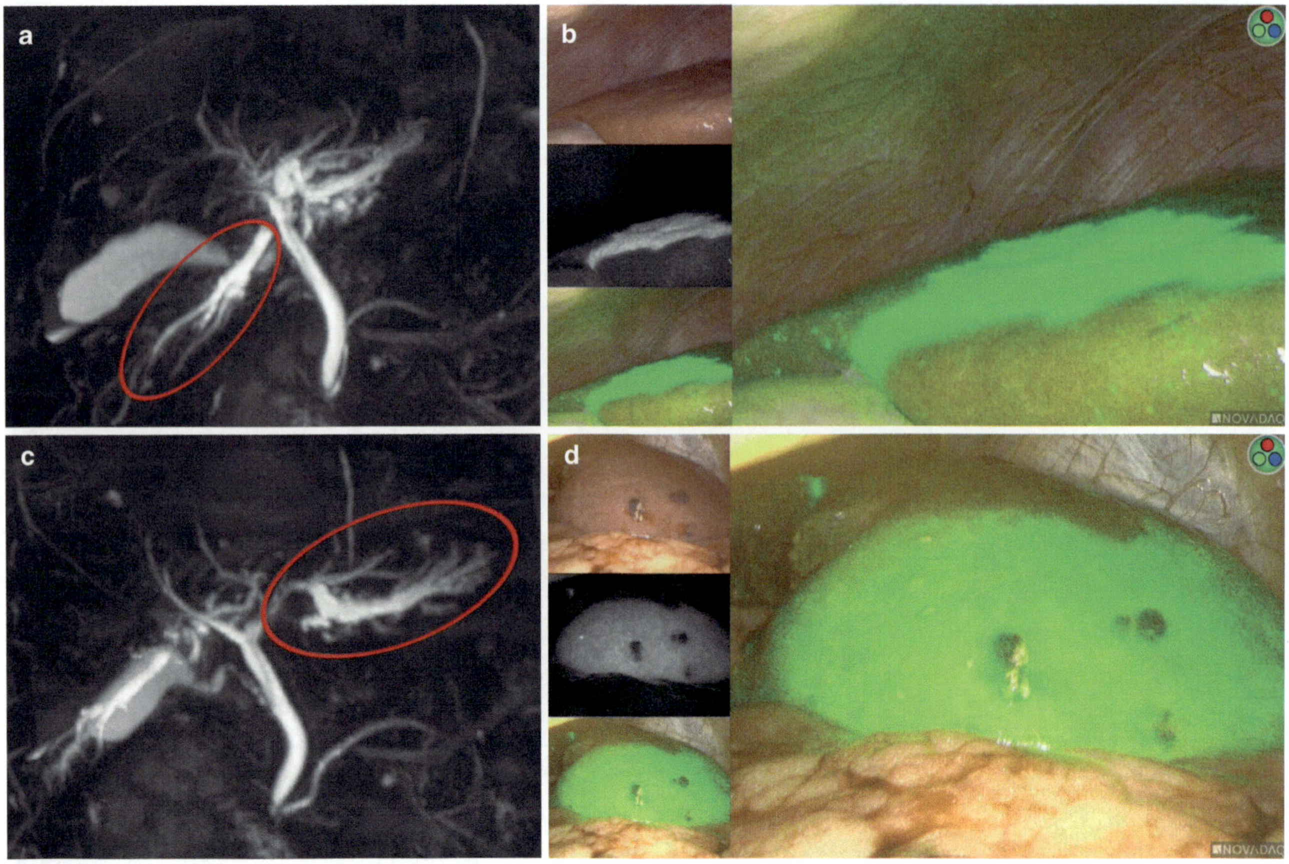

Fig. 25.4 Biliary congestion due to intrahepatic cholangiocarcinoma. A case of intrahepatic cholangiocarcinoma with two lesions localized near the portal pedicles of S5 and S3. (**a**) MRCP showed bile duct dilatation of the hepatic duct draining S5. (**b**) Fluorescence imaging following preoperative intravenous injection of ICG delineated biliary congested regions (S5) on the hepatic surfaces during surgery. (**c**) MRCP showed bile duct dilatation of the hepatic duct draining S3. (**d**) Intraoperatively, hepatic regions with biliary congestion due to the tumor in S3 were identified as fluorescing areas on hepatic surfaces

Fluorescence cholangiography can work as a navigation tool illuminating a roadmap toward the critical view of safety, which has been recommended to avoid bile duct injury during laparoscopic cholecystectomy [22].

In hepatectomy, intraoperative fluorescence cholangiography depicts the extrahepatic bile ducts and small biliary structures on raw surfaces of the liver [13, 14], enabling more precise surgery by confirming the bile duct anatomy in real time during hepatectomy procedures. In addition, fluorescence imaging following preoperative systemic injection of ICG for the purpose of assessing hepatic function can delineate the extent of obstruction in the intrahepatic bile ducts due to tumor invasion or stones (Fig. 25.4), enabling surgeons to determine which segments of the liver to be removed for complete resection [23]. Furthermore, when confirming bile leaks on hepatic raw surfaces or a biliary anastomosis during surgery, the use of ICG fluorescence imaging may enhance the detection rate, leading to reduce postoperative biliary complications.

6 Pitfalls and Limitations

6.1 Side Effects of ICG Administration

ICG is a relatively safe reagent with few side effects, but because it contains iodine, ICG fluorescence cholangiography cannot be performed in patients with a history of iodine hypersensitivity or asthma. In addition, it is recommended that ICG be carefully administered to pregnant women, nursing mothers, and children.

6.2 Near-Infrared Imaging System

At present, near-infrared imaging systems for laparoscopic and open surgery are commercially available from a lot of medical device companies. Since there are differences in the sensitivity of fluorescence signals as well as conditions of the excitation light, filter settings, and fluorescence image display methods among the systems [24], it is important to fully understand the characteristics of each model for using fluorescence cholangiography effectively. In addition, please note that the fluorescence signal identification ability of laparoscopic near-infrared observation systems tends to be lower than that of imaging systems for open surgery [24].

6.3 Appropriate ICG Dosage and Administration Timing

Since ICG is both a fluorophore and a light absorber, the fluorescence intensity reaches a saturation value at a certain ICG concentration, and as the concentration increases, it absorbs the emitted fluorescence, and the fluorescence intensity declines (photo-quenching) [25]. Therefore, it is important to perform fluorescence cholangiography using an appropriate ICG dose.

When applying the intravenous method, a dose of 2.5 mg ICG administered 1 hour before the procedure is commonly used. Some authors have reported that intravenous administration of ICG 3 hours prior to surgery would be the best to further decrease background fluorescence signals [26, 27]. When intraoperative fluorescence cholangiography is performed in patients with decreased liver function, intravenous ICG (0.5 mg/kg) administered a couple of days before the surgery may enable clear observation of the biliary tract by fluorescence imaging without additional administration of ICG during surgery.

6.4 Tissue Permeability of Fluorescence Signals

The tissue permeability of ICG fluorescence is limited up to 10 mm [10], and it is not possible to observe the bile ducts in hepatic parenchyma or in the thick connective tissues. Therefore, fluorescence imaging cannot visualize biliary structures in cases where the bile ducts are covered with thick fat or inflammatory tissues. In addition, the present technique is unsuitable for the detection of small stones in the common bile duct during cholecystectomy (impacted stones in the cystic duct can usually be visualized) [10]. In other words, fluorescence cholangiography is not a complete substitute for conventional radiographic cholangiography, and these cholangiography techniques should be selected appropriately according to the purposes and conditions of surgery. It is possible to perform radiography and fluorescence cholangiography at the same time by adding a small amount of ICG to the iodine contrast medium [28].

Point

- ICG fluorescence cholangiography is contraindicated in patients with a history of iodine hypersensitivity or asthma attacks. Its indication should be carefully considered in pregnant and lactating women.
- ICG fluorescence cholangiography is difficult to observe the biliary structure deep in the hepatic parenchyma or connective tissues and to identify small stones in the common bile duct. In such cases, conventional radiographic cholangiography should be considered.

7 Future Perspectives

Fluorescence cholangiography seems to have the potential to be a more versatile imaging method than conventional radiographic cholangiography because of its simplicity and the anatomical information obtained. In addition, the development of fluorescence imaging using novel fluorescent substances has been conducted in basic experiments [29–31], and there is a possibility of clinical application of fluorescent contrast agents that are superior to ICG in delineating bile ducts at deeper planes. In addition, a color-coded display of bile ducts and blood vessels using an imaging system that simultaneously displays near-infrared signals and visible light in color has been reported [32]. In the future, these novel technologies enabling real-time multifunction imaging may enhance the safety and efficacy of hepatobiliary surgery.

References

1. White TT, Hart MJ. Cholangiography and small duct injury. Am J Surg. 1985;149:640–3.
2. Mullock BM, Shaw LJ, Fitzharris B, et al. Sources of proteins in the human bile. Gut. 1985;26:500–9.
3. Baker KJ. Binding of sulfobromophthalein (BSP) sodium and indocyanine green (ICG) by plasma alpha-1 lipoproteins. Proc Soc Exp Biol Med. 1966;122:957–63.
4. Landsman ML, Kwant G, Mook GA, et al. Light-absorbing properties, stability, and spectral stabilization of indocyanine green. J Appl Physiol. 1976;40:575–83.
5. Kubota K, Kita J, Shimoda M, et al. Intraoperative assessment of reconstructed vessels in living-donor liver transplantation, using a novel. J Hepato-Biliary-Pancreat Surg. 2006;13:100–4.
6. Mitsuhashi N, Kimura F, Shimizu H, et al. Usefulness of intraoperative fluorescence imaging to evaluate local anatomy in hepatobiliary surgery. Hepatobiliary Pancreat Surg. 2008;15:508–14.
7. Ishizawa T, Bandai Y, Kokudo N: Fluorescent cholangiography using indocyanine green for laparoscopic cholecystectomy: an initial experience. Arch Surg 2009; 144: 381–382.
8. Yasuda D, Kusano M, Aoki T, et al. D a novel development of ICG fluorescence image-guided cholangiography. J Showa Univ Soc. 2009;69:253–62.

9. Aoki T, Murakami M, Yasuda D, et al. Intraoperative fluorescent imaging using indocyanine green for liver mapping and cholangiography. Hepatobiliary Pancreat Sci. 2010;17:590–4.
10. Ishizawa T, Bandai Y, Ijichi M, et al. Fluorescent cholangiography illuminating the biliary tree during laparoscopic cholecystectomy. Br J Surg. 2010;97:1369–77.
11. Tagaya N, Shimoda M, Kato M, et al. Intraoperative exploration of biliary anatomy using fluorescence imaging of indocyanine green in experimental and clinical cholecystectomies. J Hepatobiliary Pancreat Sci. 2010;17:595–600.
12. Dip F, LoMenzo E, Sarotto L, et al. Randomized trial of near-infrared incisionless fluorescent cholangiography. Ann Surg. 2019;270:992–9.
13. Ishizawa T, Tamura S, Masuda K, et al. Intraoperative fluorescent cholangiography using indocyanine green; a biliary road map for safe surgery. Coll Surg. 2009;208:e1–4.
14. Kawaguchi Y, Velayutham V, Fuks D, et al. Usefulness of indocyanine green-fluorescence imaging for visualization of the bile duct during Laparoscopic liver resection. J Am Coll Surg. 2015;221:e113–7.
15. Tanaka M, Inoue Y, Mise Y, et al. Laparoscopic deroofing for polycystic liver disease using laparoscopic fusion indocyanine green fluorescence. Surg Endosc. 2016;30:2620–3.
16. Kaibori M, Ishizaki M, Matsui K, et al. Intraoperative indocyanine green fluorescent imaging for prevention of bile leakage after hepatic resection. Surgery. 2011;150:91–8.
17. Mizuno S, Inoue H, Tanemura A, et al. Biliary complications in 108 consecutive recipients with duct-to-duct biliary reconstruction in living-donor. Transplant Proc. 2014;46:850–5.
18. Hong SK, Lee KW, Kim HS, et al. Optimal bile duct division using real-time indocyanine green near-infrared fluorescence cholangiography during laparoscopic donor hepatectomy. Liver Transpl. 2017;23:847–52.
19. Cherrick GR, Stein SW, Leevy CM, et al. Indocyanine green: observations on its physical properties, plasma decay, and hepatic extraction. J Clin Invest. 1960;39:592–600.
20. Hutteman M, van der Vorst JR, Mieog JS, et al. Near-infrared fluorescence imaging in patients undergoing pancreaticoduodenectomy. Eur Surg Res. 2011;47:90–7.
21. Flum DR, Dellinger EP, Cheadle A, et al. Intraoperative cholangiography and risk of common bile duct injury during cholecystectomy. JAMA. 2003;289:1639–44.
22. Strasberg SM, Hertl M, Soper NJ. An analysis of the problem of biliary injury during laparoscopic cholecystectomy. J Am Coll Surg. 1995;180:101–25.
23. Harada N, Ishizawa T, Muraoka A, et al. Fluorescence navigation hepatectomy by visualization of localized cholestasis from bile duct tumor. J Am Coll Surg. 2010;210:e2–6.
24. Kono Y, Ishizawa T, Tani K, et al. Techniques of fluorescence cholangiography during laparoscopic cholecystectomy for better delineation of the bile duct anatomy. Medicine (Baltimore). 2015;94:e1005.
25. Benson RC, Kues HA. Fluorescence properties of indocyanine green as related to angiography. Phys Med Biol. 1978;23:159–63.
26. Zarrinpar A, Dutson EP, Mobley C, et al. Intraoperative laparoscopic near-infrared fluorescence cholangiography to facilitate anatomical identification; when to give indocyanine green and how much. Surg Innov. 2016;23:360–5.
27. Boogerd LSF, Handgraaf HJM, Huurman VAL, et al. The best approach for laparoscopic fluorescence cholangiography; overview of the literature and optimization of dose and dosing time. Surg Innov. 2017;24:386–96.
28. Kawaguchi Y, Ishizawa T, Masuda K, et al. Hepatobiliary surgery guided by a novel fluorescent imaging technique for visualizing hepatic arteries. J Am Coll Surg. 2011;212:e33–9.
29. Figueiredo JL, Siegel C, Nahrendorf M, et al. Intraoperative near-infrared fluorescent cholangiography (NIRFC) in mouse models of bile duct injury. World J Surg. 2010;34:336–43.
30. Tanaka E, Choi HS, Humblet V, et al. Real-time intraoperative assessment of the extrahepatic bile ducts in rats and pigs using invisible near-infrared fluorescent light. Surgery. 2008;144:39–48.
31. van den Bos J, Al-Taher M, Hsien SG, et al. Near-infrared fluorescence laparoscopy of the cystic duct and cystic artery: first experience with two new preclinical dyes in a pig model. Surg Endosc. 2017;31:4309–14.
32. Ashitate Y, Stockdale A, Choi HS, et al. Real-time simultaneous near-infrared fluorescence imaging of bile duct and arterial anatomy. J Surg Res. 2012;176:7–13.

Hepatic Segmentation

26

Takeshi Aoki and Kazuhiro Matsuda

Summary
- Hepatic segmentation using ICG fluorescence imaging clearly visualizes hepatic segmental boundaries on the liver surfaces and hepatic raw surfaces in real time, enabling three-dimensional identification of intersegmental planes during hepatectomy.
- Hepatic segmentation by ICG fluorescence imaging includes positive staining technique and negative staining technique.
- Hepatic segmentation by fluorescence imaging can be applied to laparoscopic hepatectomy.

1 Introduction

Anatomical (sub)segmentectomy is considered to be the standard surgery for hepatic malignancies. In order to perform safe and accurate anatomical hepatectomy, it is important to understand the special relationships between the tumor and intrahepatic vessels and to identify the exact extent of hepatic segments to be removed. In this article, we describe a method of identifying boundaries of the hepatic segment (hepatic segmentation) using ICG fluorescence imaging technique.

Supplementary Information The online version contains supplementary material available at https://doi.org/10.1007/978-981-19-7372-7_26.

T. Aoki (✉) · K. Matsuda
Division of Gastroenterological and General Surgery, Department of Surgery, School of Medicine, Showa University, Tokyo, Japan
e-mail: takejp@med.showa-u.ac.jp

2 Conventional Methods of Intraoperative Hepatic Segmentation

Anatomical hepatectomy is a hepatic resection based on the anatomy of the portal vein branches as proposed by Couinaud [1]. There are two methods to identify the hepatic segment: one is to inject dye into the portal vein branch and identify the stained area, and the other is to block the Glisson sheath and identify the ischemic area.

In 1985, Makuuchi et al. introduced the use of intraoperative ultrasonography (IOUS) during hepatectomy for understanding intrahepatic structures in real time during the operation and also for identifying hepatic segmental boundaries by injecting a dye (indigo carmine) into the portal vein branch of tumor-bearing hepatic segments [2]. In 1991, Takayama et al. reported that the counterstaining method was useful for identifying hepatic segmental boundaries by puncturing the dominant portal vein branch of adjacent hepatic segments in cases where puncture of the portal branches of tumor-bearing hepatic segments was technically difficult [3]. On the other hand, in 1986, Takasaki et al. reported the Glissonean approach, in which the hepatic hilum was dissected to reach the Glisson sheath of the tumor-bearing hepatic segments and the extent of hepatic resection was determined as ischemic regions appearing on hepatic surfaces after closure of the tumor-bearing Glissonean pedicles [4].

Anatomical hepatectomy based on the above techniques for hepatic segmentation contributes greatly to the safety of hepatectomy, i.e., reduction of blood loss and prevention of biliary fistulas [3]. However, staining with indigo carmine on hepatic surfaces disappears in a couple of minutes after the injection, making it difficult to confirm hepatic segmental boundaries during parenchymal transection. Especially in cases of cirrhosis or re-hepatectomy, the stained area on hepatic surfaces is sometimes difficult to identify by naked-eye observation. The Glissonean approach also has limitations in requiring a high level of

T. Ishizawa (ed.), *Fluorescence-Guided Surgery*, https://doi.org/10.1007/978-981-19-7372-7_26

skill to reach the target Gleason sheaths from the hepatic hilum, especially in resection of the suprahepatic regions such as S7 and S8.

3 Development History of ICG Fluorescence Imaging

In recent years, the development of preoperative simulation technology using diagnostic imaging and 3D imaging has made it possible to accurately understand the local anatomy of the liver and to develop an optimal surgical plan. On the other hand, it has become necessary to establish an intraoperative navigation method that accurately reflects the actual anatomical information in order to carry out the surgical approach that has been studied in detail by preoperative simulation.

In 2008, we reported a technique for intraoperative hepatic segmentation based on the fluorescence property of ICG, in which the portal vein is punctured under IOUS guidance, ICG is injected, and the corresponding hepatic segments are identified with a near-infrared camera system (PDE-II) [5]. With this method, the boundaries of hepatic segments can be clearly observed in real time, throughout the hepatectomy procedures. In 2011, Uchiyama et al. reported a method to identify the target hepatic area as an ICG fluorescence-deficient area by administering ICG intravenously after blocking the blood flow of the dominant portal vein branch, following the conventional Glissonean approach [6]. Nowadays, these methods are widely used as a useful intraoperative navigation method for performing accurate three-dimensional anatomical hepatectomy, because the hepatic segmental boundaries can be clearly observed not only on the hepatic surface but also in detached liver sections.

In 2012, Ishizawa et al. first reported the applications of hepatic segmentation techniques using ICG fluorescence imaging to laparoscopic hepatectomy [7]. Similar to open surgery, laparoscopic hepatectomy can be performed by applying the positive staining method to identify the target liver area as an ICG fluorescence-emitting area and the negative staining method to identify the area as an ICG fluorescence-deficient area, enabling clear observation of the liver area during surgery. In the setting of laparoscopic hepatectomy, it is sometimes difficult to identify hepatic segmental boundaries by positive stating technique or negative staining technique, especially in cases of resection of suprahepatic regions such as S7 and S8. In order to overcome this problem, we developed a preoperative positive staining method in which the dominant portal vein is punctured under extracorporeal ultrasound guidance in the OR immediately after induction of anesthesia, and ICG is injected to stain the target hepatic segment [8].

Recently, the use of ICG fluorescence imaging for hepatic segmentation has expanded with the development and improvement of surgical imaging systems. Now it is expected that fluorescence imaging plays an important role as an essential navigation method for safe and accurate anatomical hepatectomy in the setting of both open and minimally invasive surgery.

4 Clinical Practice

4.1 Preoperative Simulation

In order to perform anatomical hepatectomy, various imaging tests such as abdominal ultrasonography, three-phase contrast-enhanced CT, and MRI-EOB are performed before surgery to confirm the localization of the tumor, its special relationship with surrounding vessels, and the absence of tumor invasion into major vessels. In addition, volume data from three-phase (arterial, portal, and venous phases) thin slice CT images can be converted into 3D reconstructed images on an image analysis workstation to simulate more detailed hepatic segmental resection before surgery [9]. Using the 3D reconstructed image, the dominant portal pedicle of the tumor-bearing hepatic segment is identified, and the expected extent and volumes of hepatic regions to be removed can be simulated preoperatively (Fig. 26.1). Prior to surgery, extracorporeal ultrasonography should be performed by operating surgeons to confirm whether the target portal vein branch can be identified sufficiently and whether the angle and route of puncture can be secured.

4.2 Hepatic Segmentation by ICG Fluorescence Imaging During Open Hepatectomy

In 2008, we reported a hepatic segmentation method using ICG fluorescence imaging in open hepatectomy. Recently, we selected the positive staining technique and negative staining technique according to conditions of each patient.

In the positive staining technique, the tumor-bearing portal pedicle is directly punctured under ultrasound guidance to inject ICG. When a high concentration of ICG was injected, the entire liver shows strong fluorescence signals due to ICG through the systemic circulation, making visualization of the target hepatic segment unclear. Therefore, in recent reports, 1 to 2 mL of ICG solution diluted in saline to a concentration of 0.025 to 0.25 mg/mL is mainly used for

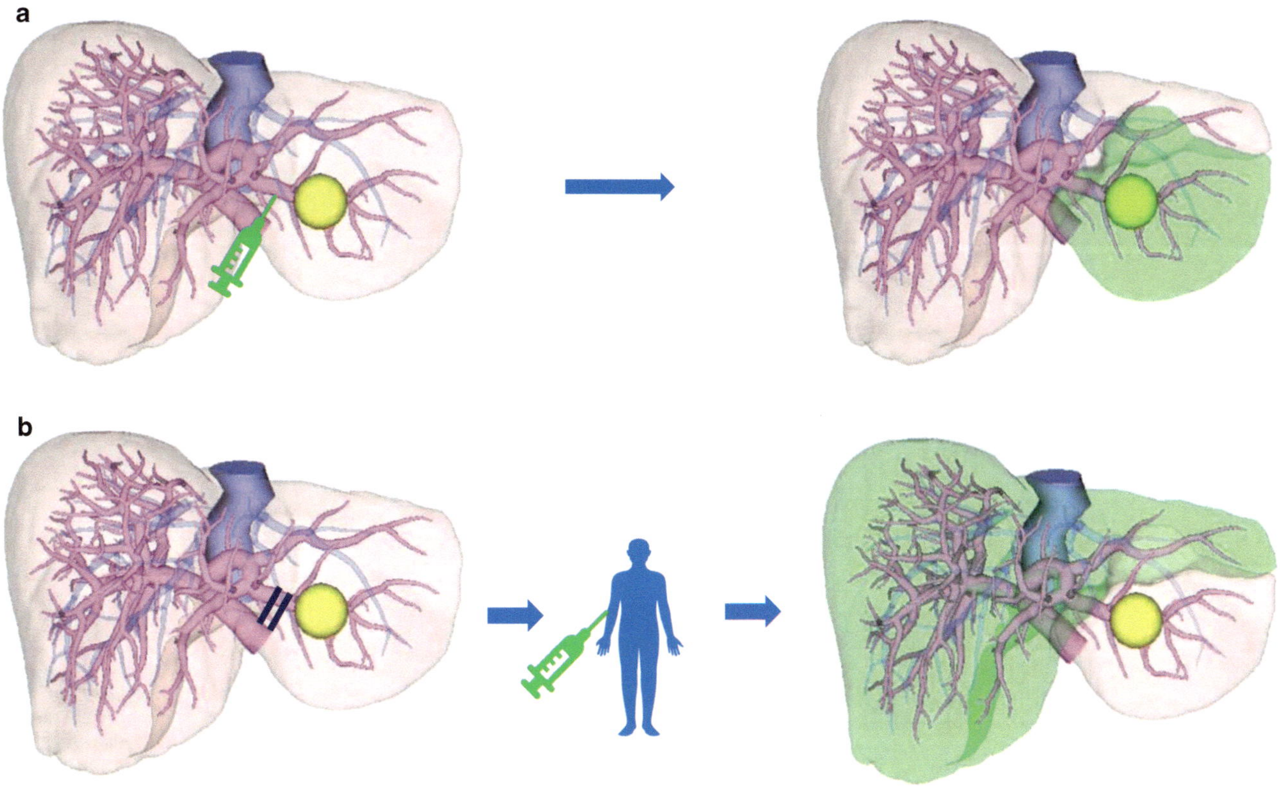

Fig. 26.1 Preoperative simulation using fluorescence imaging techniques for hepatic segmentation. (**a**) Positive staining technique. Simulation of the fluorescing regions identified after puncture and injection of ICG into the tumor-bearing portal vein branch. (**b**) Negative staining technique. Simulation of the non-fluorescing area identified when the tumor-bearing portal branch is clamped and ICG is administered intravenously

intraportal injection (Table 26.1) [5–8, 10–18]. During administration, it is important to infuse ICG slowly to avoid backflow of ICG into the adjacent hepatic portal vein. When direct puncture of the target portal branch is difficult or there are multiple dominant portal branches feeding the tumor, boundaries of hepatic segments to be removed can be confirmed by the counterstaining method [3], in which the portal branch of the adjacent hepatic segment is punctured to inject ICG solution.

The other method is to identify the target hepatic segment as an ICG fluorescence-deficient region by administering 1 mL of 2.5 mg/mL ICG solution intravenously after blocking the blood flow of the corresponding portal vein branch (negative staining technique, Fig. 26.2). In both methods, the hepatic segmental boundary can be clearly observed in real time during hepatectomy procedures, enabling anatomically accurate resection of the liver. With the use of the latest fluorescence imaging systems, we can proceed with hepatectomy procedures based on a high-resolution color image of the surgical field superimposed with the fluorescence images in real time, without turning off the surgical lights.

4.3 Hepatic Segmentation by ICG Fluorescence Imaging During Laparoscopic Hepatectomy

In 2012, Ishizawa et al. reported the negative staining method suitable for laparoscopic hepatectomy, in which the blood flow of the dominant portal vein branch is blocked as in open surgery [7]. In this method, after blocking the blood flow in the tumor-bearing portal vein, 1 mL of 2.5 mg/mL ICG solution is administered intravenously. Then, fluorescence imaging clearly visualizes hepatic segments to be removed as nonfluorescent ischemic regions on hepatic surfaces. In recent years, near-infrared laparoscopic imaging systems have been commercially available from various medical device companies. Some of these systems enable the superimposition of fluorescence signals on full-color images, enabling a three-dimensional understanding of intersegmental planes required for accurate anatomical hepatectomy.

Herein, we present a case of laparoscopic hepatectomy using the negative staining technique. In the laparoscopic hepatic S6 resection, the tumor-bearing portal vein branch

Table 26.1 Previous publications on hepatic segmentation using ICG fluorescence imaging

Researcher (year)	Staining method	Operation	Fluorescence imaging system	Route	ICG dosage	Hematologic Occlusion	Identification Rate (%)	References
Aoki (2008) [5]	Positive staining	Open	PDE	Tumor-bearing portal vein	5 mg	Pringle method	94.3	5
Uchiyama [6]	Negative staining	Open	PDE	Intravenous	0.5 mg/kg BW	Tumor-bearing portal vein	100	6
Ishizawa (2002) [7]	Positive/negative staining	Laparoscopic	Olympus	Tumor-bearing portal vein/intravenous	0.025 mg/2.5 mg	Hepatic artery/Tumor-bearing portal vein	100	7
Sakoda (2013) [10]	Positive staining	Laparoscopic	IRI	Tumor-bearing portal vein	5 mg	None	100	13
Inoue (2015) [11]	Positive/negative staining	Open	HyperEye Medical System	Tumor-bearing portal vein/intravenous	2.5 mg/2.5 mg	Hepatic artery / Tumor-bearing portal vein	95.8	14
Miyata (2015) [12]	Positive staining	Open	PDE-neo®	Tumor-bearing portal vein	0.25 mg	Hepatic artery	100	15
Mizuno (2017) [13]	Negative staining	Laparoscopic	SPY	Intravenous	Not reported	Tumor-bearing portal vein	100	16
Kobayashi (2017) [14]	Positive/negative staining	Open	PDE	Tumor-bearing portal vein/intravenous	0.25 mg/2.5 mg	None /Tumor-bearing portal vein	100	17
Terasawa (2017) [15]	Negative staining	Laparoscopic	SPY	Intravenous	1.25 mg	Tumor-bearing portal vein	100	18
Ueno (2018) [16]	Positive staining	Laparoscopic	SPY	Arterial branch	0.25 mg	Hepatic artery embolism (IVR)	100	19
Peyrat (2018) [17]	Negative staining	Open	FLUOBEAM®	Intravenous	0.625mg~1.25 mg	Tumor-bearing portal vein	80	20
Nishino (2018) [18]	Negative staining	Open	Medical Imaging Projection System	Intravenous	0.25 mg	Tumor-bearing portal vein	91.3	12
Aoki (2020) [8]	Preoperative positive staining	Laparoscopic	SPY	Tumor-bearing portal vein	0.025 mg	None	86	8

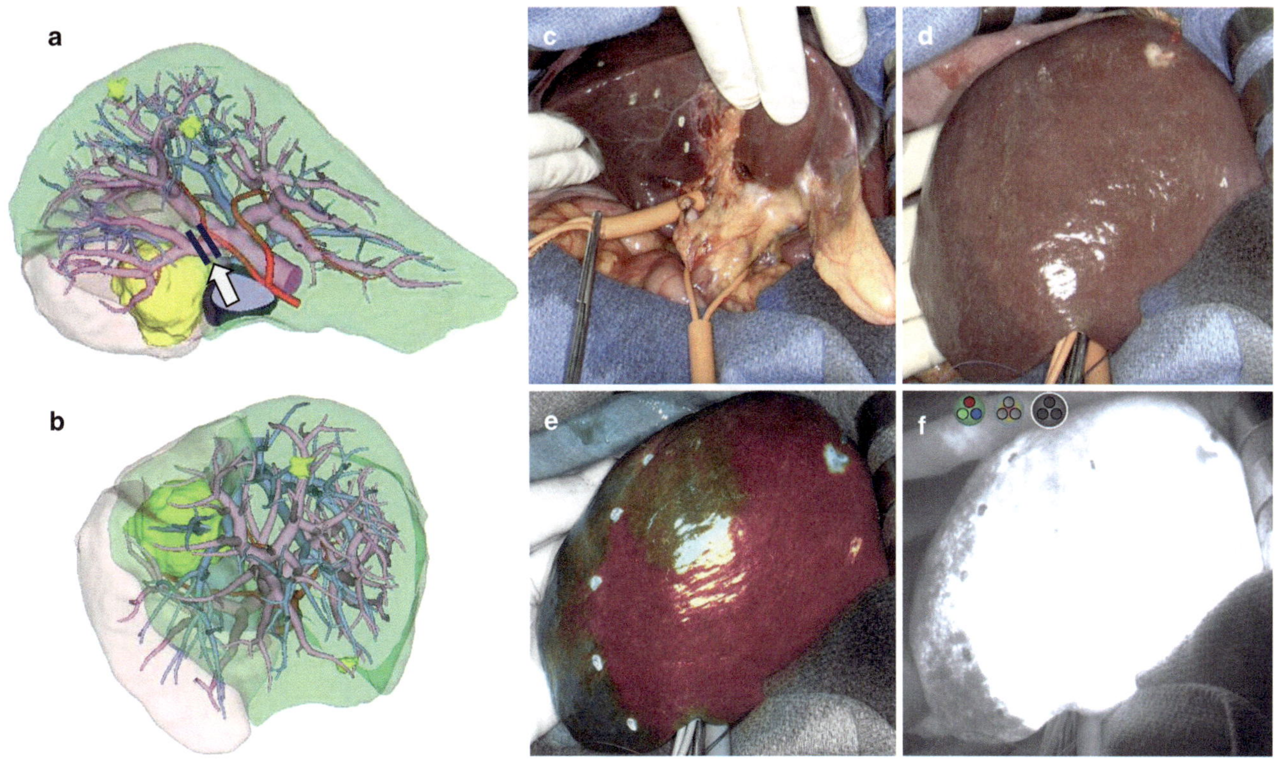

Fig. 26.2 Right posterior sectionectomy for liver metastasis in segment 7 using ICG fluorescence imaging. (**a**) Preoperative simulation identifies the tumor-bearing portal vein branch (posterior regional branch) and the corresponding hepatic segment. (**b**) Preoperative simulation of fluorescing/non-fluorescing hepatic regions. (**c**) Clamp the portal branch of the posterior section and administer 1 mL of ICG (2.5 mg) intravenously. (**d**) White light observation of hepatic surfaces. (**e**, **f**) The posterior section is identified as non-fluorescing regions by ICG fluorescence

(P6a + b) was identified by preoperative simulation, and after clipping and dissection of this vessel, 1 mL of ICG (2.5 mg/mL) was intravenously injected. Then, fluorescence imaging of the liver surfaces clearly identified the boundaries of hepatic segments fed by P6a + b as ICG fluorescence-deficient regions. During hepatic transection, another branch of the portal vein (P6c) was identified as the fluorescence-positive vasculature, and by dissecting this branch, anatomical hepatectomy of S6 was completed (Fig. 26.3) [19].

In 2012, Ishizawa et al. also reported the positive staining technique, in which the portal vein is directly punctured under IOUS guidance and ICG is injected at the same site during laparoscopic hepatectomy [7]. In laparoscopic hepatectomy, the problems of IOUS-guided puncture of the portal vein branch are difficulty in obtaining appropriate IOUS images due to the limitation of ultrasound probe operation and in integrating spatial positional information when puncturing the portal vein branch demonstrated on the IOUS monitor from the body surface. As a solution to this problem, we have devised the "preoperative positive staining method," in which the target portal branch is punctured under extracorporeal ultrasound guidance immediately after induction of anesthesia, prior to the insufflation of the abdominal cavity [8]. In this method, an 18G long needle for percutaneous transhepatic biliary drainage (PTBD) is used to inject 1 to 2 mL of ICG solution diluted to a concentration of 0.025 mg/mL in saline. Intraoperative fluorescence imaging visualized ICG-stained hepatic segments even after insufflation and throughout hepatectomy procedures.

In Fig. 26.4 [19], we present a case of laparoscopic segmentectomy of S2 for hepatocellular carcinoma using the preoperative positive staining technique. The tumor-bearing portal vein branch feeding S2 was identified preoperatively. After induction of general anesthesia, the S2 portal branch was punctured under extracorporeal ultrasound guidance using an 18G PTBD needle, and 1 mL of ICG 0.025 mg/mL was administered. Intraoperatively, the hepatic segments to be removed were identified as fluorescing regions. Boundaries of ICG fluorescent and nonfluorescent regions were clearly identified during hepatic dissection, allowing anatomical hepatectomy accurately along the intersegmental planes (Movie 26.1). If it is difficult to puncture the tumor-bearing portal branch directly, the adjacent portal branches can be punctured using the counterstaining technique.

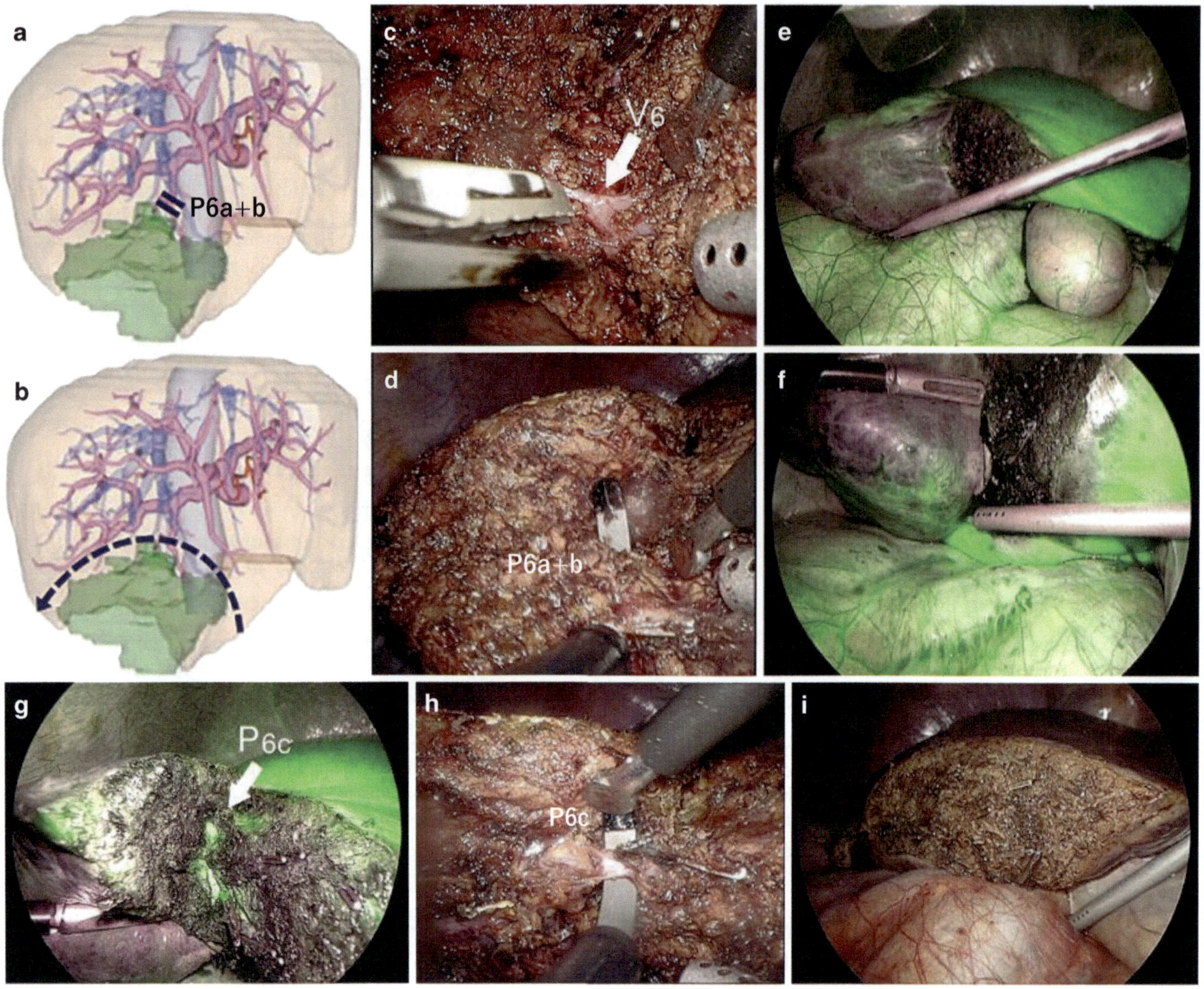

Fig. 26.3 Hepatic segmentation by ICG fluorescence imaging (negative staining technique) in laparoscopic liver resection (reproduced with permission from [19]). Laparoscopic liver resection of segment 6 for hepatic hemangioma. (**a**) Preoperative simulation confirmed the tumor-bearing portal branch (P6a + b) and the corresponding hepatic regions. (**b**) Hepatic dissection line was designed based on preoperative simulation. (**c**) Isolation of the hepatic vein (V6) appearing on the hepatic transection plane. (**d**) Tumor-bearing portal branches (P6a + b) were identified, and ICG (2.5 mg) was injected intravenously after the dissection of these vessels. (**e, f**) Segment 6 was identified as a non-fluorescing area on the liver surfaces and hepatic transection plane. (**g**) Another portal pedicle to be divided (P6c) is identified as a vasculature with ICG fluorescence. (**h**) P6c is divided. (**i**) Hepatic raw surfaces after resection of S6

When the extracorporeal puncture of the target portal branch is technically difficult, intraoperative IOUS-guided puncture may still be necessary. In such cases, a laparoscopic ultrasound probe with a hole in the probe (4-Way Laparoscopic 8666-RF; BK Medical) may be useful to make the puncture procedure easier. In this technique, a PTBD needle is inserted from the body surface through the hole in the probe, and its tip is placed in the target portal branch. Although this technique is potentially effective for identifying small regions of S7 or S8 by the positive staining technique, the puncture technique is still technically demanding because the determination of the puncture route depends on the sense of the operating surgeon (Fig. 26.5) [19].

4.4 Identification of Hepatic Subsegment for Small Hepatectomy

In patients with primary liver cancer on cirrhotic liver or metastatic liver cancer, cone unit resection is sometimes indicated [20]. Even in the case of these small hepatic resections in the distal tertiary branches, ICG fluorescence imaging can be used to identify the hepatic segmental anatomy based on the portal system, enabling anatomical hepatic resection without leaving ischemic regions on the remnant liver.

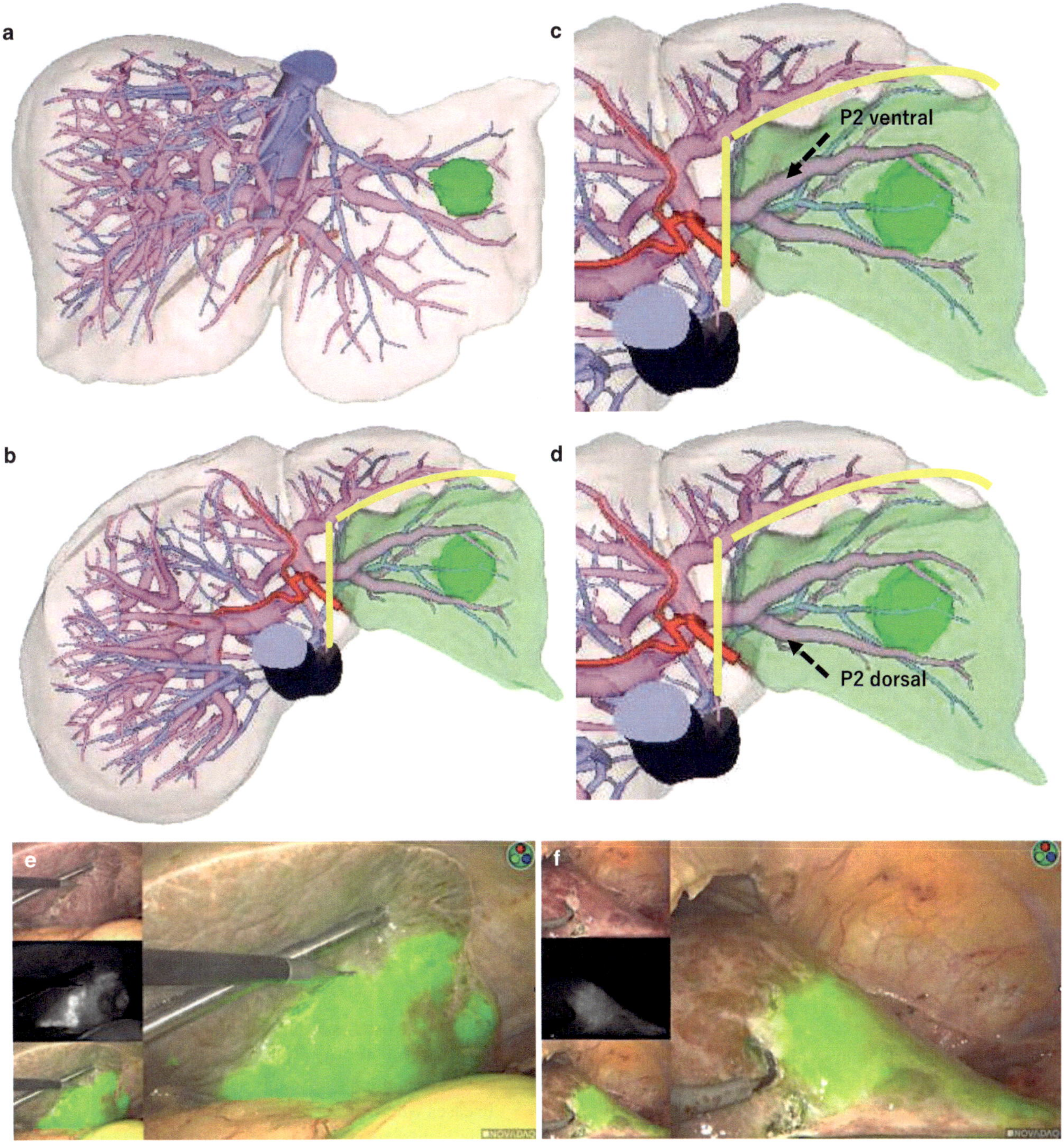

Fig. 26.4 Hepatic segmentation using ICG fluorescence imaging (preoperative positive staining method) in laparoscopic liver resection (reproduced with permission from [19]). Laparoscopic subsegmentectomy of segment 2 for hepatocellular carcinoma. (**a**) Preoperative simulation shows the tumor (S2) and the tumor-bearing portal branch (P2). (**b**) Simulation of the ICG-stained area. (**c**) Location of P2 ventral branch in preoperative simulation. (**d**) Location of P2 dorsal branch in preoperative simulation. (**e**) Hepatic S2 staining area identified on visceral surfaces of the liver after administration of 1 mL of ICG (0.025 mg/mL) into P2. (**f**) Hepatic S2 staining area identified on phrenic surfaces of the liver. (**g**) Identification of P2 ventral branch. (**h**) Division of P2 ventral branch appearing on hepatic raw surfaces. (**i**) P2 dorsal branch and V2 identified in preoperative simulation. (**j**) P2 dorsal branch and V2 were divided with a stapler. (**k**) Hepatic parenchyma of S2 to be removed was identified as fluorescing regions during hepatic transection. (**l**) Hepatic raw surfaces after anatomical resection of S2

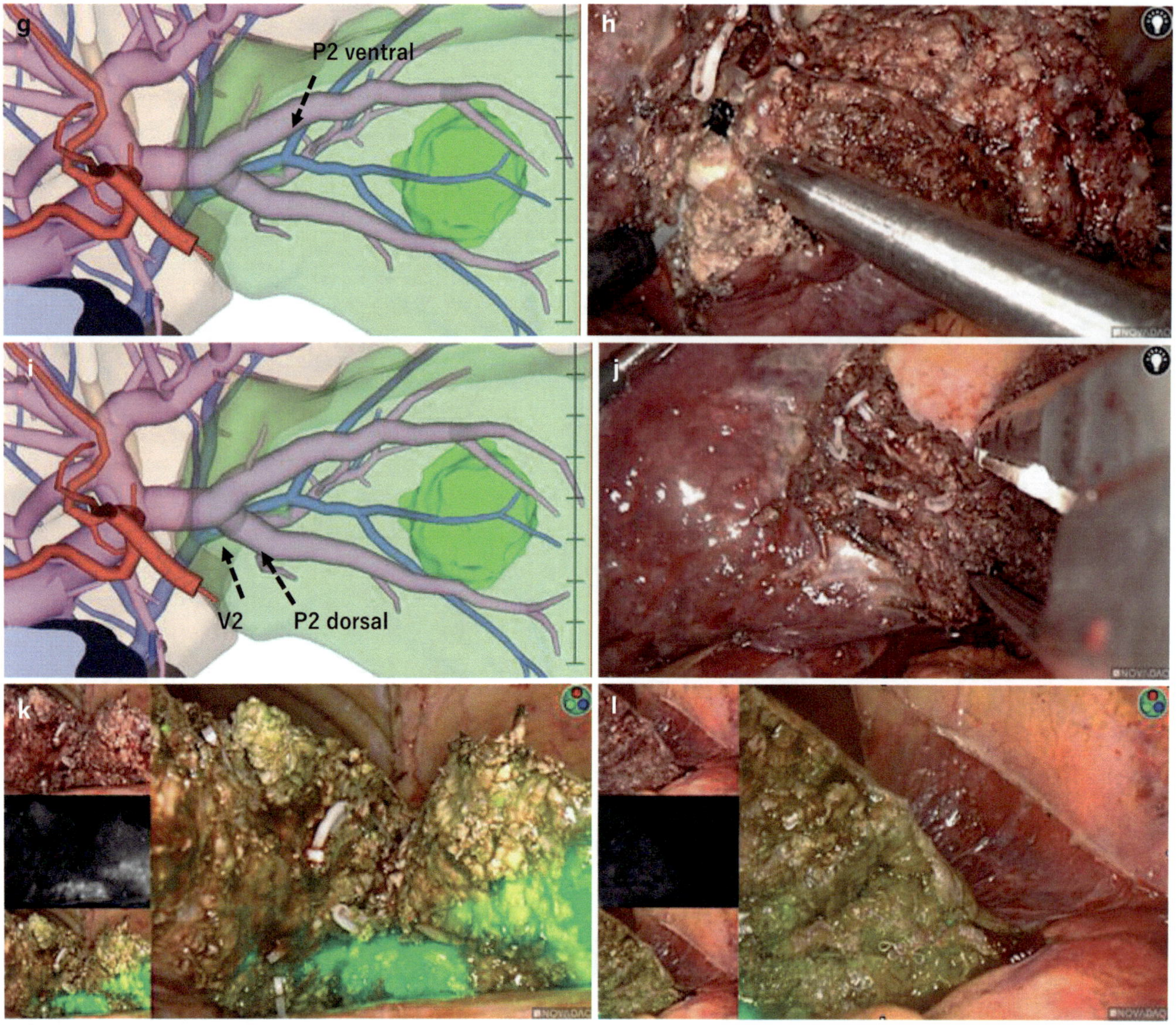

Fig. 26.4 (continued)

Point

- In the positive staining technique, the tumor-bearing portal branch is directly punctured under ultrasound guidance to inject ICG solution (0.025 mg/mL), enabling the identification of the corresponding hepatic segment as a fluorescing region.
- It is important to inject the ICG solution slowly to avoid reflux into the adjacent hepatic area.
- In laparoscopic hepatectomy, preoperative positive staining technique, in which ICG solution (0.025 mg/mL) is injected using extracorporeal ultrasound prior to insufflation, can be used as a simple method overcoming the technical difficulty of intraoperative ultrasound-guided puncture in the laparoscopic setting.
- In cases where direct puncture of the tumor-bearing portal vein is difficult, the counterstaining method can be used.
- In the negative staining technique, after the closure of blood flow into the target hepatic segment, ICG (2.5 mg) is injected intravenously, and the hepatic segment to be removed is identified as a non-fluorescing hepatic region.
- It is important to select the appropriate staining technique for each case based on detailed preoperative simulation.

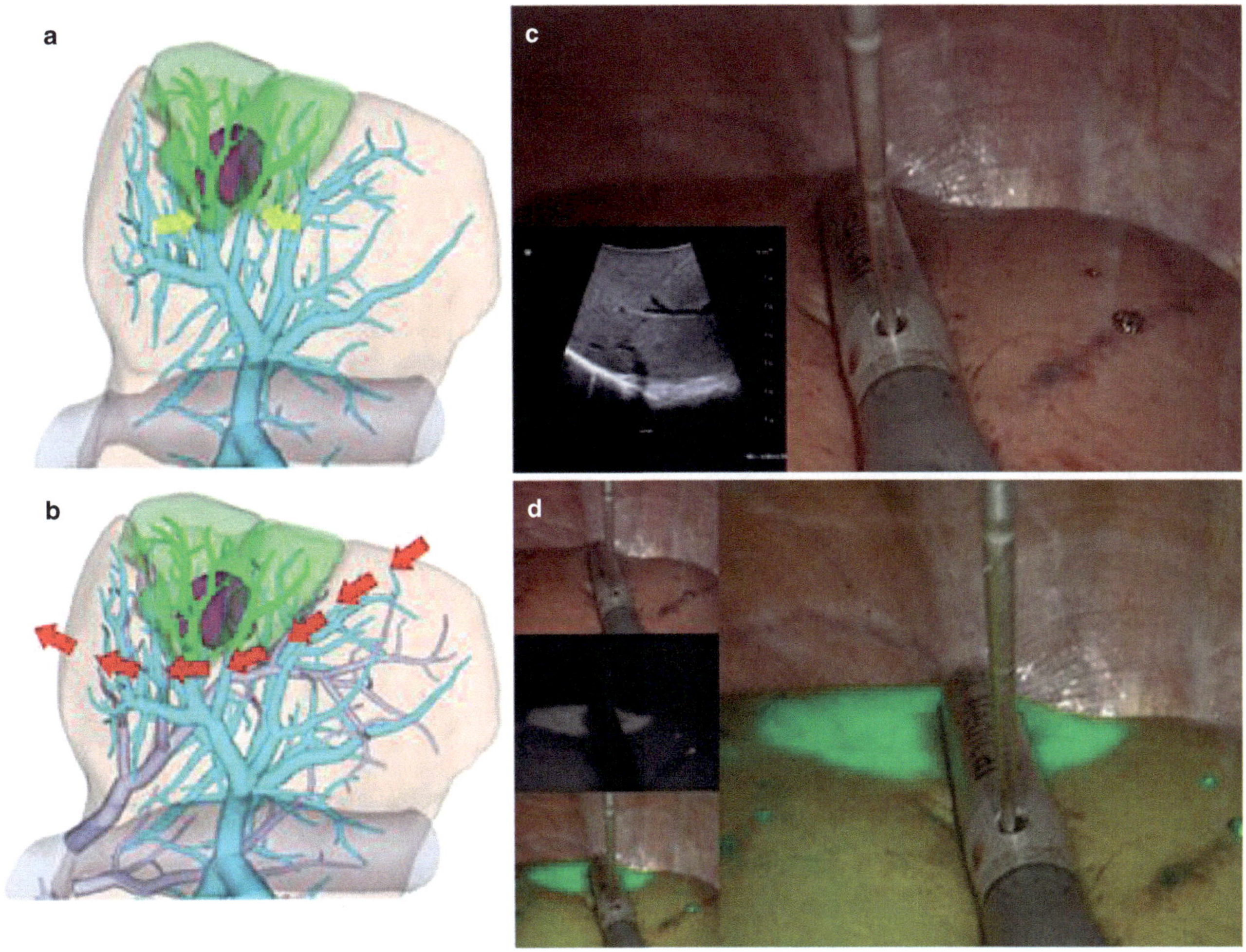

Fig. 26.5 Hepatic segmentation by ICG fluorescence imaging (intraoperative positive staining technique) in laparoscopic hepatectomy (reproduced with permission from [19]). Laparoscopic liver resection of segment 7 for liver metastasis. (**a**) The tumor-bearing portal branch was identified by preoperative simulation. (**b**) The hepatic dissection line was designed based on preoperative simulation. (**c**) Extracorporeal puncture of the target portal vein by inserting a needle through the hole made in the ultrasound probe. (**d**) Fluorescing area of segment 7 identified after injection of 1 mL of ICG (0.025 mg/mL) into the tumor-bearing portal branch

5 Expected Effects of Hepatic Segmentation by Fluorescence Imaging

The IOUS-based hepatic segmentation is still used as a golden standard method. The application of ICG fluorescence imaging to hepatic segmentation enables clearer delineation of hepatic segmental boundaries, enhancing the accuracy of anatomical hepatectomy especially in the setting of laparoscopic surgery.

Anatomical hepatic resection is known to contribute to the reduction of postoperative complications [3]. In the treatment for hepatocellular carcinoma, an indication of anatomic hepatectomy is associated with favorable overall survival as compared with nonanatomical resections [21]. Even in the case of metastatic liver cancer, it is sometimes necessary to indicate anatomic segmentectomy for complete and curative tumor resection, depending on the tumor size and localization. Therefore, the use of hepatic segmentation by ICG fluorescence imaging may enhance survival outcomes of cancer patients as well as the safety of surgery by improving the accuracy of anatomical hepatectomy, although this should be verified in large-scale multicenter studies.

6 Pitfalls and Limitations

Although ICG fluorescence is a relatively simple technique, accurate delineation of hepatic segments is not always successful in the positive staining technique because multiple branches of the innominate portal vein or small vessels of less than 2 mm are difficult to puncture. In the negative staining method, the tumor-bearing portal vein branch should be encircled first, without causing any damage to the adjacent portal pedicles. In addition, unlike indigo carmine, which is washed out in a couple of minutes, ICG fluorescence signals remain for hours, making it difficult to redo this procedure if the stained area is incorrect. Therefore, accurate preoperative simulation and planning are needed to use ICG fluorescence imaging effectively for completing anatomical hepatectomy.

Near-infrared imaging systems for open and laparoscopic surgery are currently available from various manufacturers, but there are differences in the excitation light, filter setting, fluorescence image display method, and fluorescence signal detection ability among the imaging systems [22]. In addition, it has been reported that the fluorescence signal detectability of laparoscopic imaging systems tends to be lower than that of imaging devices for open surgery [22]. In clinical installation of ICG fluorescence imaging techniques, surgeons should fully understand the characteristics of the imaging systems to be used.

Since ICG contains iodine, it should not be used in patients with a history of iodine hypersensitivity or asthma. Although the dose used for hepatic segmentation is quite small as compared with the dose used for hepatic function tests, unexpected complications may occur.

7 Future Perspectives

The fluorescence signals of ICG in the target hepatic segment can be observed on the hepatic surfaces prior to hepatectomy and raw surfaces of the liver during parenchymal procedures, due to a unique fluorescence property of ICG that remains in hepatic parenchyma for a long time. Using this fluorescence property, a more accurate and feasible real-time navigation system integrating preoperative and intraoperative information can be established in the future. In open surgery, the Medical Imaging Projection System (MIPS), which is based on projection mapping technology, has been developed, and it is now possible to proceed with hepatectomy procedures while projecting fluorescence images onto the surgical field in real time [18]. This technology is expected to be widely applied as a good example of intraoperative navigation because it follows organ deformation in real time during surgery and enables surgical participants to concentrate on the surgical field without moving their eyes. Also in laparoscopic surgery, it has become possible to superimpose the ICG fluorescent signals on the full-color images of the actual surgical field. The flexible scope system enabling near-infrared fluorescence imaging and 4 K color imaging is highly awaited. In terms of feasibility, it is necessary to establish surgical techniques and devices (ultrasound probe, needle, etc.) which make the application of the positive staining technique in the laparoscopic setting easier. With further accumulation of evidence indicating the efficacy of hepatic segmentation by ICG fluorescence imaging, this technique will be used more widely as an essential tool for completing accurate hepatectomy.

References

1. Couinaud C. Les enveloppes vasculobiliaries du foie ou capsule de Glisson:leur interet dans la chirurgie vesicularie, les resections hepatiques et l'abord du bile du foie. Lyon Chir. 1954;49:589.
2. Makuuchi M, Hasegawa H, Yamazaki S. Ultrasonically guided subsegmentectomy. Surg Gynecol Obstet. 1985;161:346–50.
3. Takayama T, Makuuchi M, Watanabe K, et al. A new method for mapping hepatic subsegment counterstaining identification technique. Surgery. 1991;109:226–9.
4. Takasaki K, Kobayashi S, Tanaka S, et al. A new systematic hepatectomy using glisson sheath processing. Surgery. 1986;40:7–14.
5. Aoki T, Yasuda D, Shimizu Y, et al. Image-guided liver mapping using fluorescence navigation system with indocyanine green for anatomical hepatic. World J Surg. 2008;32:1763–7.
6. Uchiyama K, Ueno M, Ozawa S, et al. Combined intraoperative use of contrast-enhanced ultrasonography imaging using a sonazoid and fluorescence navigation system with indocyanine green during anatomical hepatectomy. Langenbeck's Arch Surg. 2011;396:1101–7.
7. Ishizawa T, Zuker NB, Kokudo N, et al. Positive and negative staining of hepatic segments by use of fluorescent imaging techniques during laparoscopic. Arch Surg. 2012;147:393–4.
8. Aoki T, Koizumi T, Mansour DA, et al. Ultrasound-guided preoperative positive percutaneous indocyanine green fluorescence staining for laparoscopic anatomical liver resection. J Am Coll Surg. 2020;230:e7–e12.
9. Aoki T, Murakami M, Koizumi T, et al. Three-dimensional virtual endoscopy for laparoscopic and thoracoscopic liver resection. J Am Coll Surg. 2015;221:e21–6.
10. Nishino H, Hatano E, Seo S, et al. Real-time navigation for liver surgery using projection mapping with indocyanine green fluorescence: development of novel medical imaging projection system. Ann Surg. 2018;267:1134–40.
11. Sakoda M, Ueno S, Iino S, et al. Pure laparoscopic subsegmentectomy of the liver using a puncture method for the target portal branch under percutaneous ultrasound with artificial ascites. Surg Laparosc Endosc Percutan Tech. 2013;23:e45–8.
12. Inoue Y, Arita J, Sakamoto T, et al. Anatomical liver resections guided by 3-dimensional parenchymal staining using fusion indocyanine green fluorescence imaging. Ann Surg. 2015;262:105–11.
13. Miyata A, Ishizawa T, Tani K, et al. Reappraisal of a dye staining technique for anatomic hepatectomy by the concomitant use of indocyanine green fluorescence imaging. J Am Coll Surg. 2015;221:e27–36.
14. Mizuno T, Sheth R, Yamamoto M, et al. Laparoscopic Glissonean pedicle transection (Takasaki) for negative fluorescent counterstaining of segment 6. Ann Surg Oncol. 2017;24:1046–7.
15. Kobayashi Y, Kawaguchi Y, Kobayashi K, et al. Portal vein territory identification using indocyanine green fluorescence imaging: technical details and short-term outcomes. Details and short-term outcomes. J Surg Oncol. 2017;116:921–31.
16. Terasawa M, Ishizawa T, Mise Y, et al. Applications of fusion-fluorescence imaging using indocyanine green in laparoscopic hepatectomy. Surg Endosc. 2017;31:5111–8.
17. Ueno M, Hayami S, Sonomura T, et al. Indocyanine green fluorescence imaging techniques and interventional radiology during laparoscopic anatomical liver resection (with video). Surg Endosc. 2018;32:1051–5.

18. Peyrat P, Blanc E, Guillermet S, et al. HEPATOFLUO: a prospective monocentric study assessing the benefits of indocyanine green (ICG) fluorescence for hepatic surgery. J Surg Oncol. 2018;117:922–7.
19. Aoki T, et al. Laparoscopic subsegmentectomy using the ICG fluorescence technique, special issue on hepatobiliary surgery using the ICG fluorescence technique. II Surgery. 2019;73:1405–15.
20. Takasaki K. Hepatic cone unit resection (anatomical Subsegmentectomy). In: Glissonean Pedicle Transection Method for Hepatic Resection, pp. 93–143; 2007.
21. Hasegawa K, Kokudo N, Imamura H, et al. Prognostic impact of anatomic resection for hepatocellular carcinoma. Ann Surg. 2005;242:252–9.
22. Kono Y, Ishizawa T, Tani K, et al. Techniques of fluorescence cholangiography during laparoscopic cholecystectomy for better delineation of the bile duct anatomy. Medicine (Baltimore). 2015;94:e1005.

Lung Segmentation

27

Yasuo Sekine

Summary

- Fluorescence imaging of the lung segment can be done by intravenous (negative staining) or transbronchial (positive staining) injection of ICG.
- The intravenous method is very simple, while the transbronchial method is suitable for complicated lung resections.
- The combination of the two methods offers the possibility of more accurate and minimal lung resection procedures.

1 Introduction

With the development of diagnostic imaging technology, the detection of small-sized peripheral lung cancer has increased remarkably in these years. Therefore, the idea that a smaller lung resection (limited surgery) may be sufficient instead of a lobectomy, which has been considered the standard surgery, has become widespread. In accordance with this trend, the number of cases of pulmonary segmentectomy is increasing worldwide. Recently, multicenter studies in Japan clarified that the pulmonary segmentectomy prolonged overall survival more than the standard lobectomy and small partial resection was suitable in specific cases of lung cancer [1, 2].

From the anatomical standpoint, the lung can be divided into 5 lobes, 19 segments, and 42 subsegments. However, the boundaries of the pulmonary segments are not visible to the naked eye. Therefore, it is often unclear where the segmental boundaries are when performing a pulmonary segmentectomy. That is why ICG fluorescence imaging has been used as a navigation during lung resection. In this technique, ICG can be administered either intravenously (negative staining) or intra-bronchially (positive staining). The intravenous method is used more widely because of its simplicity and clarity of visualizing segmental boundaries. On the other hand, the transbronchial injection method requires a more advanced technique, but it has the advantage of being able to handle all kinds of complicated (sub)segmentectomies. In this article, we outline the characteristics and clinical applications of these techniques for intraoperative lung segmentation.

Supplementary Information The online version contains supplementary material available at https://doi.org/10.1007/978-981-19-7372-7_27.

Y. Sekine (✉)
Department of General Thoracic Surgery, Tokyo Women's Medical University Yachiyo Medical Center, Chiba, Japan
e-mail: sekine.yasuo@twmu.ac.jp

2 Conventional Methods and Problems of Lung Segmentectomy

Lung segmentectomy had been performed since the 1930s, mainly for tuberculosis. However, because of the anatomical complexity of segmental resection, it was not until the late 1940s that the technique was well established. Since then, it has become common to treat the pulmonary vessels and bronchi of the target lung segment and to transect pulmonary parenchyma along intersegmental planes. The major indication of this technique was metastatic lung tumors, benign tumors, and patients with lung cancer whose cardiopulmonary function was too impaired to undergo lobectomy (passive limited surgery). More recently, active limited surgery has become indicated for early-stage peripheral lung cancer.

The conventional method of identifying the lung segment to be removed is the inflation-deflation method, in which the air is pumped into the lungs with a ventilator and the bronchi to be resected are blocked. By collapsing the lungs, air is trapped only in the lung segment to be removed, enabling the identification of its segmental boundaries. Alternatively, air is pumped after the bronchus is blocked and the collapsed lung segment is resected. The problem with these methods is that in cases such as emphysema, alveolar destruction can cause communication of air beyond the intersegmental septum, making it difficult to determine the boundary between resected and unresected areas. In addition, in thoracoscopic

T. Ishizawa (ed.), *Fluorescence-Guided Surgery*, https://doi.org/10.1007/978-981-19-7372-7_27

surgery, inflating the lungs during surgical procedures obstructs the field of view and prolongs the operation time.

Another technique for identifying intersegmental boundaries is to insert a jet ventilation catheter into the resected area bronchus through the tracheal tube and inflate it [3]. However, this method also involves technical difficulties in inserting the jet ventilation catheter through the double-lumen tube and guiding it to the target bronchus. Fluorescence imaging using an ICG and a near-infrared camera has emerged as a promising alternative which can solve these problems in conventional techniques.

3 Development History of Fluorescence Imaging for Lung Segmentation

Fluorescence imaging in the field of pulmonary surgery has been used to identify brachial emphysema [4], sentinel lymph nodes [5], and thoracic ducts [6]. Among them, emphysema imaging is the technique to identify emphysematous regions as a defect of fluorescence following intravenous injection of ICG because of the disrupted vascular architecture, although the efficacy of this method has not yet been established.

ICG fluorescence imaging of lung segments has been developed based on the idea of emphysema imaging. Since the pulmonary artery and bronchus run parallel to each other in the periphery, intravenous injection of ICG following the closure of the segmental bronchi visualizes the target segment as a non-fluorescing region by fluorescence imaging (negative staining) [7, 8]. Recently, fluorescence imaging using negative staining has been widely used all over the world because this technique is very simple and enables clear identification of segmental boundaries in a short time.

On the other hand, the lung segment is classified based on the anatomy of the bronchi, which is the basis of fluorescence imaging using the transbronchial ICG injection method. This method is called positive staining because the target lung segment is identified as a fluorescing region following the injection of ICG through the corresponding bronchi [9, 10]. The major advantage of the positive staining technique lies in the fact that it enables accurate and clear identification of smaller subsegments of the lung. On the other hand, in this technique, ICG should be administered into the target bronchi accurately, which requires knowledge of bronchial anatomy and advanced bronchoscopic manipulation techniques of surgeons.

4 Clinical Practice of Lung Segmentation by ICG Fluorescence Imaging

4.1 Intravenous Method (Negative Staining) (Movie 27.1)

First, the resection area is determined, and the pulmonary arteries feeding the target section are divided. ICG (5–10 mL) is then administered as a one-shot intravenous infusion, which spreads throughout the body in a few tens of seconds but does not perfuse the target lung segment. As a result, a clear demarcation between the fluorescing and non-fluorescing regions of the lung is visualized, making it possible to identify the boundaries of the pulmonary segment to be removed (Fig. 27.1). ICG can penetrate the target lung segment quickly, so the time to identify the border line is about 1–2 minutes, and we need to draw the transection line on the lung surfaces with electrocautery or dyes during that time. Since ICG is washed out from the pulmonary parenchyma after a couple of minutes, this technique can be used repeatedly. The stapler is then inserted along the line, and the lung is transected to complete anatomical segmentectomy of the lung, although consistency between the results of fluorescence imaging by negative staining technique and true anatomical boundaries should be assessed in further studies [8].

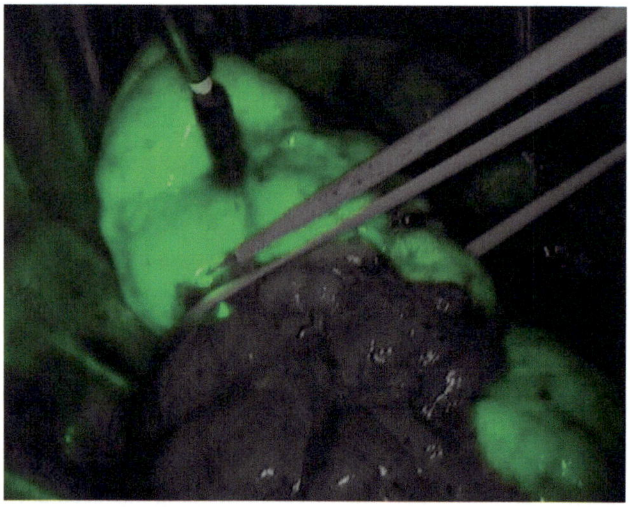

Fig. 27.1 Right S9+S10 area (unstained area) delineated by intravenous injection of ICG (negative staining). (Courtesy of Dr. Norihito Okumura, Kurashiki Central Hospital)

4.2 Transbronchial Method (Positive Staining) (Movies 27.2 and 27.3)

When performing a transbronchial infusion method, we simulate the area of the bronchus to be injected with ICG before the surgery. The image analysis software, Synapse 3D VINCENT (Fujifilm Corporation), allows us to specify the area of the bronchus and to measure the shortest distance from the tumor to the resection border (Fig. 27.2). Using this method, we can determine the optimal extent of pulmonary resection including not only by segmentectomy but also subsegmentectomy and smaller anatomical resections. The sufficient surgical margin is considered to be at least 2 cm or, in the case of tumors less than 2 cm in diameter, at least the tumor diameter [11]. Once the extent of resection is determined, the bronchus to be injected with ICG is confirmed by virtual bronchoscopy (Fig. 27.3).

In the operating room, 10 mL of ICG solution (2.5 mg/mL) + 20 mL of autologous blood +70 mL of saline solution are mixed to make 100 mL of ten-fold diluted solution. After induction of general anesthesia, artificial respiration is managed with a single lumen tube or laryngeal mask, and a fine bronchoscope (outer diameter: 5 mm) is inserted in the supine position. A catheter with a bronchoscopic balloon (disposable balloon catheter B5-2C/2LA, Olympus) was inserted into the target (sub)segmental bronchi, and the balloon is inflated to occlude the bronchial lumen (Fig. 27.4). Then, 10 mL of ten-fold diluted ICG is injected at once. Then, a total of 400 mL of air is injected with a 50 mL syringe afterward. This is to maintain a positive pressure in the injection region and to diffuse ICG to the periphery with air. After the balloon is deflated, repeat this procedure to inject ICG into all necessary bronchi. The procedure takes

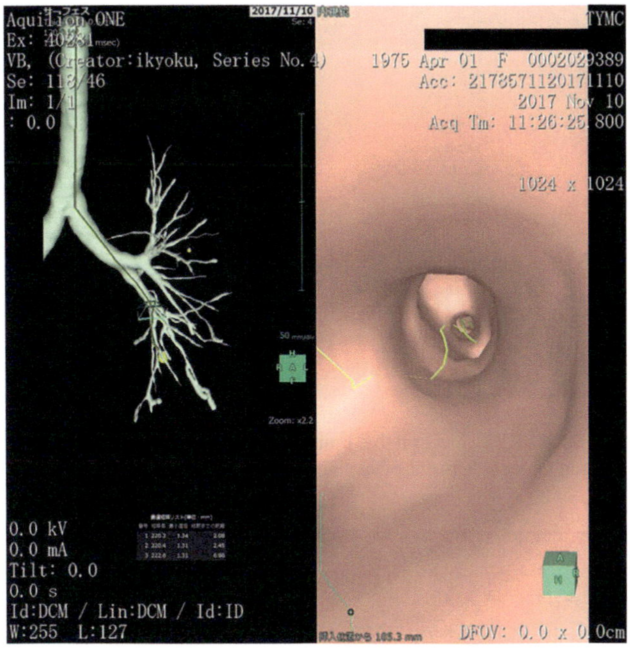

Fig. 27.3 Selection of right B10b and B10c bronchi during preoperative simulation

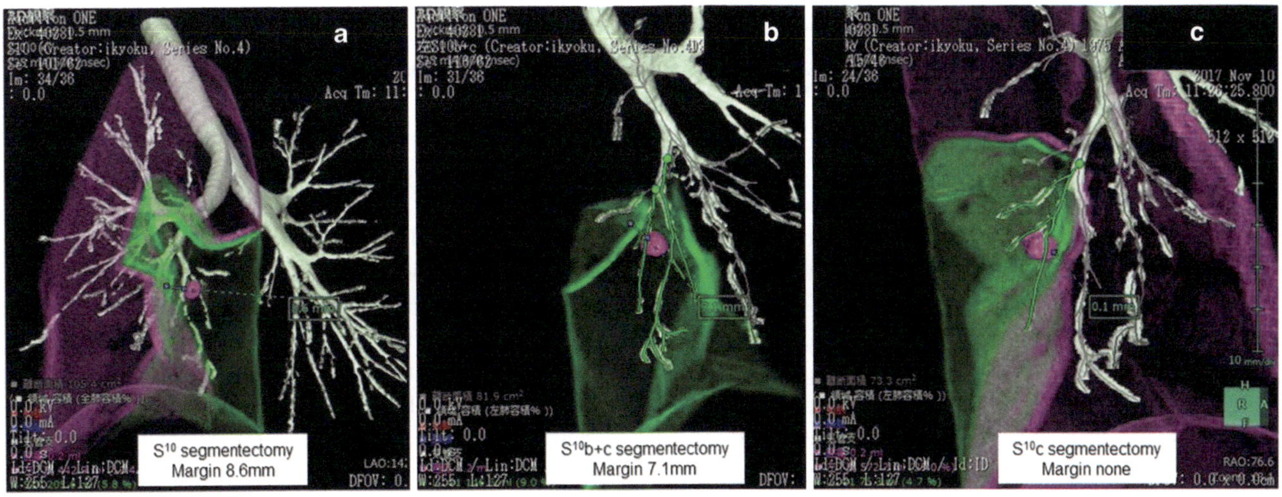

Fig. 27.2 (**a**) S10 segmentectomy, (**b**) S10b+c subsegmentectomy, (**c**) S10c subsegmentectomy

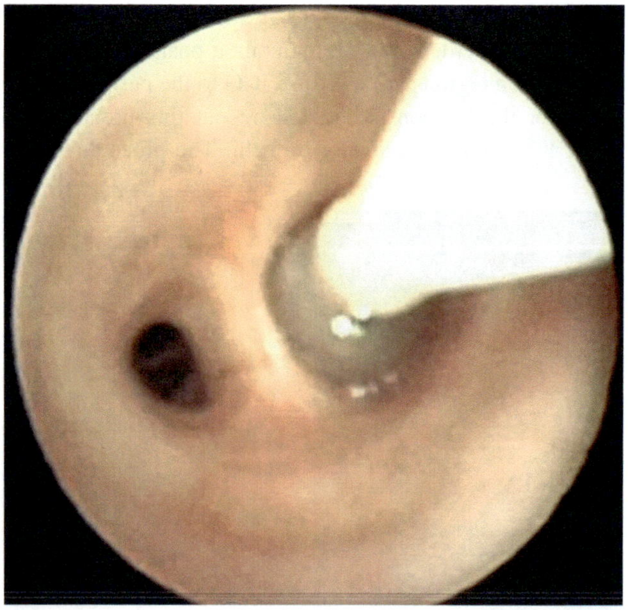

Fig. 27.4 ICG injection by transbronchial approach (a balloon is inflated to occlude the corresponding bronchus)

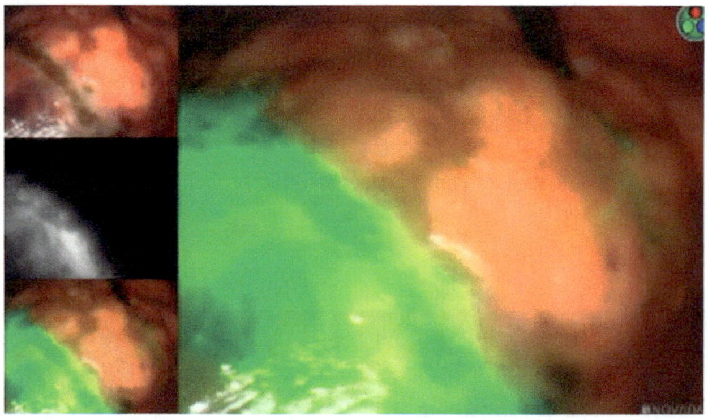

Fig. 27.5 Fluorescence imaging of right S10b+c segment following preoperative transbronchial injection of ICG

about 5–10 minutes. The patient is then asked to maintain a positive pressure of 20 cm H_2O for 20 seconds to allow ICG to diffuse more peripherally.

After switching to the double-lumen tube, the patient is repositioned, and surgery is started. Although the fluorescence of ICG can be seen immediately after the start of surgery, it takes another 15 to 30 minutes for the ICG to spread homogeneously and sufficiently. When the borderline is clearly visualized, the transection line is drawn along the segmental boundaries with electrocautery (Fig. 27.5). During the resection procedures, any vessels across the boundary into fluorescing regions can be divided safely, and the region to be removed can be detached from surrounding tissues prior to division of the corresponding bronchi. We believe that this is a new approach enabled by a positive staining

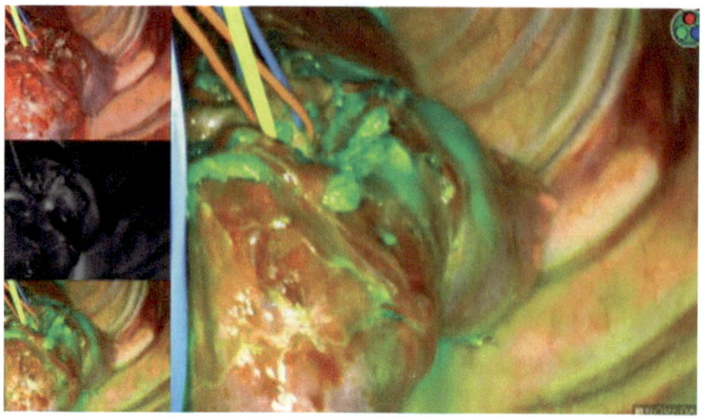

Fig. 27.6 Remaining lung after right S10b+c subsegmentectomy. The fluorescence of ICG is almost undetectable

technique that will significantly change the conventional surgical procedure. In addition, fluorescence imaging can be used to confirm whether the target pulmonary segment is successfully removed by detecting remaining fluorescence signals on raw surfaces of the lung (Fig. 27.6). As the operation time becomes longer, the ICG gradually diffuses, and its boundary becomes unclear. Therefore, it is important to determine the transection line at the early stage of surgery. It is interesting to note that the actual extent of resection indicated by the ICG is often larger than what I had initially estimated based on conventional color imaging (e.g., S6 often extends far downward, and the left S3 extends to what seems to be the lingual area).

4.3 Advantages and Disadvantages of Intravenous and Transbronchial Infusion Methods

The characteristics of the two methods are summarized in Table 27.1. The advantage of the negative staining method lies in its simplicity. By injecting ICG intravenously following the division of the corresponding pulmonary artery, the intersegmental boundaries are easily identified by fluorescence imaging. In addition, this technique can be used repeatedly when the transection line cannot be determined at once. The problem is that the borderline becomes unclear with time and even the resected area of the lung shows fluorescence signals because of the intersegmental blood perfusion. Identification of segmental boundaries by the negative staining method can decrease in patients with poor pulmonary conditions, such as emphysema, pulmonary fibrosis, strong charcoal dust deposition, and severe adhesions.

On the other hand, the positive staining method allows a very fine selection of the pulmonary segment and smaller anatomical components and can be used for a very complicated resection of the lung. In other words, this technique is

Table 27.1 Comparison between intravenous method and transbronchial methods

	Intravenous method	Transbronchial method
Ease of use	◎	△
Clarity	◎	◎ ~ ○
Stability of the procedure	◎	◎ ~ △
Minimally invasive	◎	○
Identification of vessels and bronchi to be treated	○	◎
Uniformity of ICG	◎	◎ ~ △
Duration	△	◎
Surgical time	◎	○
Repeated ICG injections	◎	×
Complex zone handling	○ ~ △	◎
COPD/IP/reop/adhesion	△	◎
Cost	○	△
Insurance coverage	○	×
Affinity for VINCENT	○	◎
Ultra-deep partial lung resection without pneumonectomy	×	◎

◎ Very good fit, ○ Suitable, △ Slightly difficult, × Not suitable

highly compatible with VINCENT and can be used to perform any type of lung resection imagined in the simulation. In addition, surgeons can safely divide vessels across the intersegmental planes into fluorescing regions based on fluorescence imaging and preoperative simulation. Even in cases of emphysema, pulmonary fibrosis, charcoal dust deposition, and adhesions, the segmental boundaries can be clearly identified by the positive staining method. This may be due to the higher local ICG concentration compared with the negative staining method. The problem is that preoperative bronchial infusion is technically difficult, requiring sufficient experience (learning curve) to obtain stable outcomes. Please note that it takes about 30 minutes for the ICG to spread sufficiently, and if the operation takes more than 3 hours, the ICG will diffuse beyond intersegmental planes, making identification of the borderline unclear.

Point

- Advantages of the intravenous method: simplicity, clarity, repeatability
- Disadvantages of the intravenous method: short contrast time, unsuitable for patients with background pulmonary diseases
- Advantages of the transbronchial method: selection of smaller (sub)segment, high affinity for preoperative simulation, long-lasting contrast, identification of the extent of resection from the beginning of surgery
- Disadvantages of the transbronchial method: technical difficulty of preoperative bronchoscopic procedures, gradual spread of ICG fluorescence, no repeatability

5 Expected Effects of Pulmonary Segmentation by Fluorescence Imaging

Fluorescence imaging using ICG can clearly identify segmental boundaries during pulmonary surgery compared with the conventional method. It can make surgical procedures for anatomical segmentectomy of the lung more accurate and straightforward, leading to shortening operation time and decreasing the incidence of postoperative complications.

Transbronchial infusion has a very wide range of applications. It can be used not only for segmental resection but also for partial lung resections including multiple sub- or sub-subsegments. In the partial resection, the resected area assuring surgical margin is simulated with VINCENT, and ICG is injected into these small subsegments to be removed (Fig. 27.7). Intraoperatively, the lung can be transected deeply along the boundaries of fluorescence signals. Unlike tumor marking (point marking) used in conventional lung resection, fluorescence imaging by transbronchial technique enables identification of resected "area" assuring sufficient margins (Movie 27.4). In contrast, conventional tumor marking only identifies the localization of the tumor, making it difficult to understand the depth of the tumor and secure surgical margins.

A double-marking technique is also available that combines the advantages of negative and positive staining techniques. The intravenous method is used for large areas such

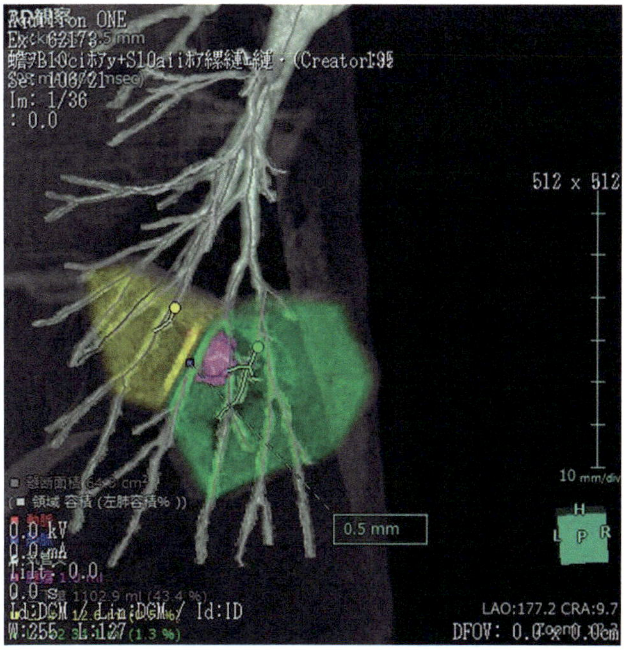

Fig. 27.7 Preoperative simulation of left S10ci α yx+S10aii α regions

as the pulmonary segment, and the transbronchial injection method is used for the identification of adjacent subsegments and smaller regions, which should also be removed to secure surgical margins. This allows for minimal resections of the lung with assuring surgical margins (Figs. 27.8 and 27.9, Movie 27.5). In addition, we developed a triple staining method in which the tumor is localized by transbronchial injection of indigo carmine, and double staining techniques based on ICG fluorescence imaging are also performed for the identification of tumor-bearing pulmonary segments. First, partial lung resection based on indigo carmine is performed to confirm the pathology of the tumor during surgery. Then, if cancer is confirmed, the minimum necessary lung area is resected for curative intent (Figs. 27.10 and 27.11). We believe that this kind of application is unique to the transbronchial approach.

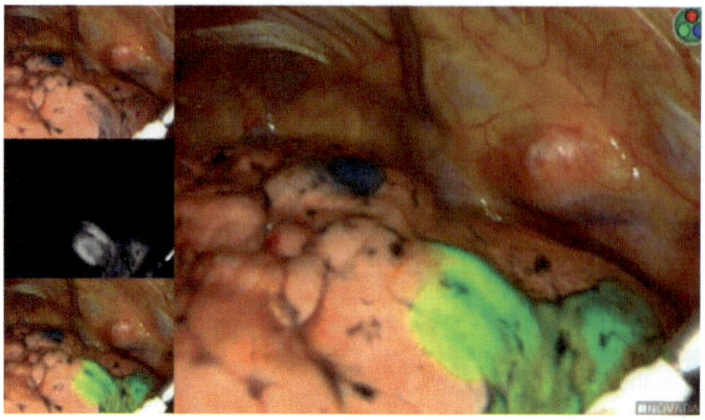

Fig. 27.10 Fluorescence imaging after indigo carmine injection into B1bii β periphery and positive staining of S3bii β during resection of right S1+S3bii β segmentectomy

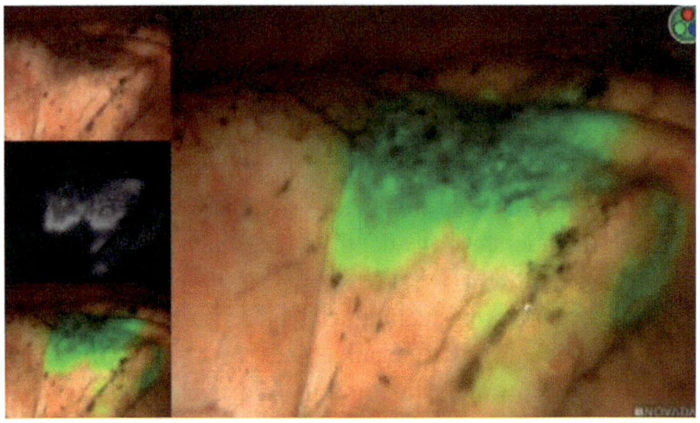

Fig. 27.8 Fluorescence imaging after positive staining of S4ai during resection of left S(1 + 2) + S3 + S4ai regions

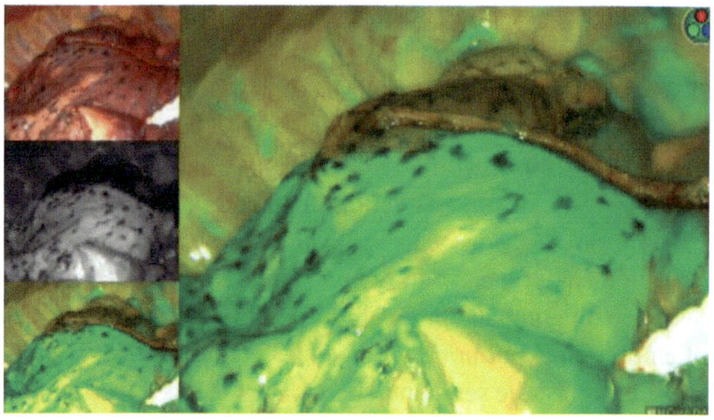

Fig. 27.11 Fluorescence imaging after positive staining of S3bii β and negative staining of S1 during right S1+S3bii β segmentectomy

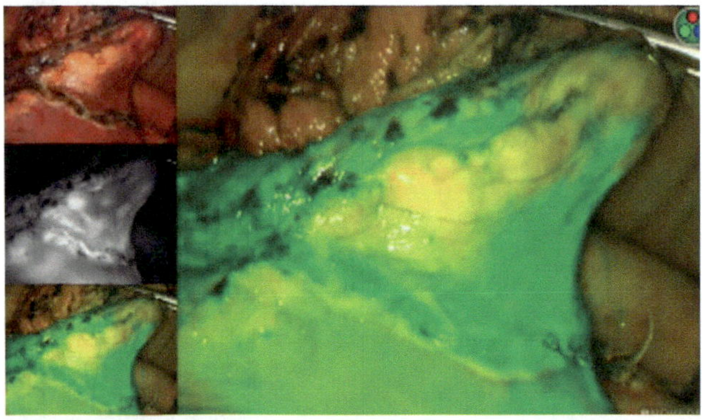

Fig. 27.9 Fluorescence imaging after positive staining of S4ai and negative staining of S(1 + 2) + S3 during resection of left S(1 + 2) + S3 + S4ai regions

Point

- The combination of intravenous and transbronchial techniques enables fine and precise anatomical resection of the lung.

6 Pitfalls and Limitations

The intravenous method is quite simple, but the challenge is to determine the transection lines and surgical approaches in a short time. If the demarcation line is not clear, surgeons will feel hesitation in the pulmonary transection and need to use the conventional inflation-deflation technique.

The most important point to note in the transbronchial method is to avoid injecting ICG into the wrong bronchus. If the injection of ICG is incorrect, this technique cannot be redone and may lead to removing the unnecessary regions of the lung. For this reason, I believe that the combined use of indigo carmine for tumor localization with lung segmentation by ICG fluorescence imaging is important. If cone beam CT or navigation bronchoscopy is available, we could inject ICG more accurately confirming the localization of the tumor at the same time.

7 Future Perspectives

In the future, marking of the tumor itself with fluorophores like ICG may become possible [12]. If cancer-specific imaging becomes available, it will enable more precise pulmonary surgery based on the information on the exact tumor localization, the presence or absence of lymph node metastasis, and the lymphatic pathways.

References

1. Saji H, Okada M, Tsuboi M, et al.; West Japan Oncology Group and Japan Clinical Oncology Group. Segmentectomy versus lobectomy in small-sized peripheral non-small-cell lung cancer (JCOG0802/WJOG4607L): a multicentre, open-label, phase 3, randomised, controlled, non-inferiority trial. Lancet. 2022;399(10335):1607–1617. https://doi.org/10.1016/S0140-6736(21)02333-3.
2. Sagawa M, Oizumi H, Suzuki H, et al. A prospective 5-year follow-up study after limited resection for lung cancer with ground-glass opacity. Eur J Cardiothorac Surg. 2018;53(4):849–56. https://doi.org/10.1093/ejcts/ezx418.
3. Okada M, Mimura T, Ikegaki J, et al. A novel video-assisted anatomic segmentectomy technique: selective segmental inflation via bronchofiberoptic jet followed by cautery cutting. J Thorac Cardiovasc Surg. 2007;133:753–8.
4. Gotoh M, Yamamoto Y, Igai H, et al. Clinical application of infrared thoracoscopy to detect bullous or emphysematous lesions of the lung. J Thorac Cardiovasc Surg. 2007;134:1498–501.
5. Yamashita S, Tokuishi K, Anami K, et al. Video-assisted thoracoscopic indocyanine green fluorescence imaging system shows sentinel lymph nodes in non-small-cell lung cancer. Thorac Cardiovasc Surg. 2011;141:141–4.
6. Ashitate Y, Tanaka E, Stockdale A, et al. Near-infrared fluorescence imaging of thoracic duct anatomy and function in open surgery and video-assisted. J Thorac Cardiovasc Surg. 2011;142:31-8.e1-2.
7. Misaki N, Chang SS, Gotoh M, et al. A novel method for determining adjacent lung segments with infrared thoracoscopy. J Thorac Cardiovasc Surg. 2009;138:613–8.
8. Sun Y, Zhang Q, Wang Z, et al. Is the near-infrared fluorescence imaging with intravenous indocyanine green method for identifying the intersegmental plane concordant with a modified inflation-deflation method in lung segmentectomy? Thorac Cancer. 2019;10:2013–21.
9. Sekine Y, Ko E, Oishi H, et al. A simple and effective technique for identification of intersegmental planes by infrared thoracoscopy after. J Thorac Cardiovasc Surg. 2012;143:1330–5. (8) Sekine Y, Ko E, Oishi H et al: A simple and effective technique for identification of intersegmental planes by infrared thoracoscopy after transbronchial injection of indocyanine green
10. Sekine Y, Itoh T, Toyoda T, et al. Precise anatomical sublobar resection using a 3D medical image analyzer and fluorescence-guided surgery with transbronchial instillation of indocyanine green. Semin Thorac Cardiovasc Surg. 2019;31:595–602.
11. Cao C, Chandrakumar D, Gupta S, et al. Could less be more?-A systematic review and meta-analysis of sublobar resections versus lobectomy for non-small lung cancer according to patient selection. Lung Cancer. 2015;89:121–32.
12. Predina JD, Newton AD, Xia L, et al. An open label trial of folate receptor-targeted intraoperative molecular imaging to localize pulmonary squamous cell carcinomas. Oncotarget. 2018;9:13517–29.

Visualization of the Ureter

28

Toshihiko Nishidate, Koichi Okuya, Kenji Okita, and Ichiro Takemasa

Summary

- The ureter can be identified intraoperatively by fluorescence imaging following retrograde ureteral injection of indocyanine green (ICG) or intravenous injection of methylene blue (MB).
- Fluorescent catheters can also be inserted for intraoperative visualization of the ureter.

1 Introduction

Recently, intraoperative fluorescence imaging has been used for real-time identification of cancer tissues, blood vessels, and lymph nodes. In this article, we demonstrate the methods for ureter identification by fluorescence imaging and discuss future prospects of this novel technique.

2 Conventional Methods for Intraoperative Ureter Identification

Intraoperative ureteral injury can develop with the incidence ranging from 0.007% to 1.8% [1], mainly during gynecological surgery (50%) followed by urological surgery (30%) and colorectal surgery (5–15%) [2, 3].

When ureteral injury is suspected intraoperatively, indigo carmine or methylene blue (MB) can be injected to identify the injured site based on leakage of dye from the ureter, although this technique is not useful in the case of complete obstruction of the ureter by ligation or coagulation. Placement of the ureteral stent is very effective for the identification of the ureter and prevention of ureteral injury especially in endoscopic surgery for high-risk patients, such as those with severe inflammation/abscess formation due to Crohn's disease or diverticulitis, highly invasive cancer, pelvic surgery, or reoperation due to recurrence. However, since the ureteral catheter should be inserted under cystoscopy, catheter placement itself involves a risk of complications such as hematuria (98.4%), oliguria/anuria (0.5–6.1%), and ureteral perforation (1.1%). Cost and prolonged operation time (10–25 minutes for catheter placement) are the other problems in the use of prophylactic ureteral catheters. Therefore, intraoperative imaging techniques enabling the identification of the ureter are highly awaited.

3 Ureter Identification with the Use of Fluorescent Dye

Even if a conventional single J-shaped ureteral catheter is placed, it is difficult to confirm the exact location of the ureter by visual inspection and palpation especially in the laparoscopic setting. That is why intraoperative imaging techniques using fluorescent agents such as indocyanine green (ICG), methylene blue (MB), and ZW801 enabling real-time visualization of the ureter have been developed (Table 28.1) [4].

The use of ICG has been reported only in a small number of patients, in whom fluorescence images of the ureter are obtained following retrograde injection of ICG through ureteral catheters. MB can be administered intravenously (0.25–1.0 mg/kg) during surgery, as reported in previous studies. In the MB technique, the anatomy of the ureter is visualized for up to 2 hours after intravenous injection, although the duration varies depending on patient factors. Visualization rates ranged from 20% to 100%, and there were no adverse events or urinary tract infections associated with MB (Table 28.1) [4]. ZW801 is a small zwitterionic substance of 943 Da. In a previous study, ZW801 was administered intravenously in 28 patients, enabling visualization of the ureter within 10 min without adverse events [11].

T. Nishidate (✉)
Department of Surgery and Gastroenterological surgery, Muroran City General Hospital, Muroran, Hokkaido, Japan
e-mail: nisidate@sapmed.ac.jp

K. Okuya · K. Okita · I. Takemasa
Department of Surgical Oncology and Science, Sapporo Medical University, Sapporo, Hokkaido, Japan

T. Ishizawa (ed.), *Fluorescence-Guided Surgery*, https://doi.org/10.1007/978-981-19-7372-7_28

Table 28.1 Fluorescent dyes used for visualization of the ureter (原著 p.255)

	Dye	E_x (nm)	E_m (nm)	Renal clearance	Ext. coeff. (M^{-1} cm^{-1})	Quantum yield (%)
Currently available dyes	ICG	807	822	–	121,000	9.3
	MB	670	690	+	71,200	3.8
Experimental dyes	CW800-CA	786	800	+	237,000	14.2
	CW800-BK	774	790	NR	NR	NR
	ZW800-1	772	788	++	249,000	15.1
	cRGD-ZW800-1	NR	NR	++	NR	NR
	Fluorescein	494	512	+	92,300	95.0
	Liposomal ICG	NR	NR	+	NR	NR
	Genhance 750	750	775	NR	240,000	NR
	UL-766	766	789	++	229,000	9.5
	UreterGlow	800	830	+	NR	NR

From Slooter MD, et al. Tech Coloproctol. 2019;23:305–313 [4]
NR not reported

4 Ureter Identification with the Use of Fluorescent Catheter

Recently, infrared illumination system (IRIS U-Kit®, Stryker) and near-infrared ray catheter (NIRC™, Cardinal Health™) have become commercially available. The IRIS is a single-use fiber catheter that is inserted into the ureter during lower abdominal or pelvic surgery to make its anatomy visible using near-infrared light generated by a light source. The NIRC fluorescent ureteral catheter uses a fluorescent material with excitation and fluorescence wavelengths similar to those of ICG, which enables visualization of the ureter by fluorescence imaging using near-infrared camera systems. These devices are designed to assist the surgeon by revealing anatomical structures during laparoscopic surgery. By using near-infrared light for illumination, the risk of heat damage to the ureter can be minimized.

At the time of catheter insertion, the ureteral orifice is confirmed using a cystoscope, and after insertion of a guidewire, an external catheter for IRIS is placed. After the removal of the cystoscope, a luminescent fiber is inserted into the external catheter to enable intraoperative visualization. Since only the 20 cm tip of the fiber emits light, it should be adjusted to the optimal position (Figs. 28.1 and 28.2).

We herein demonstrate operative pictures after the placement of IRIS during robot-assisted surgery (Fig. 28.3): IRIS was used for an 80 mm rectal cancer with suspected retroperitoneal invasion). Figures 28.4, 28.5, and 28.6 demonstrate the second case of surgical treatment for retroperitoneal abscess due to diverticulitis, in which IRIS was useful for dissection around the ureter. The third case is lateral lymph node dissection for postoperative recurrence of rectal cancer. Since this was the second surgery, the use of NIRC™ was effective for identifying the ureter during the dissection of the adhesion (Fig. 28.7).

Point

- The luminescent catheter is effective for identifying the ureter especially in laparoscopic surgery, where palpation of the ureteral catheter is limited.
- In the use of the luminescent catheter, it is important to recognize the place end length of the light-emitting portion of the catheter.

a : Outer cylinder, luminous fiber

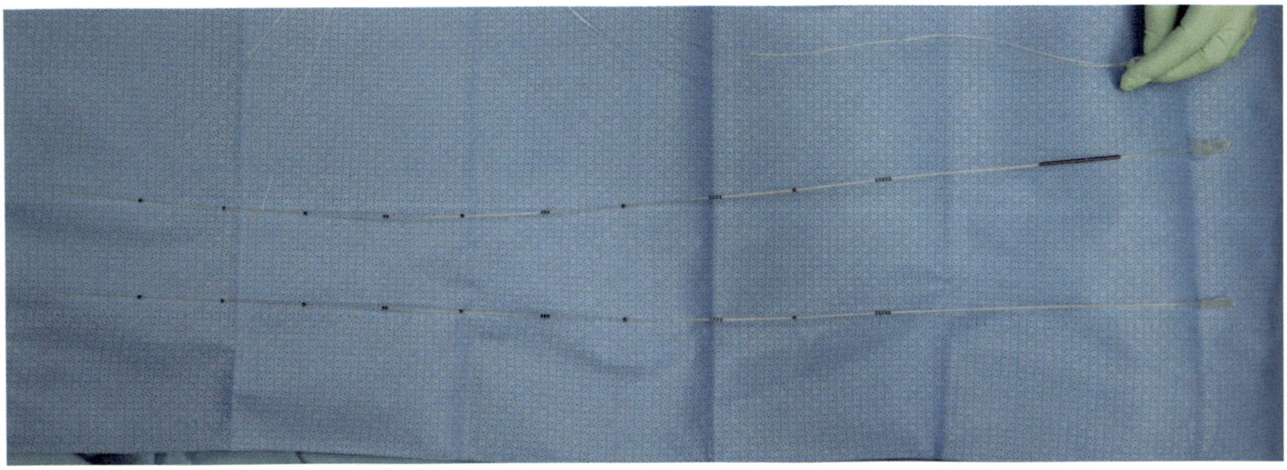

b : System for ureteral illumination

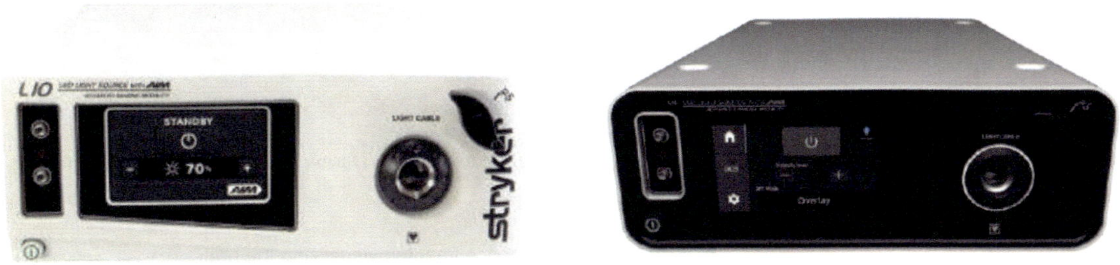

Fig. 28.1 IRIS U-kit®. (**a**) Outer cylinder, luminous fiber. (**b**) System for ureteral illumination

a : Identification of the ureteral orifice using a cystoscope

b : Guidewire Insertion

c : After placement of outer cylinder and luminescent fiber

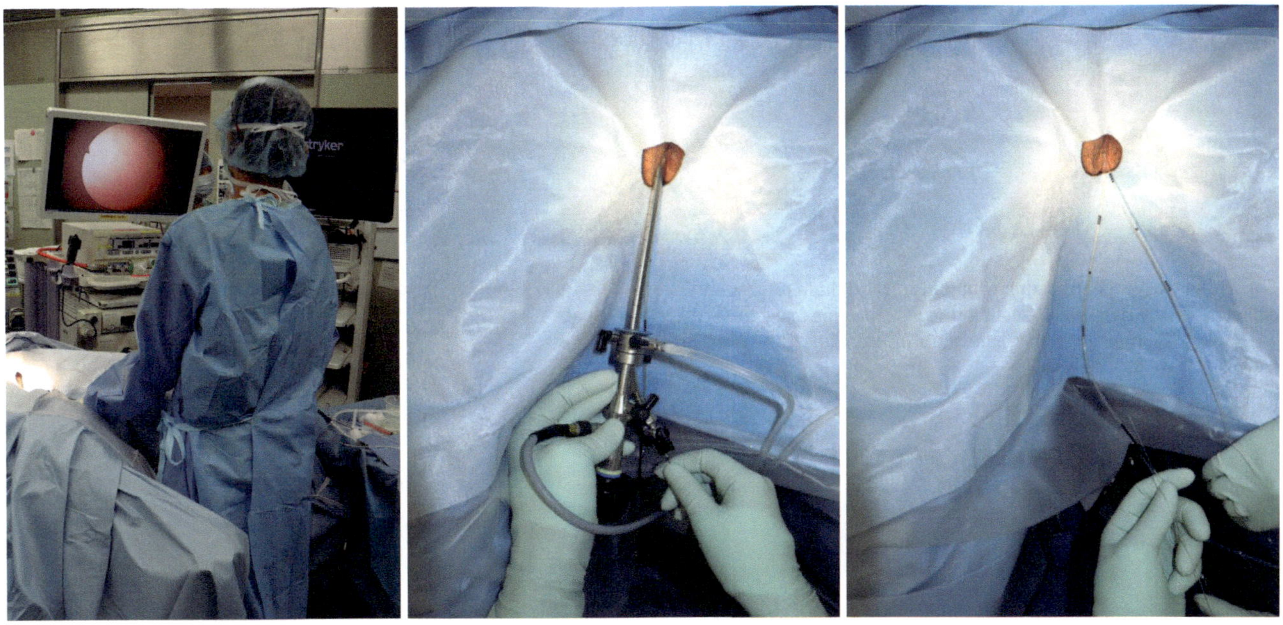

Fig. 28.2 Light-emitting fiber insertion method. (**a**) Identification of the ureteral orifice using a cystoscope. (**b**) Guidewire insertion. (**c**) After placement of outer cylinder and luminescent fiber

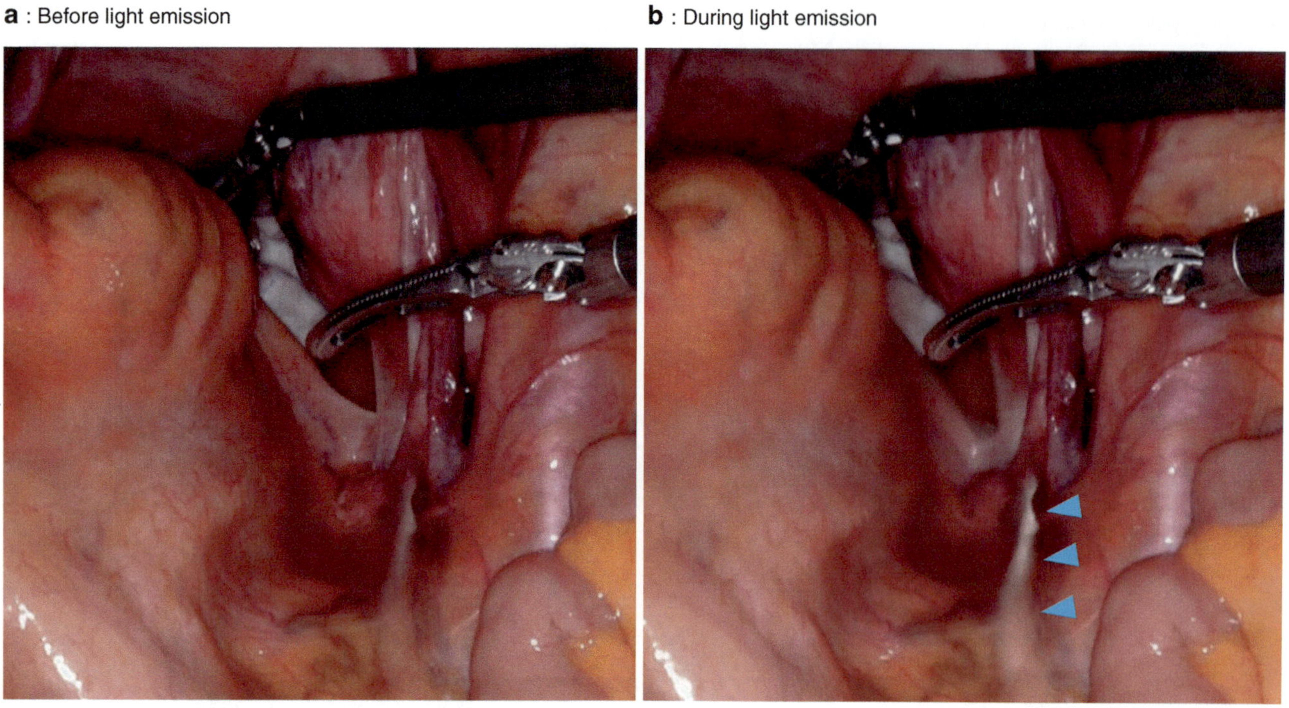

Fig. 28.3 Luminescent fiber placement during robot-assisted surgery. (**a**) Before light emission. (**b**) During light emission

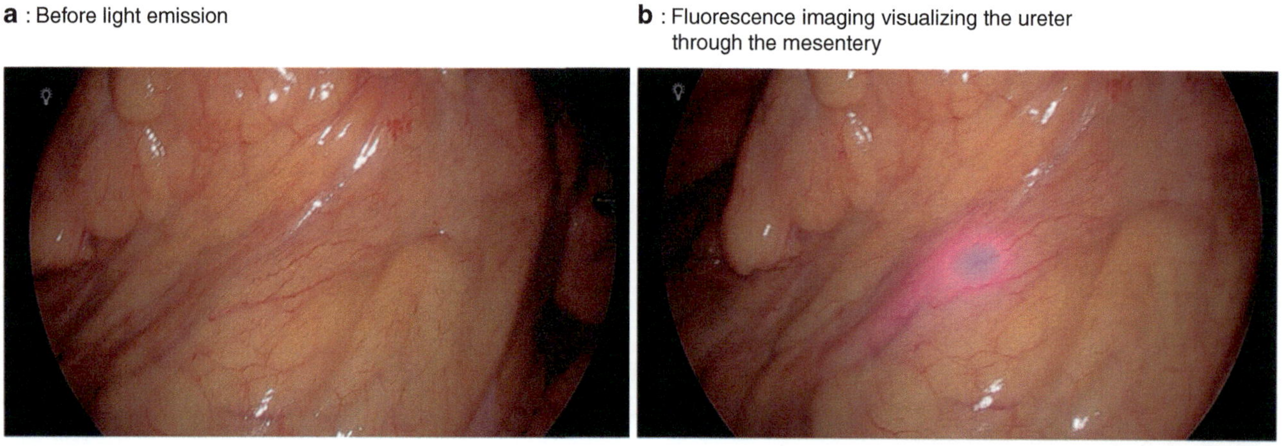

Fig. 28.4 Ureter identification with placement of luminescent ureteral catheter into the right ureter and fluorescence imaging using the Stryker 1588 system during surgical treatment for diverticulitis. (**a**) Before light emission. (**b**) Fluorescence imaging visualizing the ureter through the mesentery

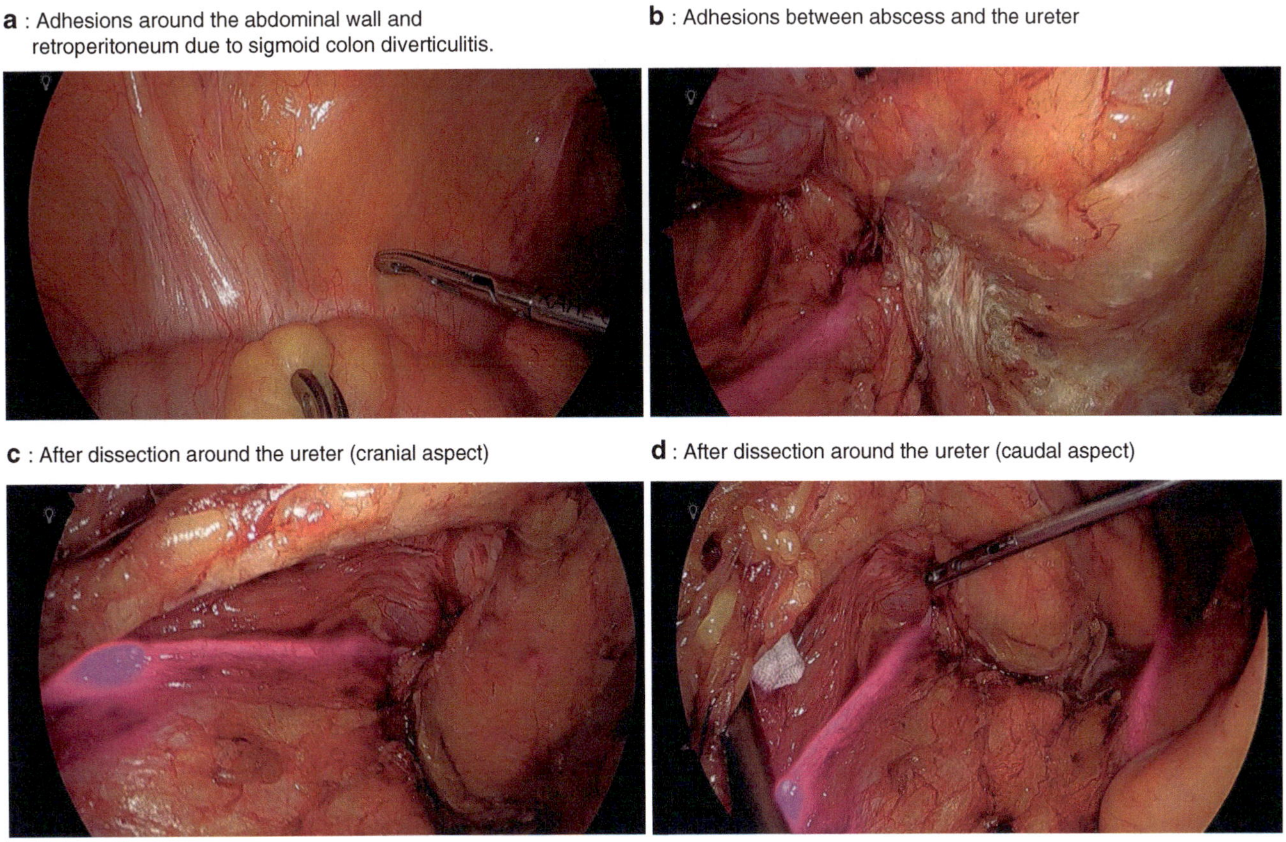

Fig. 28.5 Ureter identification by fluorescence imaging during surgical treatment for diverticulitis. (**a**) Adhesions around the abdominal wall and retroperitoneum due to sigmoid colon diverticulitis. (**b**) Adhesions between the abscess and the ureter. (**c**) After dissection around the ureter (cranial aspect). (**d**) After dissection around the ureter (caudal aspect)

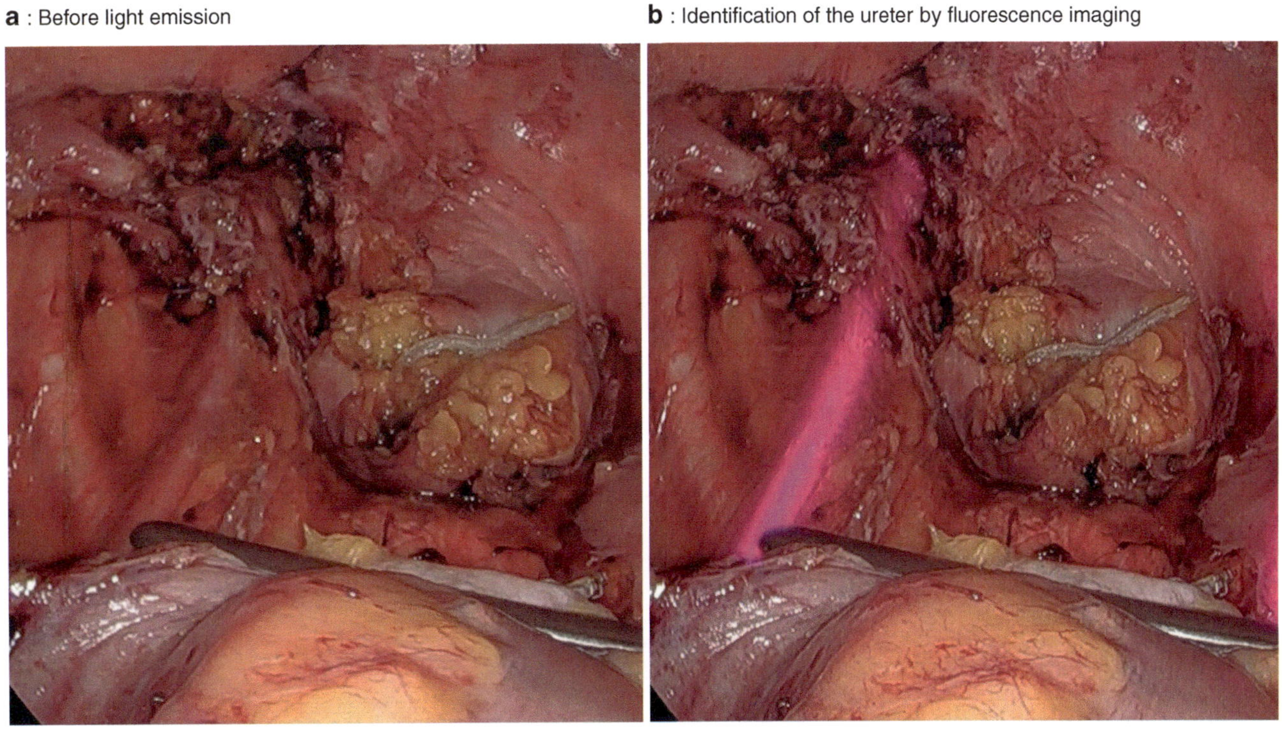

Fig. 28.6 Dissection of adhesions after sigmoidectomy. (**a**) Before light emission. (**b**) Identification of the ureter by fluorescence imaging

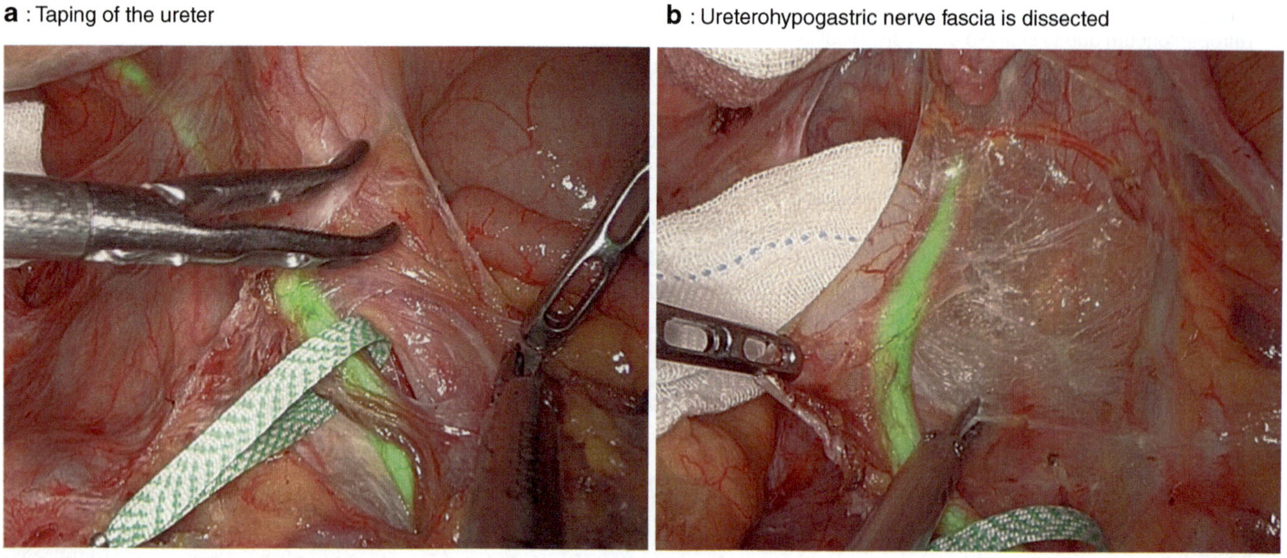

Fig. 28.7 Ureter identification by fluorescence imaging using the Stryker 1688 and luminescent ureteral catheter during lateral lymph dissection for recurrent lymph node metastasis. (**a**) Taping of the ureter. (**b**) The ureterohypogastric nerve fascia is dissected

5 Pitfalls

The use of a cystoscope requires the cooperation of a urologist. There can be cases of difficult stent placement, difficulty in deep reinsertion during surgery, and inability to advance the inner tube. Forcible manipulation may lead to ureteral injury during stent placement.

6 Future Perspective

In order to visualize the ureter without catheter placement, fluorescence imaging techniques using various fluorescent agents in the near-infrared region have been developed and evaluated in clinical studies (Table 28.2) [4–11]. For example, Tanaka et al. [12] reported that intravenous injection of near-infrared fluorescent dye CW800-CA enables real-time visualization of the ureter in the animal model. The CW800-CA is excreted into the kidney and thus accumulates in high concentrations in the ureter, resulting in selective identification of the urinary system. Other fluorophores under development include CW800-BK, cRGD-ZW800-1, fluorescein, liposomal ICG, Genhance 750, UL-766, and UreterGlow (Table 28.3) [4, 12–23]. In the future, the intraoperative ureteral identification technique will become widely used, enhancing the accuracy and safety of surgical treatments in the pelvic cavity.

Table 28.2 Previous reports on the use of ICG, MB, and ZW800-1

	Article	Surgical specialism	Laparoscopic/ open	Number of patients	Administration	Dose	Ureters identified	Complications	References
Indocyanine green	Siddighi et al. (2014)	Gynecologic	Laparoscopic	>10	Ureteral catheter	25 mg in 10 mL/ ureter	All	None	[5]
	Lee et al. (2015)	Urologic	Laparoscopic	25	Ureteral catheter and/or PNT	25 mg in 10 mL/ ureter	All	Yes, in one patient[a]	[6]
Methylene blue	Verbeek et al. (2013)	Abdominal	Open	12	Intravenous	0.25–1 mg/kg	24/24	None	[7]
	Yeung et al. (2016)	Colorectal	Both	8	Intravenous	0.25–1 mg/kg	10/11	None	[8]
	Al-Taher et al. (2016)	Colorectal	Laparoscopic	9	Intravenous	0.125–1.0 mg/kg	6/9	None	[9]
	Barnes et al. (2018)	Colorectal	Both	40	Intravenous	0.25–1 mg/kg	63/69	None	[10]
ZW800-1	de Valk et al. (2019)	Abdominal	Laparoscopic	44	Intravenous	3.9 mg/kg		None	[11]

From Slooter et al. [4]

Table 28.3 Previous studies of intraoperative visualization of the ureter

Dye	Article	Number of animals	Duration of visualization	Doses (IV[a], mg/kg)	Toxicity in rats (mg/kg)	Ongoing clinical trials	References
CW800-CA	Tanaka et al. (2007)	12 rats, 6 pigs	120 min	0.0015–0.015	>20	NCT03387410[c] NCT03106038[c]	[12]
	Schols et al. (2014)	2 pigs		0.007–0.086			[13]
	Korb et al. (2015)	6 pigs		0.030–0.12			[14]
CW800-BK	Al-Taher et al. (2018)	3 pigs		0.08–0.3			[15]
cRGD-ZW800-1	Verbeek et al. (2014)	3 rats	<7.5 h	0.25–30 nmol	> 24.5	2017-001954-32[c]	[16]
Fluorescein	Dip et al. (2014)	9 rats	<12 h	7	LD_{50} = 600	Unknown	[17]
	Meershoek et al. (2018)	3 pigs		5 ml, 100 mg/ml SC/IM[b]			[18]
Liposomal ICG	Portnoy et al. (2015)	25 mice	>90 min	8	Liposomes: 10 ICG: LD_{50} = 87	Unknown	[19]
	Friedman-Levi et al. (2018)	> 12 mice, 2 pigs		4–16			[20]
Genhance 750	Rowe et al. (2012)	10 swine	>20 min	0.5	NR	Unknown	[21]
UL-766	Cha et al. (2018)	8 rats	>60 min	0.09	NR	Unknown	[22]
UreterGlow	Mahalingam et al. (2018)	5 pigs	<6 h	0.1	NR	Unknown	[23]

From Slooter et al. [4]

References

1. Matsui A, Tanaka E, Choi HS, et al. Real-time, near-infrared, fluorescence-guided identification of the ureters using methylene blue. Surgery. 2010;148:78–86.
2. Selzman AA, Spirnak JP. Iatrogenic ureteral injuries: a 20-year experience in treating 165 injuries. J Urol. 1996;155:878–81.
3. Burks FN, Santucci RA. Management of iatrogenic ureteral injury. Ther Adv Urol. 2014;6:115–24.
4. Slooter MD, Jansen A, Bemelman WA, et al. Tech Coloproctol. 2019;23:305–13.
5. Siddighi S, Yune JJ, Hardesty J. Indocyanine green for intraoperative localization of ureter. Am J Obstet Gynecol. 2014;211:436.e1–2.
6. Lee Z, Moore B, Giusto L, et al. Use of indocyanine green during robot-assisted ureteral reconstructions. Eur Urol. 2015;67:291–8.
7. Verbeek FP, Vorst JR, Schaafsma BE, et al. Intraoperative near-infrared fluorescence guided identification of ureters using low dose methylene blue: a first in human experience. J Urol. 2013;190:574–9.
8. Yeung TM, Volpi D, Tullis ID, et al. Identifying ureters in situ under fluorescence during laparoscopic and open colorectal surgery. Ann Surg. 2016;263:e1–2.
9. Al-Taher M, van den Bos J, Schols RM. Fluorescence ureteral visualization in human laparoscopic colorectal surgery using methylene blue. J Laparoendosc Adv Surg Tech A. 2016;26:870–5.
10. Barnes TG, Hompes R, Birks J, et al. Methylene blue fluorescence of the ureter during colorectal surgery. Surg Endosc. 2018;32:4036–43.
11. de Valk KS, Handgraaf HJ, Deken MM, et al. A zwitterionic near-infrared fluorophore for real-time ureter identification during laparoscopic abdominopelvic surgery. Nat Commun. 2019;10:3118.
12. Tanaka E, Ohnishi S, Laurence RG, et al. Real-time intraoperative ureteral guidance using invisible near-infrared fluorescence. J Urol. 2007;178:2197–202.
13. Schols RM, Lodewick TM, Bouvy ND, et al. Application of a new dye for near-infrared fluorescence laparoscopy of ureters: demonstration in a pig model. Dis Colon Rectum. 2014;57:407–11.
14. Korb ML, Huh WK, Boone JD, et al. Laparoscopic fluorescent visualization of the ureter with intravenous IRDye800CW. J Minim Invasive Gynecol. 2015;22:799–806.
15. Al-Taher M, van den Bos J, Schols RM, et al. Evaluation of a novel dye for near-infrared fluorescence delineation of ureters during laparoscopy. BJS Open. 2018;2:254–61.
16. Verbeek FP, van der Vorst JR, Tummers QR, et al. Near-infrared fluorescence imaging of both colorectal cancer and ureters using a low-dose integrin targeted probe. Ann Surg Oncol. 2014;21:S528–37.
17. Dip FD, Nahmod M, Anzorena FS, et al. Novel technique for identification of ureters using sodium fluorescein. Surg Endosc. 2014;28:2730–3.
18. Meershoek P, KleinJan GH, van Oosterom MN, et al. Multispectral fluorescence imaging as a tool to separate healthy and disease-related lymphatic anatomies during robot-assisted laparoscopic procedures. J Nucl Med. 2018;59:1757–60.
19. Portnoy E, Nizri E, Golenser J, et al. Imaging the urinary pathways in mice by liposomal indocyanine green. Nanomedicine. 2015;11:1057–64.
20. Friedman-Levi Y, Larush L, Diana M, et al. Optimization of liposomal indocyanine green for imaging of urinary pathways and a proof of concept in a pig model. Surg Endosc. 2018;32:963–70.
21. Rowe CK, Franco FB, Barbosa JA, et al. A novel method of evaluating ureteropelvic junction obstruction: dynamic near infrared fluorescence imaging compared to standard modalities to assess urinary obstruction in a swine model.
22. Cha J, Nani RR, Luciano MP, et al. A chemically stable fluorescent marker of the ureter. Bioorg Med Chem Lett. 2018;28:2741–5.
23. Mahalingam SM, Dip F, Castillo M, et al. Intraoperative ureter visualization using a novel near-infrared fluorescent dye. Mol Pharm. 2018;15:3442–7.

Imaging of the Parathyroid Gland

29

Akihiro Nakajo and Yoshiaki Shinden

Summary

- ICG fluorescence imaging is useful for intraoperative identification of small biological structures like parathyroid glands, although it can be interfered by nonspecific fluorescence signals.
- Since the discovery of autofluorescence in parathyroid glands in 2008, the usefulness of autofluorescence imaging without using specific fluorescent agents has been widely reported.
- Autofluorescence imaging of the parathyroid glands is a highly sensitive technique that can be easily performed using commercially available near-infrared imaging systems, and its widespread use is expected in the future.

1 Introduction

The parathyroid glands are endocrine organs that surround the thyroid and usually consist of four glands. Each gland is about the size of a grain of rice and secretes parathyroid hormones. Parathyroid hormones increase the calcium concentration in the blood by indirectly increasing calcium absorption from the digestive tract and kidneys, in addition to transferring calcium stored in the bones to the bloodstream.

Calcium metabolism is important for the human body, and abnormalities in parathyroid gland function can cause a variety of symptoms. For example, hyperparathyroidism is associated with symptoms such as thirst, fatigue, and nausea due to hypercalcemia and urinary tract stone disease. On the other hand, if all of the parathyroid glands are removed during thyroid surgery or if they become dysfunctional due to decreased blood flow, hypoparathyroidism occurs. Hypoparathyroidism is characterized by numbness in the fingertips and lips due to hypocalcemia and tetany, a spasm of the limbs. On the other hand, the function of the parathyroid glands can often be restored by intramuscular autotransplantation of the removed parathyroid glands. In this sense, accurate intraoperative identification of parathyroid tissues is very important for assuring the safety of thyroid and parathyroid surgery. For example, normal parathyroid glands must be preserved in surgery for benign thyroid disease. In malignant thyroid disease, intramuscular autotransplantation of parathyroid glands removed during lymph node dissection is indicated whenever possible to avoid permanent postoperative hypoparathyroidism. In addition, in surgery for hyperparathyroidism, it is essential but often difficult to identify relatively small parathyroid tissues for preservation.

Conventionally, the identification of parathyroid glands has relied on the surgeon's eye. Although experienced endocrine surgeons have been able to identify most of the normal parathyroid glands and perform parathyroid preservation or autotransplantation, inexperienced surgeons often fail to detect parathyroid glands. Even experienced surgeons may overlook parathyroid glands buried in fatty tissue. In recent years, fluorescence imaging has been clinically applied to accurately identify parathyroid glands using fluorescently labeled agents, but there were some problems such as difficulty in obtaining sufficient contrast between the glands and surrounding tissues. In 2008, the discovery that parathyroid glands have autofluorescence led to the focus of attention on autofluorescence imaging, which could be a simple method to identify parathyroid glands without the use of fluorescent labeling agents.

Supplementary Information The online version contains supplementary material available at https://doi.org/10.1007/978-981-19-7372-7_29.

A. Nakajo (✉) · Y. Shinden
Department of Digestive Surgery, Breast and Thyroid Surgery, Kagoshima University Graduate School of Medical and Dental Sciences, Kagoshima, Japan
e-mail: anakajo@m.kufm.kagoshima-u.ac.jp

T. Ishizawa (ed.), *Fluorescence-Guided Surgery*, https://doi.org/10.1007/978-981-19-7372-7_29

2 Conventional Methods of Identifying the Parathyroid Glands

For intraoperative identification of parathyroid glands, fluorescence imaging using fluorescent dyes such as methylene blue, 5-ALA, and ICG has been reported [1–5]. Although methylene blue retains in adenomas and hyperplasia, it does not accumulate in normal parathyroid glands, and its use has been limited due to toxicity issues [6, 7]. Optical coherence tomography, confocal reflectance microscopy, and Raman spectroscopy have been clinically applied to identify parathyroid glands without using fluorescent dyes. Although these methods showed some usefulness, they were not widely used because of the complexity of imaging procedures [8–10].

3 Development History of Autofluorescence Imaging of the Parathyroid Glands

Around 2008, researchers at Vanderbilt University in the USA discovered that parathyroid glands have endogenous autofluorescence in the near-infrared region [11] Autofluorescence refers to the spontaneous emission of light that occurs when biological structures such as mitochondria and lysosomes absorb light and is distinct from light emitted from artificially added fluorophores. They found that the intensity of autofluorescence of excited parathyroid glands peaks at a wavelength in the infrared region around 822 nm (Fig. 29.1, upper left graph) and succeeded in identifying parathyroid glands intraoperatively using a dedicated excitation device and an infrared observation camera [12–14] This method is safe because it does not require the administration of specific fluorescent dyes or contrast agents. In addition, autofluorescence imaging is innovative in that it is not susceptible to nonspecific accumulation, which remains a problem with fluorescence imaging using fluorescent dyes. Furthermore, since the peak wavelength of parathyroid glands is similar to that of ICG (around 810–830 nm), autofluorescence of parathyroid glands can be easily detected with high sensitivity using near-infrared camera systems that are already commercially available for ICG fluorescence imaging [15–17]. Thus, autofluorescence imaging of parathyroid glands has attracted wide attention and has been applied clinically.

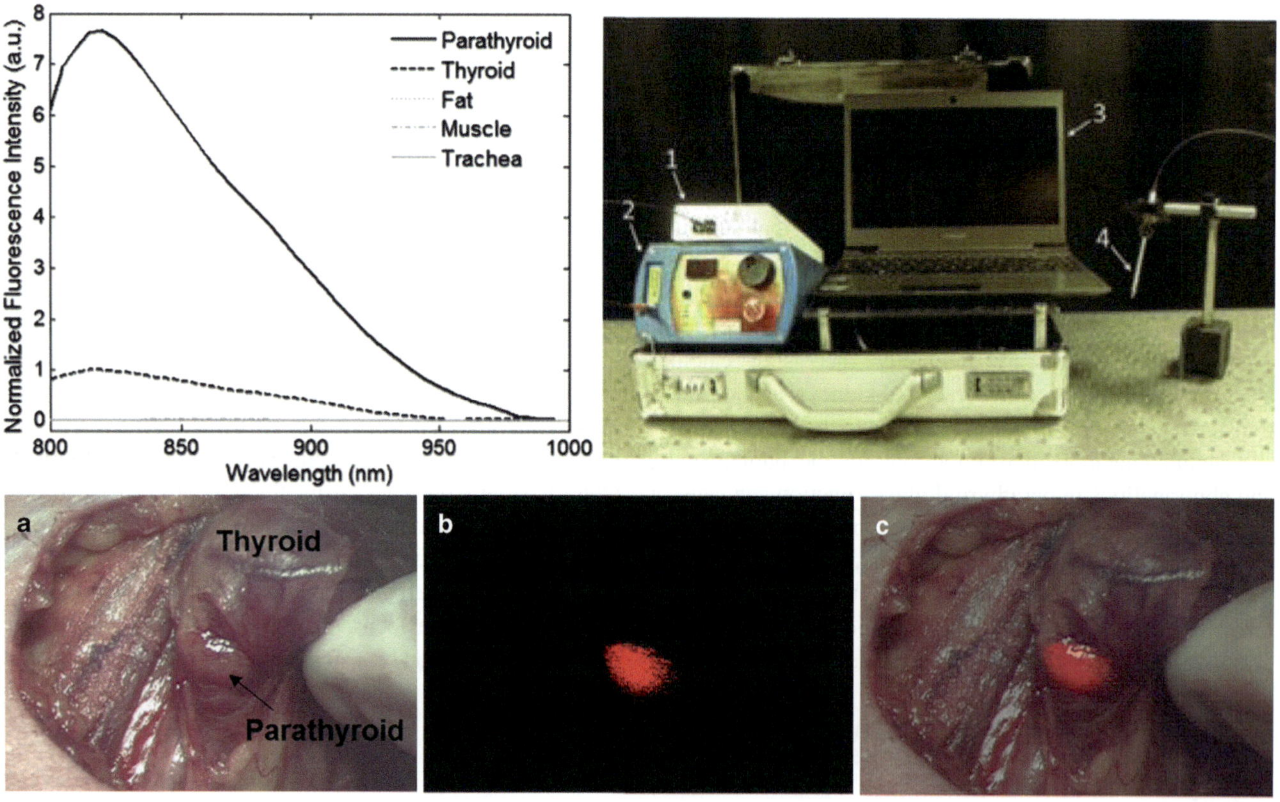

Fig. 29.1 (**a**–**c**) Discovery of autofluorescence in parathyroid glands. Intraoperative fluorescence images of the parathyroid gland. (**a**) Bright-field image of parathyroid and thyroid from patient undergoing thyroidectomy. (**b**) NIR fluorescence image shows parathyroid in red. (**c**) Bright-field and fluorescence images overlaid show parathyroid fluorescence. (Adapted from McWade et al [12, 13])

4 Clinical Practice of Autofluorescence Imaging of Parathyroid Glands

Since the autofluorescence of parathyroid glands and the fluorescence spectrum of ICG are similar, a commercially available near-infrared camera system that detects ICG fluorescence can be used for this method without any modification. In this article, we introduce autofluorescent imaging using the PDE-neo®, a near-infrared imaging system for open surgery.

First, the imaging parameters of PDE-neo® were set to the following values: brightness, minimum; contrast, maximum; excitation light, maximum; and imaging mode, fluorescence mapping mode.

For intraoperative real-time identification of the parathyroid glands (in vivo imaging), a camera unit with a special sterile cover is fixed at a distance of 5–30 cm from the observation position, focused, all lights in the operating room are turned off, and autofluorescence is observed on a monitor. The location of the fluorescing parathyroid gland is identified by turning the surgical lights on and off several times to integrate fluorescence images and visual inspection of surgical fields (Fig. 29.2).

To identify the parathyroid glands in the removed thyroid gland or fat tissues including lymph nodes (ex vivo imaging), observation is performed using a shielding box on the operating table. When autotransplanting parathyroid glands in thyroid cancer surgery, the removed tissue is observed in a sterilized condition.

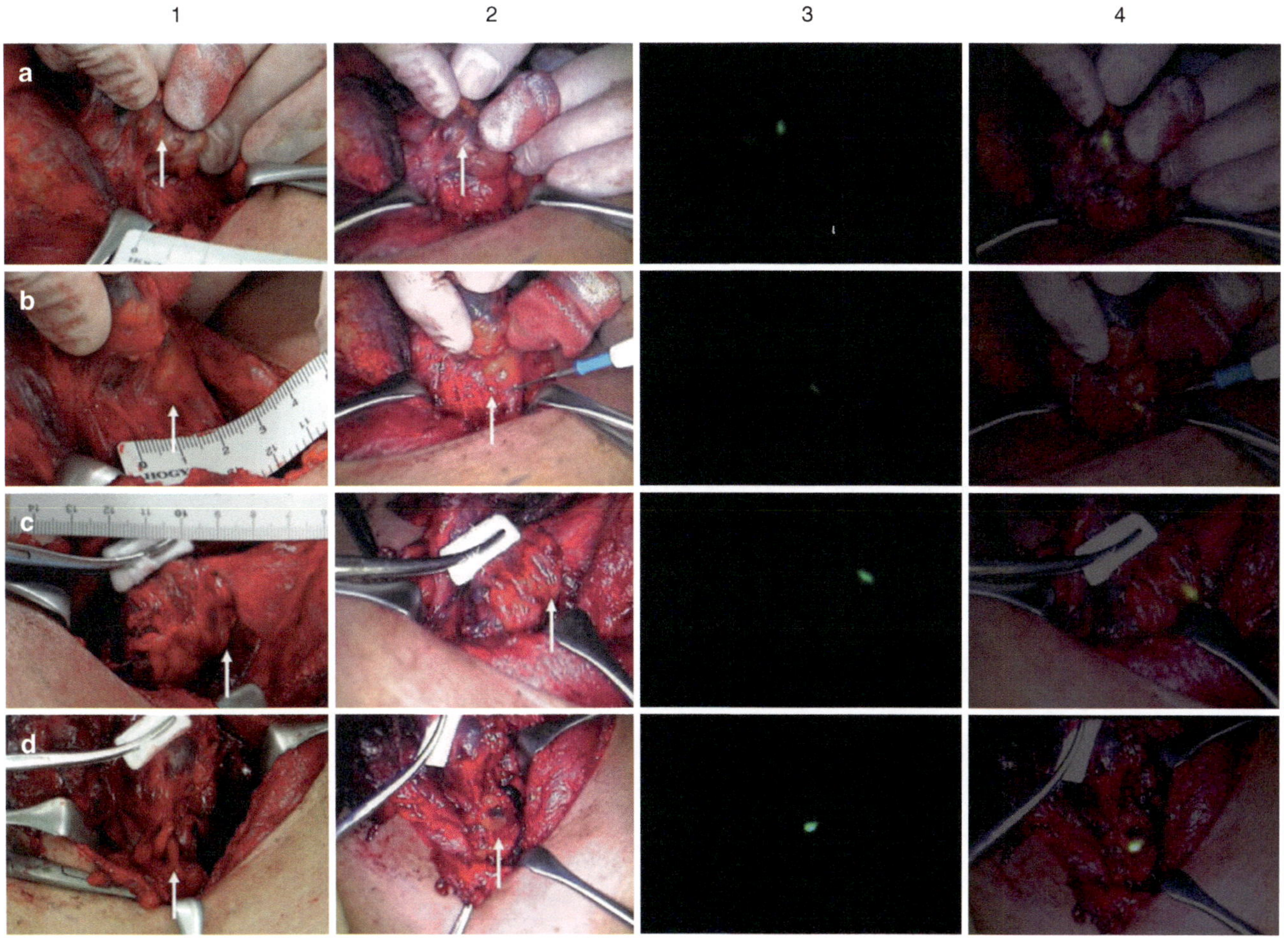

Fig. 29.2 (**a**–**d**) In vivo identification of the parathyroid gland using the intrinsic fluorescence method. Four parathyroid glands; upper right (**a**), lower right (**b**), upper left (**c**) and lower left (**d**) from one patient are shown. The first column shows high-resolution pictures, and the second column shows pictures taken with the PDE system under white light. White arrows indicate the parathyroid glands. Under white light, some tissues were difficult to identify as parathyroid tissue. In the third column, fluorescent pictures are shown. Here, image qualities were adjusted for printing. In the fourth column, merged images from the second column and the third column are shown. All of these glands were identified as parathyroid tissue pathologically. (Adapted from Shinden et al. [16])

Point

- Identification of parathyroid glands using autofluorescence detected with PDE-neo®.
  1. Intraoperative real-time identification (in vivo imaging)
     - (i) Fluorescence mapping mode is used under the following condition (brightness, minimum value; contrast, maximum value; excitation light, maximum value).
     - (ii) Attach the sterile cover to the camera head, hold the camera at a distance of approximately 5 cm from the observation position of the surgical field, and focus the camera.
     - (iii) Turn off all lights in the operating room, and observe autofluorescence on a monitor. The location of the parathyroid gland is determined by integrating autofluorescent images and visual inspection of the operation field.
     - (iv) Remove the identified parathyroid gland, turn off the light, and confirm autofluorescence signals.
  2. Identification of parathyroid glands in the removed tissues (ex vivo imaging)
     - (i) Observe fluorescence signals on the resected specimen, using a shielding box on the table outside the surgical field.
     - (ii) When autotransplanting parathyroid glands in thyroid cancer surgery, the removed tissue should be kept sterilized.

Table 29.1 Autofluorescence intensity and detection rate for each tissue

Tissue	Mean value of fluorescence intensity	Fluorescent positive rate
Parathyroid	3.53 (0.17 ~ 15.3)	100% (34/34)
Lymph node	0.01 (0 ~ 2.0)	0.5% (1/198)
Fat	0.15 (0 ~ 2.1)	12% (2/17)
Thyroid	0.45 (0 ~ 3.71)	27% (3/11)

Adapted from Shinden et al. [16]

We measured the fluorescence intensity of autofluorescence in surgically removed parathyroid glands ($n = 34$), thyroid glands ($n = 11$), lymph nodes ($n = 198$), and adipose tissue ($n = 17$) using PDE-neo® in our department. When the cutoff value was set at a fluorescence intensity that could detect all of the excised parathyroid tissue, it was possible to differentiate most lymph nodes and fat tissues (Table 29.1 and Fig. 29.3). Autofluorescence was also found in about 30% of the thyroid glands, but in the setting of in vivo imaging, autofluorescence of parathyroid could be discriminated from that of the thyroid. This demonstrates that PDE-neo® autofluorescence imaging is clinically applicable, simple, and extremely useful for intraoperative parathyroid identification (Movie 29.1).

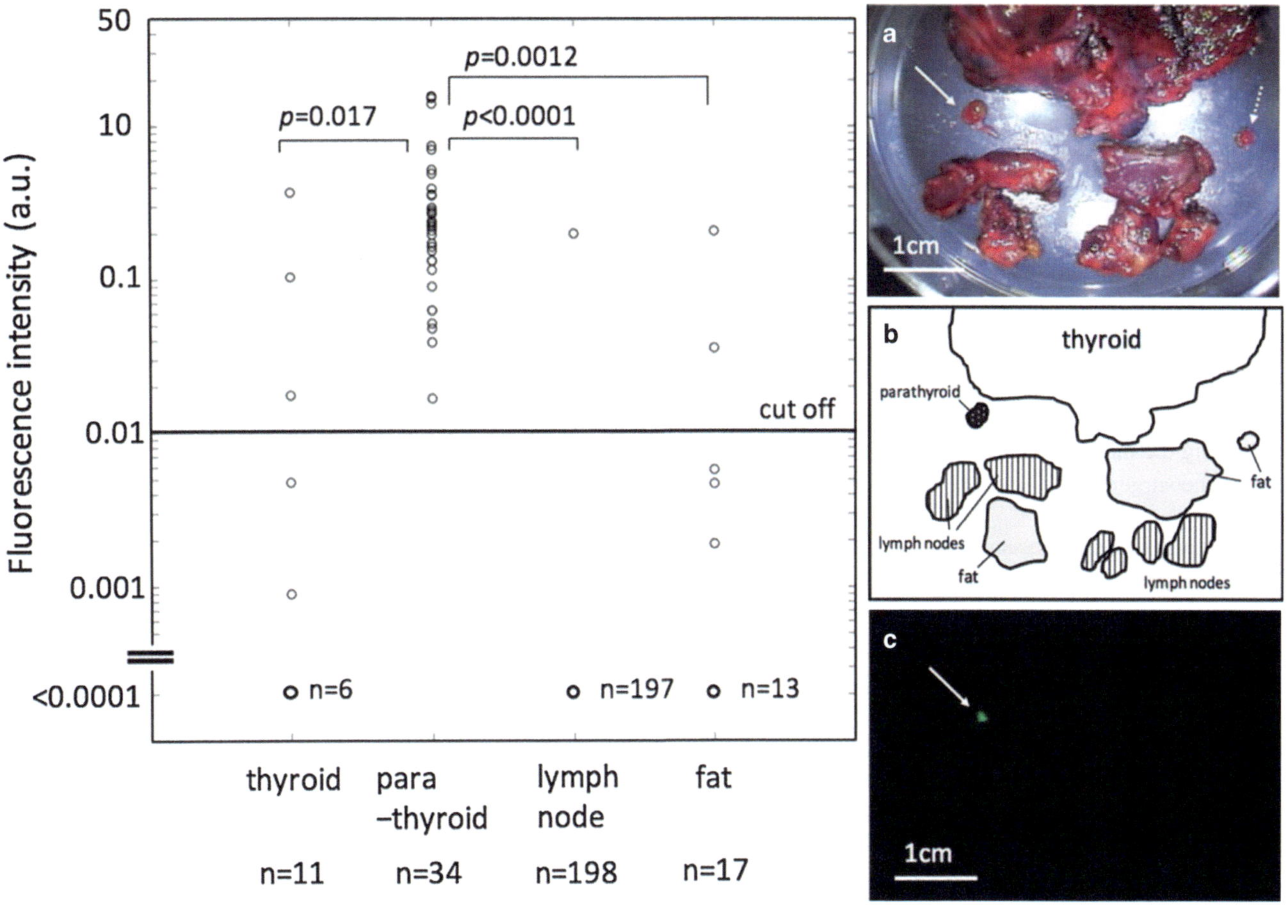

Fig. 29.3 (**a**–**c**) Autofluorescence intensity of parathyroid glands and neighboring tissue (ex vivo imaging by PDE-neo®). An example of ex vivo measurement of the resected parathyroid gland. Pictures of parathyroid glands and other tissues with the PDE system taken under white light (**a**) and excitation light (**c**) are shown. Fluorescence of the parathyroid gland can be seen as green light (**c**, white arrow), and it was identified as normal parathyroid tissue pathologically. The scheme of each tissue was shown in (**b**). The dotted white arrow in a indicated fat tissue which was identified as parathyroid by surgeon. However, it had no fluorescence and it was revealed to be fat tissue after peeling surrounding tissue. (Adapted from Shinden et al. [16])

5 Expected Effects of Autofluorescence Imaging of the Parathyroid Glands

Autofluorescence imaging of the parathyroid glands is most valuable in identifying small, difficult-to-detect parathyroid glands. Specifically, autofluorescence imaging of the parathyroid glands is useful in the search for small responsible lesions in primary hyperparathyroidism, ectopic parathyroid lesions, and normal parathyroid glands in thyroid cancer surgery (autotransplantation). In particular, if the detection rate of small normal parathyroid glands is increased, the frequency of postoperative permanent hypoparathyroidism may be significantly reduced. Autofluorescence imaging does not require any specific fluorescence labeling agent or contrast medium and is therefore nontoxic and easy to perform. The detection rate of parathyroid by autofluorescence imaging was reported to be higher than that by ICG fluorescence imaging [18].

6 Pitfalls and Limitations

This method is characterized by low nonspecific background fluorescence and contrast with surrounding tissues such as the thyroid gland, but there are some points to be noted. First of all, it may be difficult to identify the parathyroid glands in patients with strong autofluorescence in the thyroid gland. Fluorescence signals are rarely detected in fat tissues, but it may lead to misidentification. Some surgical sutures (e.g., Vicryl®) also show strong autofluorescence signals.

Application to endoscopic surgery remains a problem. Although it has been reported that autofluorescence imaging could be used with the KARL STORZ endoscopic surgery system [19], our experience is that the current surgical endoscopic systems available for near-infrared fluorescence imaging do not have the high detection rate as achieved in the use of open surgical imaging systems. In the future, it will be necessary to optimize the imaging conditions (excitation lights, filter settings, camera sensitivity, etc.) and develop endoscopic imaging systems dedicated to parathyroid autofluorescence imaging.

What actually causes autofluorescence in the parathyroid glands remains unclear. Calcium-sensing receptors and other receptors have been suggested as candidates for the origin of autofluorescence, but the exact mechanism is unknown. At present, we can say that autofluorescence is often weak in parathyroid glands with secondary hyperparathyroidism associated with renal failure (hyperplasia) and strong in parathyroid glands with a lot of eosinophils [7]. As the mechanism of autofluorescence is elucidated in the future, it will lead to developing a more sensitive imaging system and extending the indication of autofluorescence imaging during endocrine surgery.

7 Future Perspectives

Autofluorescence imaging of parathyroid glands is a highly sensitive method for parathyroid gland detection that can be used with existing near-infrared imaging systems. In addition, this technique has clinical advantages in identifying parathyroid glands without using fluorescent dyes such as ICG or other contrast agents. As the mechanism of autofluorescence is elucidated, the technique will be brushed up to be more sensitive and effective in endoscopic surgery as well as open surgery. If this method becomes widely used in the future, it will not only shorten the time required for the identification of parathyroid glands but also have the potential to reduce the incidence of postoperative permanent hypoparathyroidism.

References

1. Inoue A, Inokuchi A. Parathyroidectomy using methylene blue. Otorhinolaryngol Clin. 2008;101:652–3.
2. Majithia A, Stearns MP. Methylene blue toxicity following infusion to localize parathyroid adenoma. J Laryngol Otol. 2006;120:138–40.
3. Takeuchi S, Shimizu K, Shimizu K Jr, et al. Identification of pathological and normal parathyroid tissue by fluorescent labeling with 5- aminolevulinic acid during endocrine neck surgery. J Nippon Med Sch. 2014;81:84–93.
4. Zaidi N, Bucak E, Yazici P, et al. The feasibility of indocyanine green fluorescence imaging for identifying and assessing the perfusion of parathyroid glands during total thyroidectomy. J Surg Oncol. 2016;113:775–8.
5. Sound S, Okoh A, Yigitbas H, et al. Utility of indocyanine green fluorescence imaging for intraoperative localization in reoperative parathyroid surgery. Surg Innov. 2019;26:774–9.
6. Rudin AV, Berber E. Impact of fluorescence and autofluorescence on surgical strategy in benign and malignant neck endocrine diseases. Best Pract Res Clin Endocrinol Metab. 2019;33:101311.
7. Solórzano CC, Thomas G, Baregamian N, et al. Detecting the near infrared autofluorescence of the human parathyroid: hype or opportunity? Ann Surg. 2019;272:973.
8. Ladurner R, Hallfeldt KK, Al Arabi N, et al. Optical coherence tomography as a method to identify parathyroid glands. Lasers Surg Med. 2013;45:654–9.
9. White WM, Tearney GJ, Pilch BZ, et al. A novel, noninvasive imaging technique for intraoperative assessment of parathyroid glands: confocal reflectance microscopy. Surgery. 2000;128:1088–101.
10. Das K, Stone N, Kendall C, et al. Raman spectroscopy of parathyroid tissue pathology. Lasers Med Sci. 2006;21:192–7.
11. Paras C, Keller M, White L, et al. Near-infrared autofluorescence for the detection of parathyroid glands. J Biomed Opt. 2011;16:067012.
12. McWade MA, Paras C, White LM, et al. A novel optical approach to intraoperative detection of parathyroid glands. Surgery. 2013;154:1371–7.
13. McWade MA, Paras C, White LM, et al. Label-free intraoperative parathyroid localization with near-infrared autofluorescence imaging. J Clin Endocrinol Metab. 2014;99:4574–80.
14. McWade MA, Sanders ME, Broome JT, et al. Establishing the clinical utility of autofluorescence spectroscopy for parathyroid detection. Surgery. 2016;159:193–203.
15. De Leeuw F, Breuskin I, Abbaci M, et al. Intraoperative near-infrared imaging for parathyroid gland identification by autofluorescence: a feasibility study. World J Surg. 2016;40:2131–8.
16. Shinden Y, Nakajo A, Arima H, et al. Intraoperative identification of the parathyroid gland with a fluorescence detection system. World J Surg. 2017;41:1506–12.
17. Ladurner R, Al Arabi N, Guendogar U, et al. Near-infrared autofluorescence imaging to detect parathyroid glands in thyroid surgery. Ann R Coll Surg Engl. 2018;100:33–6.
18. Kahramangil B, Berber E. Comparison of indocyanine green fluorescence and parathyroid autofluorescence imaging in the identification of parathyroid glands during thyroidectomy. Gland Surg. 2017;6:644–8.
19. Ladurner R, Sommerey S, Al Arabi N, et al. Intraoperative near-infrared autofluorescence imaging of parathyroid glands. Surg Endosc. 2017;31:3140–5.

Part VI

Intraoperative Fluorescence Imaging in Practice [Development]

Takeaki Ishizawa

Although ICG is an excellent fluorophore that can be used for many applications, it was not originally designed for in vivo fluorescence imaging (developed mainly as testing drug of cardiac output and liver function). Therefore, we are waiting for the development of a new fluorescent probe that can "delineate only the target such as cancer with high sensitivity." Imaging devices have evolved remarkably in the past 20 years or so, but there is still room for improvement not only in the calculational sensitivity and resolution but also in the "display method" of the images in the operating room, in order to make fluorescence guided surgery "more visible than the naked eye."

In terms of cancer therapy, the ultimate goal of fluorescence imaging should be to use the accumulation of probes not only for diagnosis but also for treatment. In fact, we are surprised to find that photodynamic therapy using fluorescent probes has already demonstrated visible tumor reduction effects in animal clinical settings. We cannot help but hope that this technology will contribute to the treatment of cancer patients in the near future.

30 Development of Novel Fluorescent Probes: Rapid Intraoperative Visualization of Microcarcinoma by Local Application of Chemical Fluorescence Probes

Yasuteru Urano

Summary

- Activatable fluorescent probes are molecular tools that emit fluorescence only when they bind to and react with target molecules.
- In recent years, we have developed a number of fluorescent probes that detect exo-type protease activity and have succeeded in the rapid detection of microcarcinoma by topical administration of these probes to fresh clinical specimens.
- The fluorescent probes for detecting aminopeptidase GGT and DPP4 activity enable rapid detection of breast and esophageal cancers within a few minutes of probe application.
- A fluorescent probe detecting PSMA, a carboxypeptidase, enables rapid detection of prostate cancer.
- For some probes, clinical performance tests and first-in-human tests have been started toward early clinical installation.

1 Introduction

Fluorescence probes are molecular tools that change their fluorescent property dramatically by specific reactions to the target molecule. In recent years, fluorescent probes have been widely used to observe the dynamics of various bioactive substances in living biological samples in real time. Fluorescent probes can be broadly classified into those based on fluorescent proteins such as GFP and those based on synthetic organic small molecules. We have actively developed the latter organic small molecules and achieved novel in vivo cancer imaging.

2 Features of Cancer Imaging with Activatable Fluorescent Probes

Organic small molecule fluorescent probes are nonfluorescent molecules that emit strong fluorescence when they react with or bind to target molecules. In this paper, we refer to such fluorescent probes as activatable fluorescent probes, which are characterized by their ability to observe target molecules in real time without the need to remove unreacted probes. Figure 30.1 shows the contrast between the always on-type probe, which has fluorescence from the beginning and requires washing of unreacted probes to image the cancer site, and the activatable-type probe. The advantage of the latter probe as a cancer imaging probe is clear. In addition, these probes can be rapidly introduced into all cells in a controlled concentration by simply administering them to the cancer site environment, which enables selective and sensitive imaging of target molecules. We have established a number of logical design methods for small organic activatable fluorescent probes and have successfully developed probes for various target molecules. In the following, we will focus on the fluorescent probes for visualization of biomarker enzymatic activities that characterize cancer cells and describe their application to rapid intraoperative cancer imaging.

Y. Urano (✉)
Graduate School of Pharmaceutical Sciences, Faculty of Pharmaceutical Sciences, The University of Tokyo, Tokyo, Japan
e-mail: uranokun@m.u-tokyo.ac.jp

T. Ishizawa (ed.), *Fluorescence-Guided Surgery*, https://doi.org/10.1007/978-981-19-7372-7_30

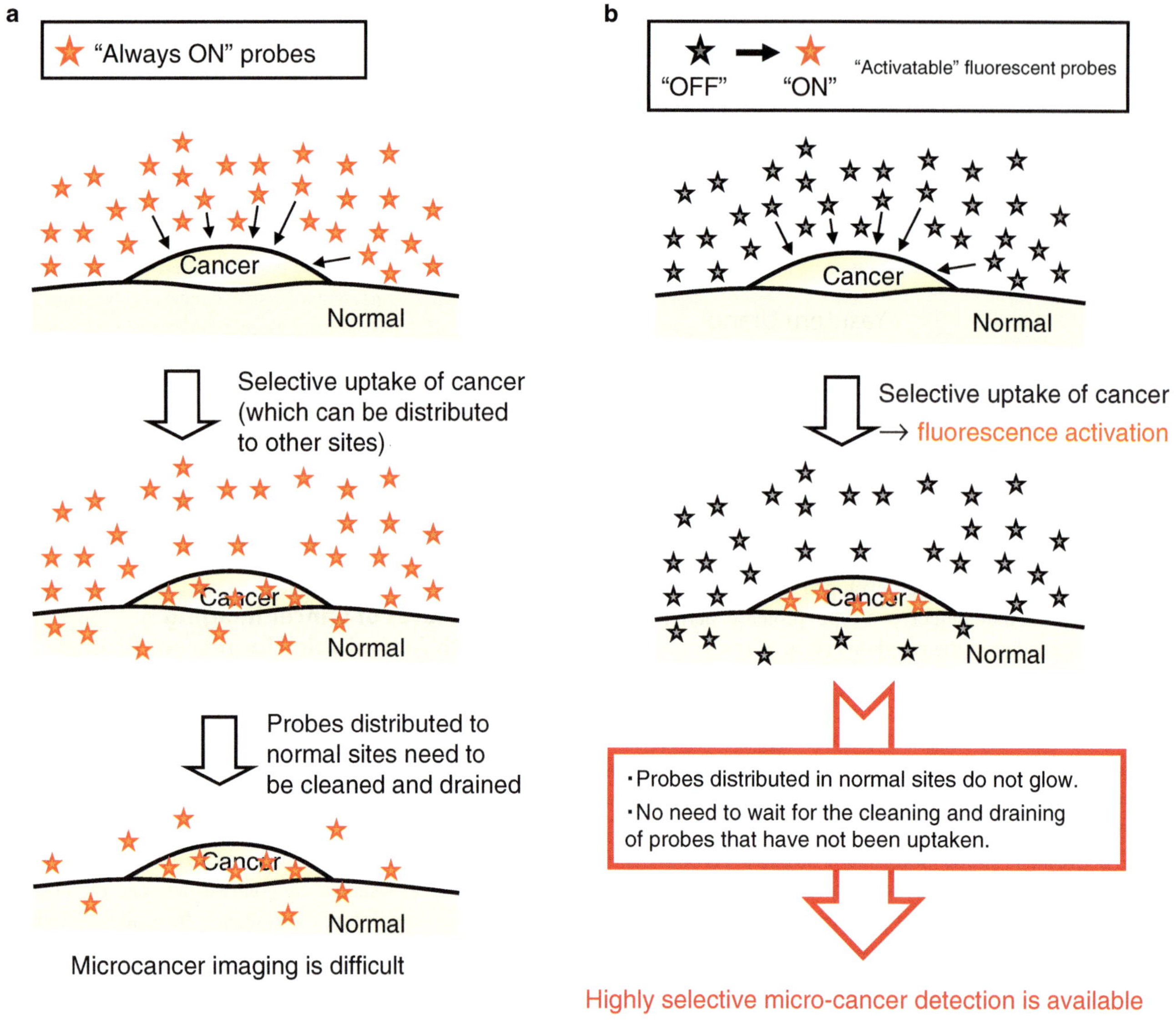

Fig. 30.1 Conventional cancer imaging and our newly developed highly selective fluorescence imaging technique. (**a**) Cancer imaging methods using "always on" probes such as PET and MRI. (**b**) Microcancer imaging method based on the precise design of "activatable" fluorescent probes established in this study

3 Development of Aminopeptidase Fluorescent Probes for Rapid Visualization of Cancer

If we consider cancer imaging as a practical surgical guidance using a biomarker characterizing cancer cells, one possible method is to administer a probe before surgery and perform surgical treatments at a time when the probe accumulated only in cancer tissues. The almost only clinical example of this method is brain tumor imaging using 5-aminolevulinic acid (5-ALA). 5-ALA itself is an amino acid that exists naturally in our body, but it has no aromatic rings, no color, and no fluorescence. In vivo, 5-ALA is converted to protoporphyrin IX, a precursor of hemoglobin, by a multistep biochemical reaction. 5ALA tends to accumulate in some cancer cells, and visualization of cancer sites has been achieved. Thus, 5-ALA is an example of an activatable fluorescent probe. However, its application is limited in terms of cancer types; it takes several hours for cancer to fluoresce after administration, so a relatively large amount (about 1 gram) of 5-ALA needs to be administered systemically, and there are problems with side effects such as photosensitivity. The amount of protoporphyrin produced decreases with prolonged surgery, and it is also susceptible to photobleaching.

Therefore, we have started to develop a completely new cancer imaging technique, in which cancer cells start to fluoresce immediately after a few minutes of local application instead of systemic administration. Specifically, we focused on proteolytic enzymes (peptidases), which are reported to be up-regulated in cancer cells, in order to target enzymes with sufficiently fast reaction rates for imaging. When we started our research, several activatable fluorescent probes had already been developed to detect endo-type peptidase activity, which recognizes the middle of a long peptide chain and hydrolyzes it. However, because the activity of endopeptidases is generally not high, the fluorescence enhancement rate at the cancer site after administration of fluorescent probes is slow, and there have been no examples of cancer imaging achieved by local dissemination. Then, we aimed to develop a fluorescent probe that detects the activity of exo-type peptidase, which has a faster reaction rate. Exo-type peptidases include aminopeptidases, which recognize and hydrolyze amino-terminal amino acids (Fig. 30.2a), and carboxypeptidases, which recognize and hydrolyze carboxy-terminal amino acids (Fig. 30.2b). Since there was no method for designing a highly sensitive fluorescent probe that functions in the visible light region for either of these peptidases, it was essential to establish original methods. In the following, we briefly introduce one of our molecular design methods, which is an aminopeptidase fluorescent probe design method based on intramolecular spirocyclization.

Rhodamine green is a conventional fluorophore that always shows strong fluorescence signals independent of pH. In contrast, hydroxymethyl rhodamine green (HMRG), in which the carboxy group is replaced by a hydroxymethyl group, is colorless and nonfluorescent in alkaline environments stronger than pH 9 due to intramolecular spirocyclization (Fig. 30.3, upper panel). In alkaline environments stronger than pH 9, HMRG becomes colorless and nonfluorescent (Fig. 30.3, middle panel). Furthermore, AcHMRG, in which one amino group of HMRG is amidated by acetic acid, gives preference to the spirocyclic form even in more acidic environments, emitting color and fluorescence only in acidic environments below pH 6 and becoming colorless and nonfluorescent in neutral pH environments at pH 7.4 (lower part of Fig. 30.3). If the amide group of this colorless, nonfluorescent AcHMRG is cleaved by a hydrolytic enzyme; the change indicated by the orange box in the figure occurs under neutral pH conditions to produce HMRG, which emits strong fluorescence. Therefore, by replacing the acetyl group of AcHMRG with various amino acids, it became possible to design and develop fluorescent probes to detect various aminopeptidase activities (bottom row of Fig. 30.3) [1, 2].

First, we designed and developed a fluorescent probe, gGlu-HMRG, to detect the activity of γ-glutamyltranspeptidase (GGT), which recognizes and hydrolyzes the amino-terminal γ-glutamyl group and whose activity has been reported to be upregulated in various types of cancer cells (Fig. 30.4a) (1,2) The imaging mechanism of cancer cells by gGlu-HMRG is shown in Fig. 30.4b. gGlu-HMRG, which is nearly nonfluorescent in a neutral pH environment, remains nearly nonfluorescent in a normal cell environment due to its low GGT activity, and thus background fluorescence is extremely low. However, in the presence of cancer cells, HMRG is efficiently hydrolyzed by GGT, which is highly expressed on the surface of cancer cells, and converted to highly fluorescent HMRG. The generated HMRG is immediately taken up into the cancer cells due to its high hydrophobicity, and thus the cancer site emits strong fluorescence.

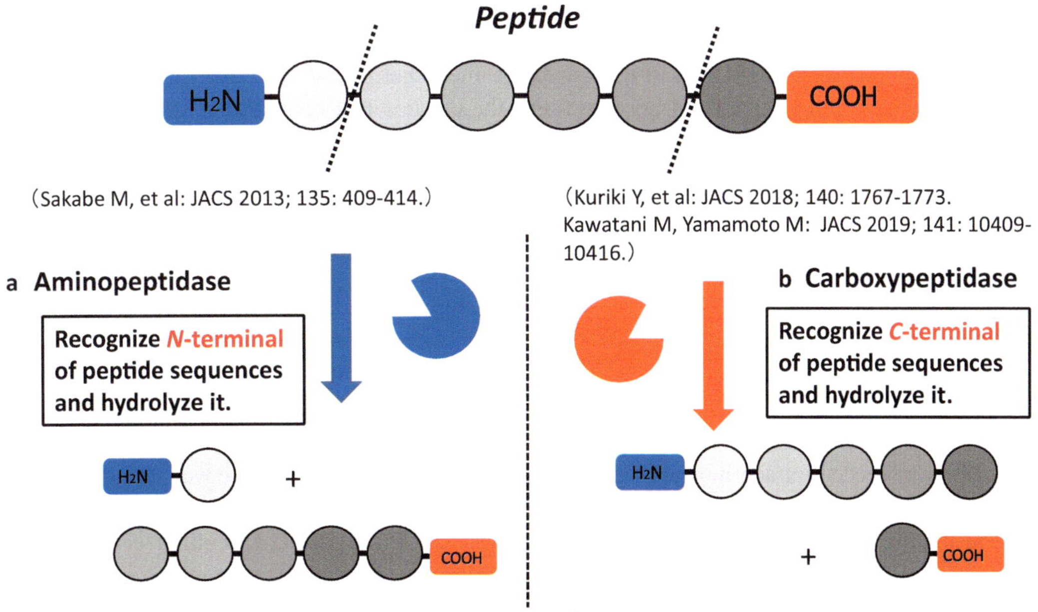

Fig. 30.2 Exo-type peptidase as a target for rapid fluorescence visualization. (**a**) Aminopeptidase and (**b**) carboxypeptidase are present

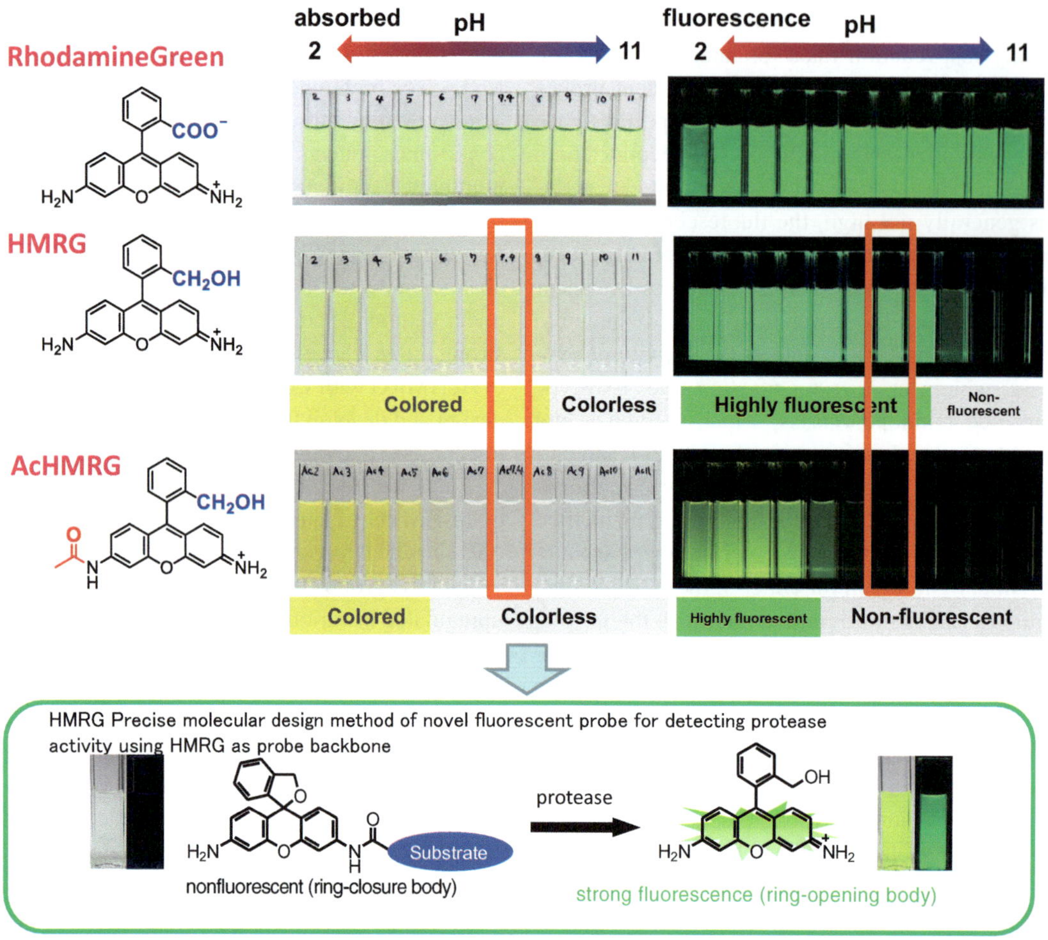

Fig. 30.3 Establishment of precise molecular design method of novel fluorescent probe for detecting protease activity using HMRG as probe backbone

We applied the developed gGlu-HMRG to various types of cultured cancer cells and normal cells (normal human umbilical vein endothelial cells; HUVECs) and compared their enzymatic activities. We found that GGT activity was higher in ovarian, lung, liver, and cholangiocarcinoma cells and lower in HUVECs (Fig. 30.4c). On the other hand, there were some cancer cells whose GGT activity was not so high. The number of cancer cells with significantly increased GGT activity was about 60% of the total. Although not all cancer cells can be detected by gGlu-HMRG, we believed that our technique would have value in detecting small cancer cells during surgery and performed in vivo fluorescence imaging using a mouse model of cancer.

We prepared a cancer model mouse in which various ovarian cancer cells were seeded intraperitoneally and intraperitoneally administered a PBS solution of gGlu-HMRG to the mouse. As a result, even after 5 min of probe administration, the cancer site emitted extremely strong fluorescence, and even small cancers of less than 1 mm were clearly detected (Fig. 30.4d, e) [1]. The photographs in Fig. 30.4e were not taken with any special equipment but through a 515-nm long-pass filter with a commercially available digital camera that I personally own. This is due to the turnover of the enzyme that converts the probe into highly fluorescent products one after another, which could not be achieved by any other fluorescence imaging techniques.

Next, as a model experiment of cancer site detection and treatment under endoscopy, we made a small hole in the peritoneum of anesthetized cancer model mice, inserted a fluorescent endoscope (in collaboration with Olympus Corporation) through the hole and locally sprayed a probe with an applier through the forceps hole to examine whether it was possible to detect small cancer sites. As a result, it was found that the cancerous area gradually started to glow immediately after the probe was sprayed and that it was possible to clearly visualize the microcancerous area, which was not distinguishable with a conventional white-light endoscope, by fluorescence within only several tens of seconds to several minutes after the probe was sprayed (Fig. 30.4f) [1].

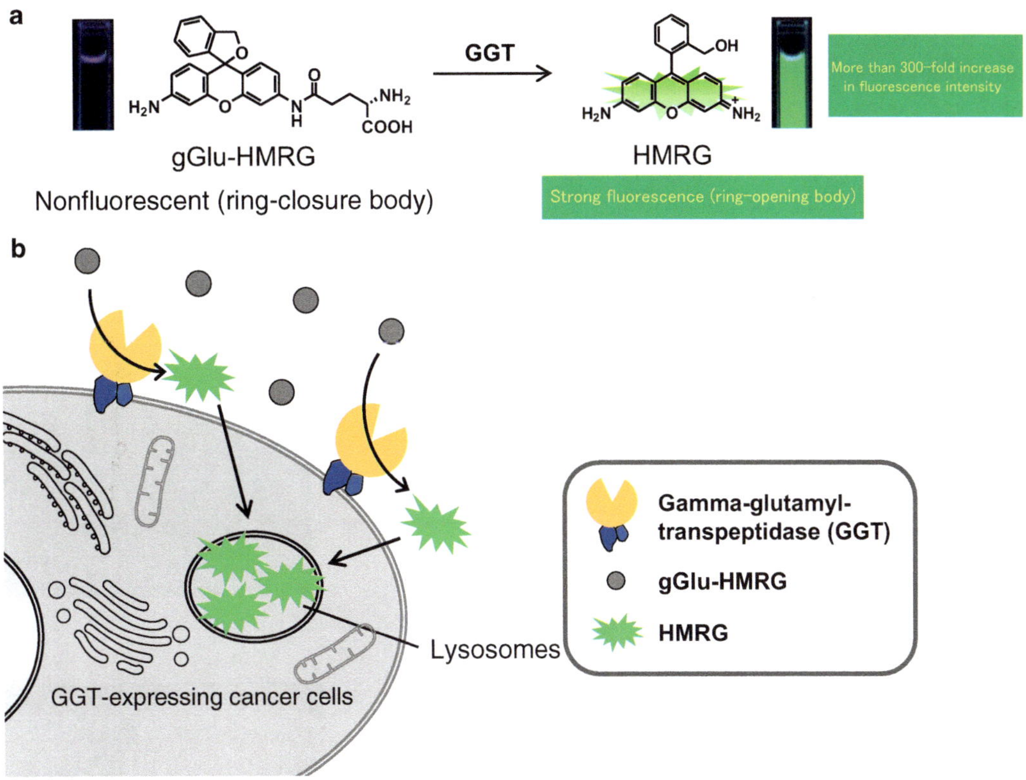

Fig. 30.4 Development of a group of fluorescent probes for detecting aminopeptidase activity, enabling rapid detection of small cancers after their local spraying. (**a**) Development of a novel fluorescent probe for detection of GGT activity, gGlu-HMRG. (**b**) Cancer cell imaging mechanism by GGT activity detection fluorescent probe gGlu-HMRG. (**c**) Comparison of GGT activity of various cancer cells and normal cells by gGlu-HMRG. (**d**) Intraperitoneal administration of gGlu-HMRG for fluorescence imaging of mesenteric dissemination (10 min after probe administration). (**e**) Fluorescence imaging of peritoneal dissemination using gGlu-HMRG (imaging with a commercially available digital camera). (**f**) Detection of tiny cancerous regions by local administration of gGlu-HMRG under fluorescence endoscopy (left, normal white light image; right, fluorescence image [after 5 min of probe spraying]). (**g**) Rapid visualization of microcancer by gGlu-HMRG spraying on a human breast cancer partial resection sample and EP-HMRG spraying on a human esophageal cancer sample. The microcancer sites were visualized within 1–3 min after the probe administration, and the fluorescence became strong enough to be seen by the surgeon's eyes within a few minutes. The fluorescing areas and postoperative histopathological staining images (H&E) were in good agreement. Top, noninvasive ductal carcinoma of the breast; middle, invasive ductal carcinoma of the breast; bottom, esophageal carcinoma

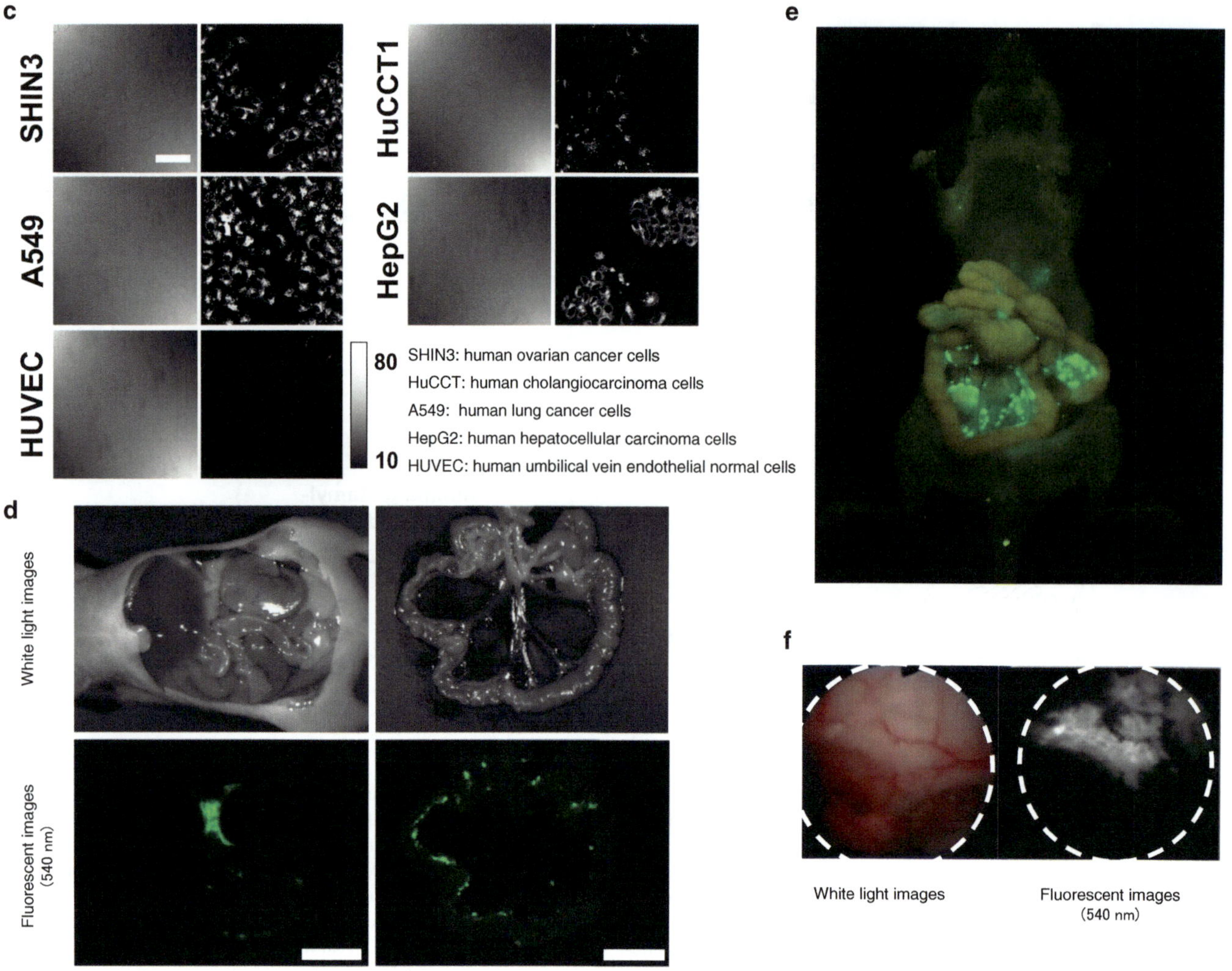

Fig. 30.4 (continued)

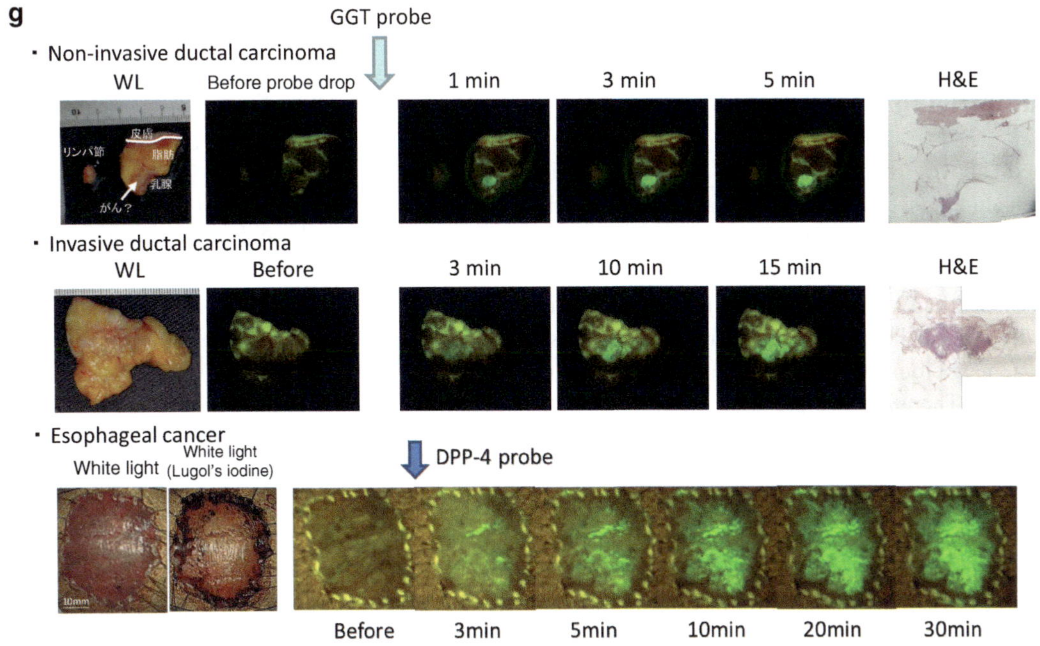

Fig. 30.4 (continued)

4 Rapid Cancer Visualization in Fresh Human Clinical Specimens Using Fluorescent Probes for Detecting GGT Activity

The technology for visualizing cancer sites in a short time achieved above is a highly practical technology that can be inserted into the timeframe of cancer surgery and endoscopic resection and is therefore strongly expected to be applied to actual clinical practice. On the other hand, cancer is a highly heterogeneous disease, and it cannot be said to be effective as a new cancer medical technology without verifying whether gGlu-HMRG can detect cancer in actual patients.

As described previously, the efficacy of conventional cancer diagnostic agents must be verified by introducing them into the patient's body by injection, and this is not easy to achieve. However, the gGlu-HMRG-based rapid cancer imaging is the first method that can be validated ex vivo using human clinical specimens. Therefore, this technique has attracted the interest of many clinical surgeons, and many validation experiments using freshly removed human clinical cancer specimens are now underway in Japan and overseas. The following is an example of verification in breast cancer.

In collaboration with Prof. Koji Mimori of Kyushu University Beppu Hospital and Dr. Hiroaki Ageo of Ageo Breast Surgery, we performed imaging of clinical breast cancer specimens removed from actual human patients. In a partial mastectomy, which accounts for about half of all surgeries for breast cancer, it is essential to diagnose whether cancer cells are contained in the margins of the resected specimen during surgery in order to confirm curability. However, it is impossible to perform a pathological examination of the entire resection surface (only a few representative areas are usually examined), so the possibility of missing cancer has been pointed out, which may lead to local recurrence after surgery. In this study, we examined the usefulness of gGlu-HMRG by topical administration on actual fresh clinical specimens removed during breast cancer surgery. As a result, we found that gGlu-HMRG can detect various types of mammary tumors, such as noninvasive ductal carcinoma (upper part of Fig. 30.4g) and invasive ductal carcinoma (middle part of Fig. 30.4g), in a few minutes. In a validation study using more than 100 clinical specimens, it was found that even small cancers of 1 mm or less could be detected with high sensitivity within 5 min after application of the probe and that detection with unusually high sensitivity and specificity (>90%) was achieved even for cancer, which is potentially a highly heterogeneous disease [3]. Since this technique can image the entire resection margin, it is expected to dramatically reduce the frequency of local recurrence by enabling clear detection of microcancerous lesions left at the margin that are not detectable visually. Furthermore, this method may also be effective for detecting lymph node metastasis in breast cancer. In fact, it was shown that the presence or absence of lymph node metastasis can be determined by rapid fluorescence imaging using excised lymph nodes during breast surgery [4].

5 Development of a Rapid Imaging Probe for Esophageal Cancer by Creating a Fluorescent Probe Library and Applying it to Clinical Specimens

As mentioned above, the heterogeneity of cancers is extremely high, and in fact, gGlu-HMRG does not react with all cancer types. Esophageal cancer is one such example, and it was difficult to visualize the fluorescence rapidly with high sensitivity and specificity by localized gGlu-HMRG distribution. In order to visualize esophageal cancer, it is necessary to find a new biomarker enzyme that is specific to esophageal cancer. Therefore, we prepared a library of 400 probes for the detection of aminopeptidase and dipeptidyl peptidase enzyme activities by introducing one or two amino acids into HMRG as a fluorescent probe matrix. Next, in collaboration with Prof. Yasuyuki Seto and his colleagues at the Department of Gastroesophageal Surgery, University of Tokyo Hospital, we applied this library to fresh clinical specimens of esophageal cancer and began searching for fluorescent probes with high esophageal cancer specificity. As a result, it was found that the fluorescence intensity of several HMRG probes increased more in the tumor area than in the non-tumor area, and among these probes, GP-HMRG showed the greatest difference in fluorescence intensity. It was predicted from the amino acid sequence that dipeptidylpeptidase-4 (DPP-4) is the enzyme responsible for the hydrolysis of this probe, and in fact, the coadministration of the DPP-4 inhibitor suppressed the increase in fluorescence. In addition, immunostaining of fresh clinical specimens revealed that cancer cells in the luminal surface of the esophagus showed strong DPP-4 expression.

DPP-4 is known to recognize substrates whose amino-terminal second residue is Pro or Ala. Therefore, we applied various candidate HMRG probes containing this sequence to fresh clinical specimens to identify probes that show maximum fluorescence enhancement in cancer sites and suppress fluorescence in normal sites. As a result, EP-HMRG was found to give the best results. We applied EP-HMRG to fresh endoscopically resected ESD specimens of esophageal cancer and found that the boundary between tumor and non-tumor areas could be clearly visualized within a few minutes after the probe was applied, as shown in Fig. 30.4g below. We applied this method to more than 70 fresh clinical specimens and analyzed the fluorescence behavior of tumor and non-tumor areas. As a result, the sensitivity, specificity, and positive detection rate were 96.9%, 85.7%, and 90.5%, respectively, indicating that this method has sufficient performance as a rapid intraoperative imaging method [5]. We are currently working on the development of novel fluorescent probes for various types of cancers using a similar approach and in collaboration with many surgeons.

Point

We have developed about 400 kinds of fluorescent probes for detecting aminopeptidase activity.

The GGT-detecting fluorescent probe enables rapid intraoperative identification of tiny (< 1 mm) breast cancer foci on surgical margins of resected specimens.

We found that DPP4 activity was enhanced in esophageal cancer and achieved rapid cancer imaging by topical administration of the noel fluorophores.

6 Prostate Cancer Imaging by Developing a Fluorescent Probe for Detecting Carboxypeptidase Activity

Exo-type peptidases include carboxypeptidases (CPs) that hydrolyze C-terminal amino acids (Fig. 30.2b), in addition to the aminopeptidases described in the previous section, and their activity has been reported to be enhanced in certain cancers. However, there are no practical activatable fluorescent probes for detecting CP activity.

In this context, we have recently succeeded in establishing several methods for designing fluorescent probes for detecting CP activity [6, 7] Here, we introduce the development of PSMA (prostate-specific membrane antigen, which has CP activity to recognize and hydrolyze carboxy-terminal glutamate) probes, which are attracting attention as prostate cancer biomarkers. First of all, we synthesized a variety of candidate substrates to determine the substrate structure of glutamate that PSMA recognizes and hydrolyzes. In this study, we found that when the substrate amino acid of phenyl azoformyl (AF) derivatives, which have been reported as substrates for carboxypeptidase A and carboxypeptidase B, was converted to glutamic acid (Ph-AF-Glu), it could be a good substrate for PSMA (Fig. 30.5a) [7]. The AF chain has a property of degradation through spontaneous decarboxylation and denitrogenation following hydrolysis of the substrate amino acids, evoking big changes in electrodensity before and after the reaction with the enzyme. In order to visualize this change, we designed a molecule using photoinduced electron transfer (PeT) (Fig. 30.5b). We synthesized a fluorescein derivative with glutamic acid introduced via the AF group from the benzene ring moiety of the fluorescein derivative and found that the fluorescence was quenched by the PeT from the xanthene ring moiety to the benzene ring moiety, but a fluorescent dye with strong fluorescence was generated by the reaction with PSMA, and the fluorescence increased significantly (Fig. 30.6a). In addition, we developed a membrane-permeable probe by converting the fluorescein moiety to the TokyoGreen skeleton (Fig. 30.6b) and applied it to PSMA-expressing cancer cells, showing that it enables live staining of PSMA-expressing cells. Finally,

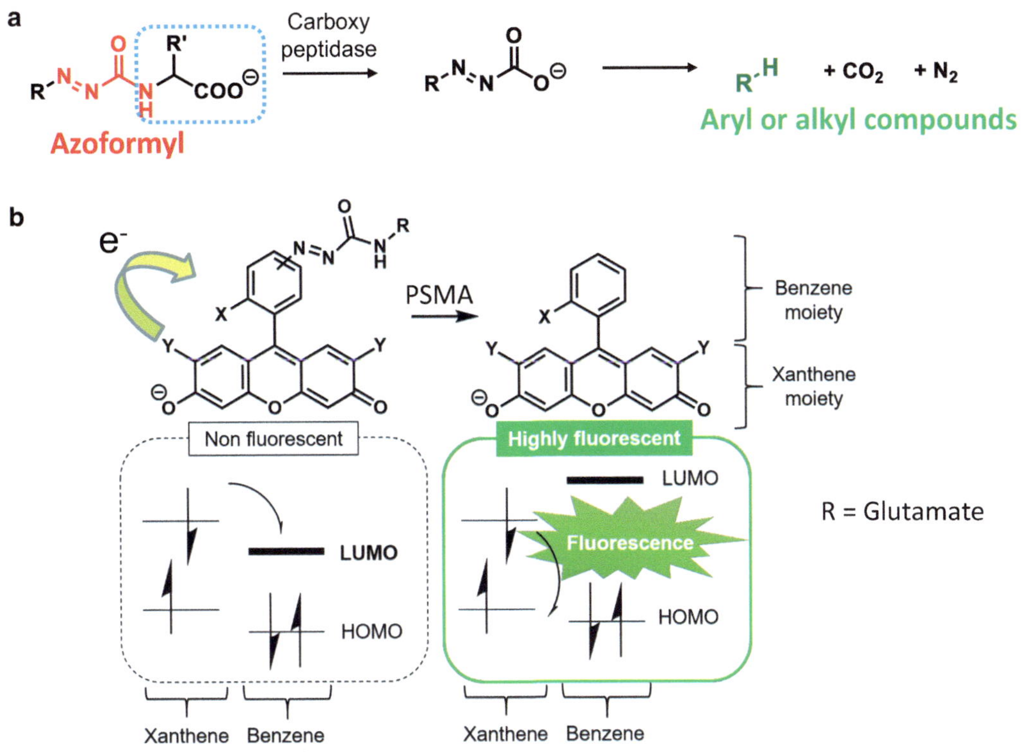

Fig. 30.5 Development of activatable PSMA activity detection fluorescent probe based on photoinduced electron transfer. (**a**) Hydrolysis of azoformyl derivatives by PSMA and subsequent decarboxylation and denitration reactions to form allyl and alkyl compounds. (**b**) Principle of operation of an activatable PSMA activity-detecting fluorescent probe based on photoinduced electron transfer

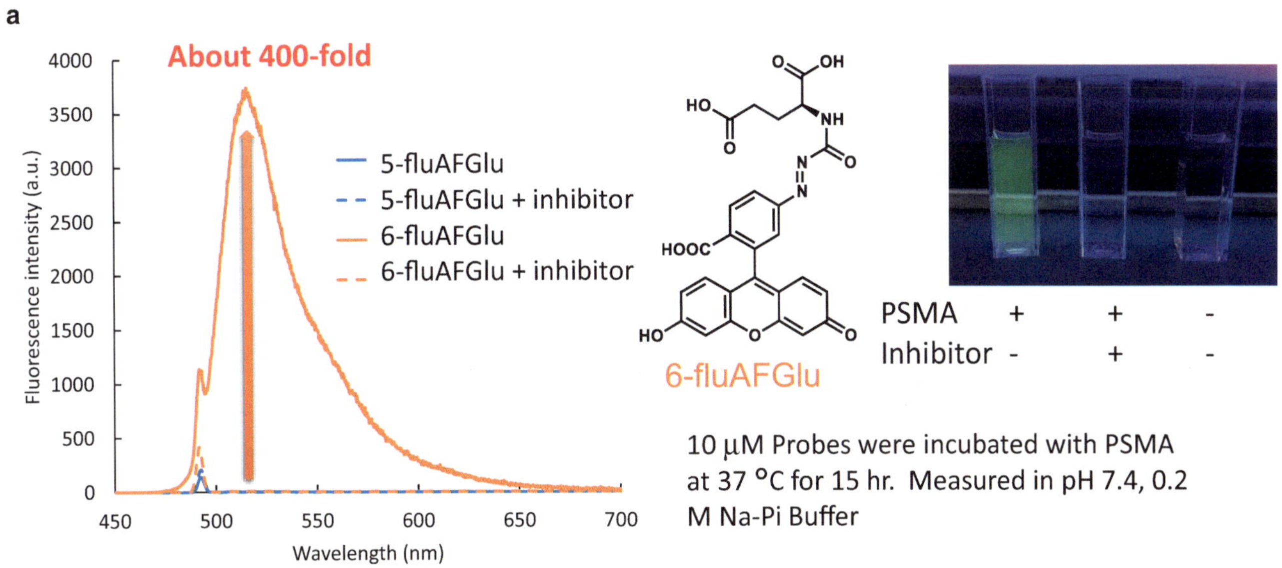

Fig. 30.6 Fluorescent detection of prostate cancer site in a fresh human specimen by a topical spray of the developed PSMA probe. (**a**) Reactivity of the successfully developed fluorescent probe for PSMA activity detection with PSMA. (**b**) Microprostate cancer imaging in a fresh specimen by dropping a transmembrane PSMA activity-detecting fluorescent probe

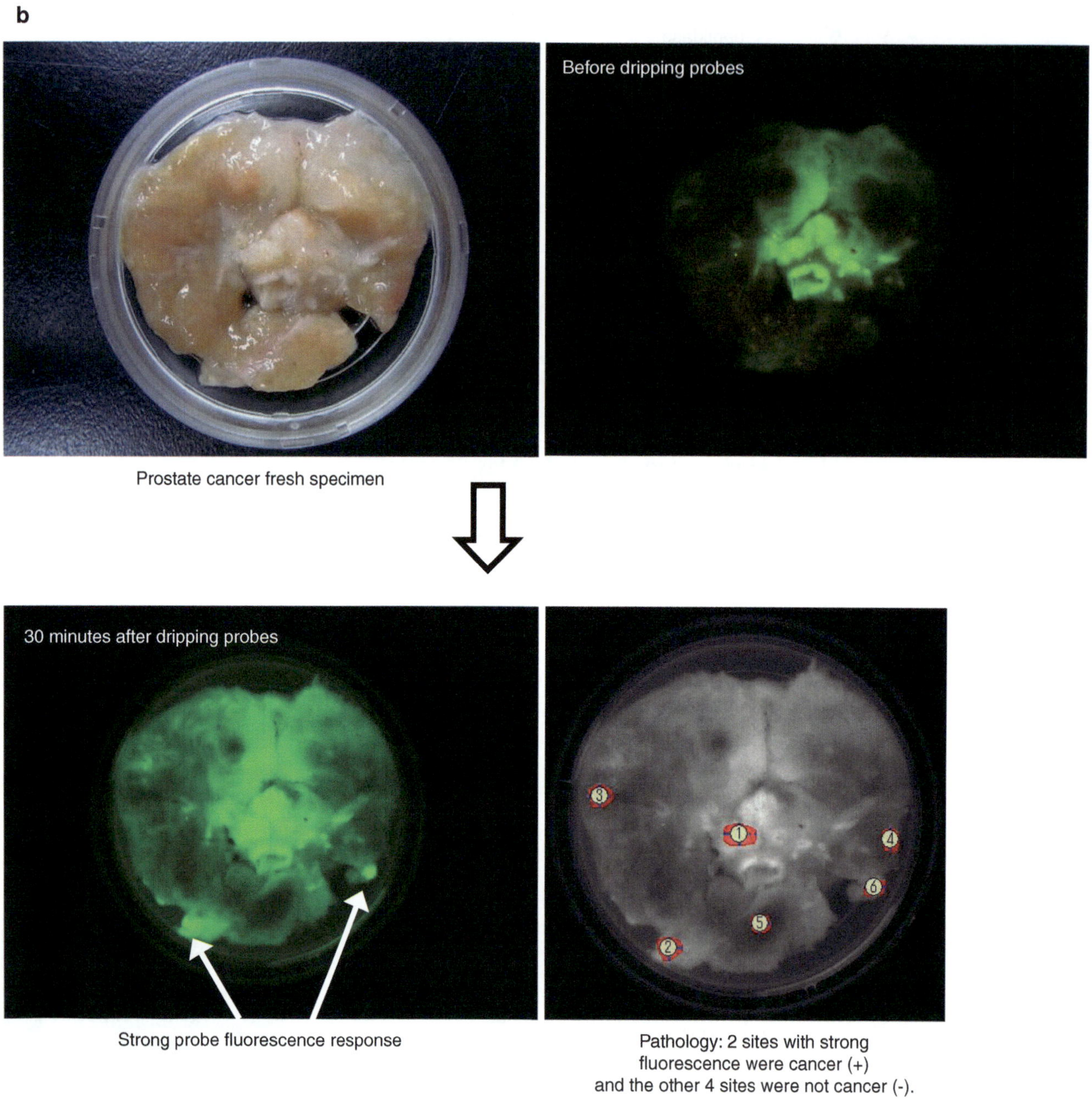

Fig. 30.6 (continued)

when the probe was applied to resected specimens from prostate cancer patients, it was found that even small cancers of several mm in diameter could be detected, and the world's first rapid fluorescence visualization of microscopic prostate cancer sites was achieved by targeting the carboxypeptidase activity of PSMA [7].

Point

- We have succeeded in developing the world's first fluorescent probe for detecting PSMA activity.
- The use of this probe is expected to enable intraoperative detection of prostate cancer, which is difficult to detect by conventional imaging modalities.

7 Future Perspectives

In this paper, we have introduced a number of examples of the development of activatable-type rapid fluorescent probes whose fluorescence characteristics are significantly altered based on cancer-specific enzymatic activities, utilizing our original fluorescent probe precision design methods. Previous studies have indicated that novel fluorophores designed by our technologies are sufficiently effective for cancer localization in actual patients beyond the level of experiments using animal models, and are practical enough to be used in clinical settings. Following expectations from many surgeons and endoscopists, we have started preclinical trials of several probes in cooperation with a domestic chemical venture, an in vitro fluorescent imager manufacturer, and an endoscope manufacturer, with investment from venture capitalists. Among them, fluorescent probes for ex vivo breast cancer imaging and endoscopic identification of esophageal cancer have recently passed nonclinical studies such as toxicity tests after face-to-face advice with the Japanese FDA, and the first-in-human study is about to start after the review by the hospital ethics committee. I am very much looking forward to the day when surgeons will be able to clearly determine the location of cancer to be removed with the use of novel fluorescence imaging techniques.

References

1. Urano Y, Sakabe M, Kosaka N, et al. Rapid cancer detection by topically spraying a γ-glutamyltranspeptidase-activated fluorescent probe. Sci Transl Med. 2011;3:110ra119.
2. Sakabe M, Asanuma D, Kamiya M, et al. Rational design of highly sensitive fluorescence probes for protease and glycosidase based on precisely. J Am Chem Soc. 2013;135:409–14.
3. Ueo H, Shinden Y, Tobo T, et al. Rapid intraoperative visualization of breast lesions with γ -glutamyl hydroxymethyl rhodamine green. Sci Rep. 2015;5:12080.
4. Shinden Y, Ueo H, Tobo T, et al. Rapid diagnosis of lymph node metastasis in breast cancer using a new fluorescent method with γ -glutamyl hydroxymethyl rhodamine green. Sci Rep. 2016;6:27525.
5. Onoyama H, Kamiya M, Kuriki Y, et al. Rapid and sensitive detection of early esophageal squamous cell carcinoma with fluorescence probe targeting Dipeptidylpeptidase IV. Sci Rep. 2016;6:26399.
6. Kuriki Y, Kamiya M, Kubo H, et al. Establishment of a molecular design strategy to obtain activatable fluorescent probes for carboxypeptidases. J Am Chem Soc. 2018;140:1767–73.
7. Kawatani M, Yamamoto K, Yamada D, et al. Fluorescence detection of prostate cancer by an activatable fluorescence probe for PSMA carboxypeptidase activity. J Am Chem Soc. 2019;141:10409–16.

Development of a New Imaging System

31

Satoru Seo and Etsuro Hatano

Summary

- Although preoperative simulation has become essential for hepatectomy, it cannot follow intraoperative movement and deformation of the liver. That's why intraoperative navigation has to be developed.
- ICG fluorescence imaging is useful for real-time visualization of hepatic segments and tumor locations, but it is not practical to use conventional handheld devices throughout surgical procedures.
- We developed the Medical Imaging Projection System (MIPS) using an industry-academia collaboration framework, which projects ICG fluorescence images directly onto the patient's organs by applying the projection mapping technology that has been used in the field of entertainment.
- The system has recently been launched in Japan with pharmaceutical approval.

1 Introduction

In recent years, the development of intraoperative navigation using fluorescent dyes has progressed. Among the intraoperative navigation techniques, fluorescence imaging using ICG and a near-infrared camera system has attracted particular attention. This technique is based on a mechanism in which ICG bound to plasma proteins emits fluorescence signals in the near-infrared range, which is outside of the absorption wavelengths of hemoglobin and water [1]. In the field of liver surgery, it has been reported to be useful for the visualization of hepatic segments [2], liver tumors [3], and bile ducts [4].

Since the development of near-infrared camera systems for open surgery, ICG fluorescence imaging has also been incorporated into imaging systems for laparoscopic or robot-assisted surgery. In the setting of laparoscopic or robot-assisted surgery where surgeons are always watching a monitor, it has become possible to continue surgical procedures while checking ICG fluorescence images overlayed on the full-color images in real time. In contrast, the use of a hand-held camera during open surgery has other problems: frequent eye movement between the operative field and the monitor is required to see the fluorescence images displayed on the monitor, the camera needs to be held by a surgeon, the image can be blurred by the movement of the hand-held device, and the surgical lights need to be turned off during fluorescence imaging to remove the light interference (Fig. 31.1). Because of these problems, continuous real-time navigation using ICG fluorescence imaging is difficult to perform in the setting of open surgery, and in reality, only intermittent navigation is usually applied.

Supplementary Information The online version contains supplementary material available at https://doi.org/10.1007/978-981-19-7372-7_31.

S. Seo (✉) · E. Hatano
Department of Surgery, Kochi Medical School, Nankoku, Kochi, Japan
e-mail: rutosa@kuhp.kyoto-u.ac.jp

T. Ishizawa (ed.), *Fluorescence-Guided Surgery*, https://doi.org/10.1007/978-981-19-7372-7_31

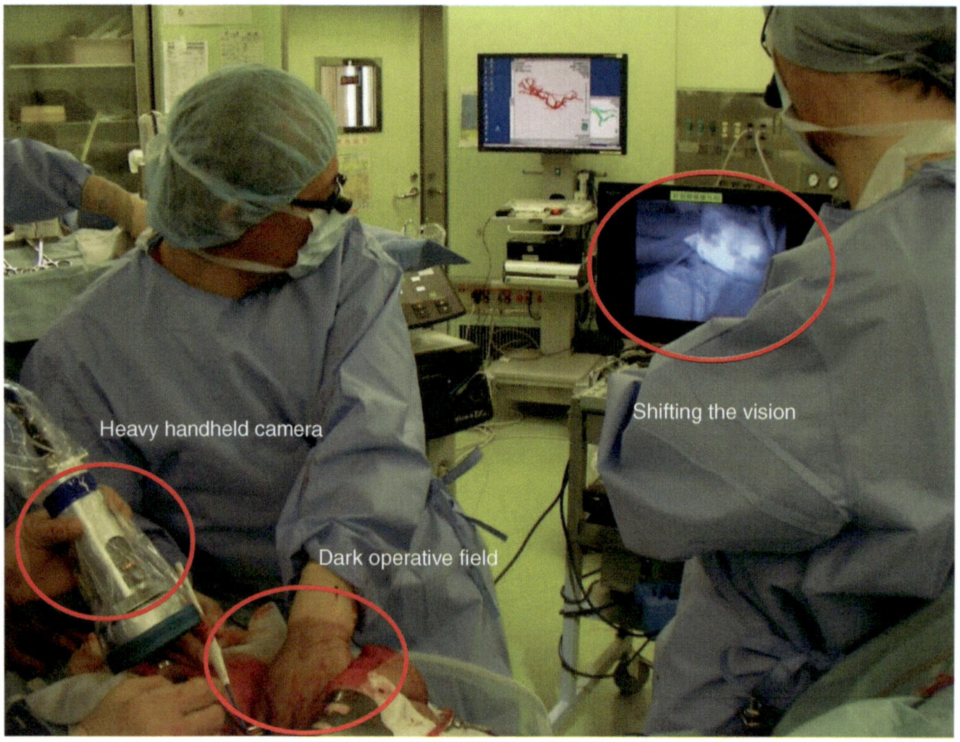

Fig. 31.1 Problems of the conventional fluorescence imaging

2 Motivation to Develop a Novel ICG Fluorescence Imaging System for Open Abdominal Surgery

Using software such as Synapse Vincent (Fujifilm), the shape of the liver, vasculature, and tumor are reconstructed in 3D based on preoperative CT images. In clinical settings, a preoperative simulation is becoming an indispensable tool for determining the resection line with consideration of the balance between curability and safety. However, simulation images cannot reflect real-time information because they do not correspond to intraoperative positional movement and organ deformation. Technological innovation in preoperative simulation images has enabled detailed surgical procedure planning, but in order to complete surgery as simulated, it is necessary to establish real-time navigation that follows intraoperative positional movement and deformation.

The technology we focused on is the projection mapping technique, which has been widely used in the field of entertainment. By applying this technology, we have started to develop a system that can project ICG fluorescence images directly onto a patient's organs during surgery.

3 Launch of the MIPS Project

We started joint development with Panasonic Corporation as an industry-academia collaboration project and first made a prototype of the principle (Fig. 31.2). In experiments using this prototype, we confirmed that ICG injected into the pig's liver could be projection-mapped without any positional shift, and we then developed a prototype for clinical use (Fig. 31.3). We named the prototype the Medical Imaging Projection System (MIPS), and the project was named the MIPS Project. In animal experiments using the prototype, we confirmed that the system could be used in vivo without misalignment and could follow movement. At the same time, we identified the problem that the excitation light illumination was close to the surgical field and interfered with surgical operations. Then, we improved the design to incorporate the projector and excitation light illumination into the imaging head (Fig. 31.4a). We have used this updated system in clinical trials for breast cancer and liver cancer surgery, achieving the <2 mm misalignment and the <0.2 s time difference between projections. In addition, by using an algorithm

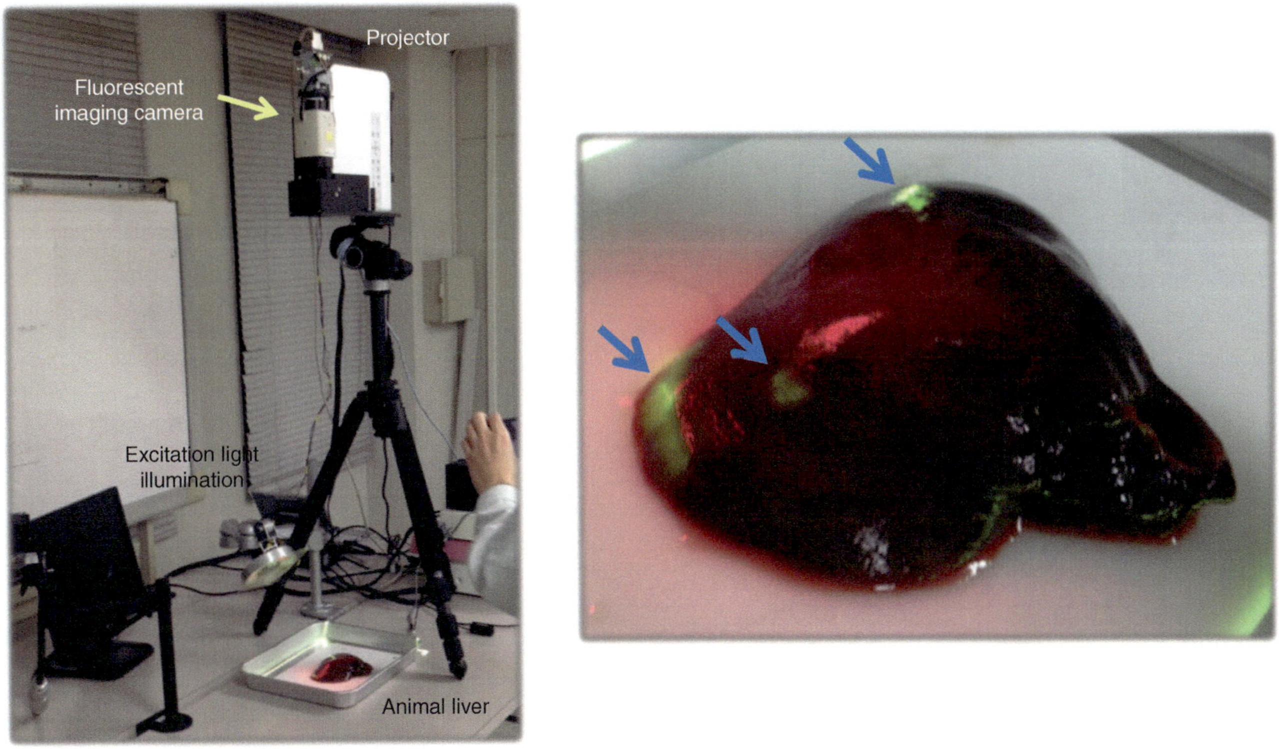

Fig. 31.2 Principle of the prototype imaging system

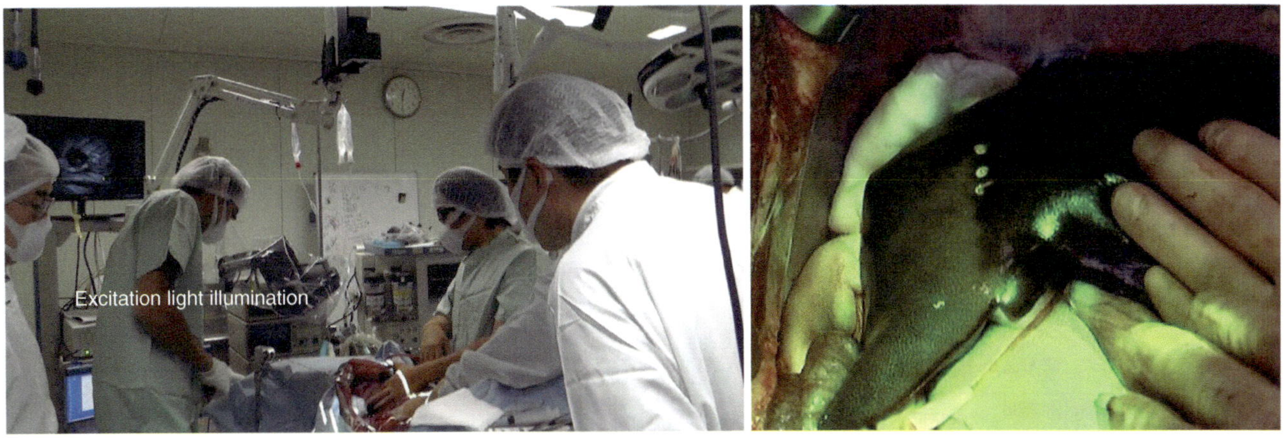

Fig. 31.3 Animal experiments using the prototype imaging system

that converts fluorescent areas to specific colors and non-fluorescent areas to white, we were able to provide a sufficiently bright surgical field even in the circumstance where the surgical lights were turned off. With these improvements, we were able to overcome all of the aforementioned problems (Fig. 31.5).

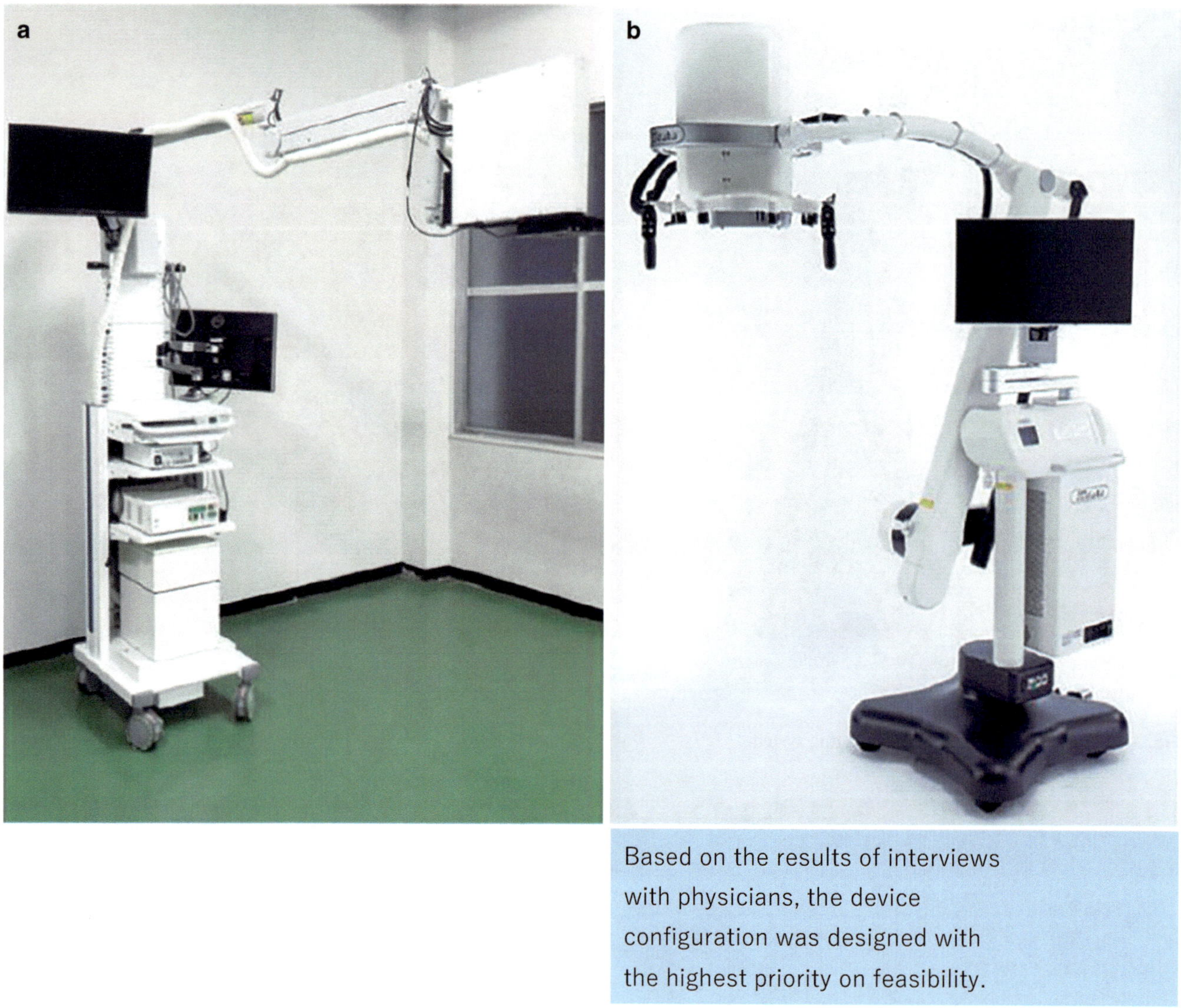

Fig. 31.4 Updates of the MIPS system. (**a**) First machine. (**b**) Final machine. Based on the results of interviews with physicians, the device configuration was designed with the highest priority on feasibility

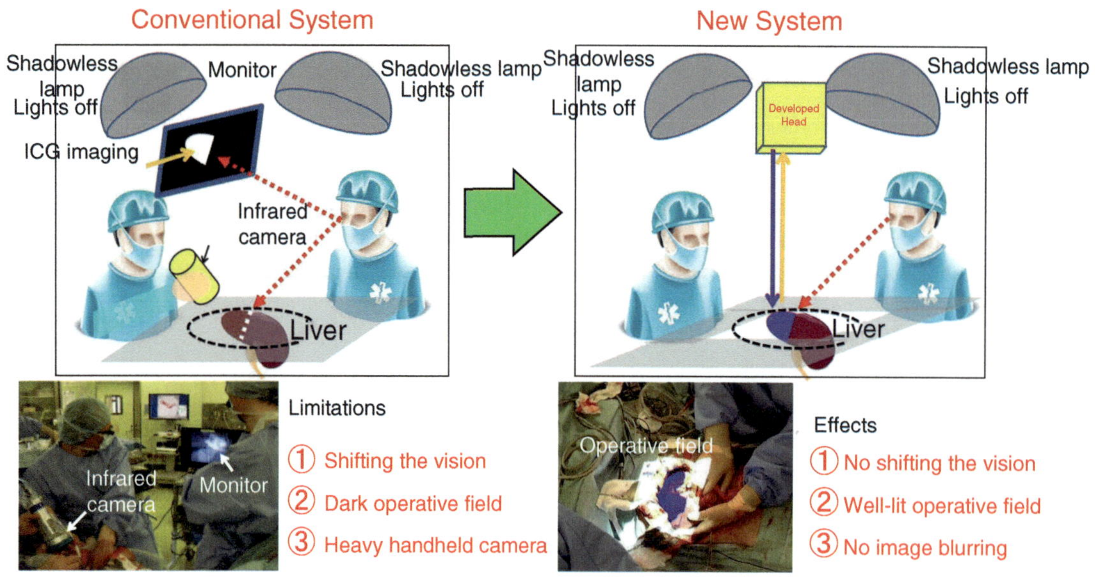

Fig. 31.5 Advantages of MIPS

4 Challenges in MIPS Development

One of the challenges the project faced at the time of its inception was that Japanese insurance coverage for the intraoperative use of ICG was limited to sentinel node identification (breast cancer and malignant melanoma). Although ICG was subsequently approved for use in the "evaluation of blood flow in blood vessels and tissues," the companies acknowledged concerns about potential complications and liability for inaccurate MIPS navigation. This concern was also an issue with ICG fluorescence imaging itself. These were discussed within the project and resolved by emphasizing at subsequent conferences and in publicity that MIPS is a medical device to be used by specialists to perform more precise surgery. In order to perform accurate ICG fluorescence imaging during surgery, it was necessary to adjust the ICG dose and fluorescence gain based on individual patient information. For this reason, we added an image superimposing function to the monitor and improved it so that we can always check whether the settings are appropriate.

Another issue was how to raise the costs for research and development. Fortunately, we were selected for research support by the Japan Agency for Medical Development (AMED), which has offered various advantages to our project. For example, the academia side was able to bear the medical expenses for expensive clinical trials using unapproved medical devices, and we were able to actively present the obtained data at domestic and international conferences. The results of the initial clinical trials were published in the *Annals of Surgery*, and this provided an opportunity to raise awareness of MIPS [5]. The company was not only able to bear the cost of developing and fabricating an expensive prototype, but the fact that the project was selected encouraged the company to obtain internal development funds, which enabled the project to be transferred from the research department to the business department. By holding progress meetings every 2 months for AMED site visits and annual reports, academia and the company were able to work closely together and establish a clear goal and roadmap. This was one of the factors that enabled the project to be successful.

5 Key Factors for Successful Medical-Industrial Collaboration

The key to success in medical-engineering collaboration is the matching of seeds and needs, and in our case, we had two successful matches: the first was between the seeds of Panasonic Corporation's optical technology and image processing technology, and our need to perform ICG fluorescence imaging in open surgery in the same way as in endoscopic surgery. We also matched the need for doctors to move the projection head freely, which arose in the MIPS project, with the seeds of arm control technology that Mitaka Koki had already commercialized (Fig. 31.4b, Movie 31.1). Through this matching, MIPS has grown into a medical device that enables surgeon-friendly and continuous color-coded surgery (Movie 31.2).

6 Unexpected Difficulties and Solutions

Industry-academia framework is a collaboration between a company and an academia for the purpose of R&D of new technology and creation of new business. However, there are often cases where communication fails due to differences in position and culture, even if the members share the same purpose. At Kyoto University, the Integrated Center for Clinical Research (iACT) acted as a bridge between the two parties, supporting the preparation of communication tools (e.g., internet conference, teleconference, opening of a shared server) and the holding of progress confirmation meetings at Kyoto University every 2 months. In addition, we received support for confidentiality agreements, joint research agreements, and advice for regulatory approval and were able to deal with all steps quickly.

Initially we were aiming to obtain Class 2 regulatory approval in March 2019, but the schedule has been significantly delayed. One of the reasons is the recent tightening of the laws and regulations for clinical research in Japan. In our case, it took 8 months for the ethical review after the amendment, and we could not conduct the clinical study using the actual device during that period, which delayed the schedule significantly. Even with these unexpected difficulties, we were able to share information and consider solutions within the project and successfully obtain Class 2 regulatory approval.

7 Future Perspectives

The number of laparoscopic liver resections has been increasing in Japan since the insurance coverage was introduced in 2010. In the course of the MIPS project, we sometimes heard the opinion, "What is the point of developing a medical device for open surgery in the era of endoscopic surgery?". Especially in the field of hepatobiliary surgery, however, there remain many cases that can only be treated by open surgery, such as complicated repeated liver resections and hepatectomy for huge tumors. In addition, the ICG fluorescence method has many problems such as administration route and dosage. Therefore, we believe that it is time to overcome the problems of ICG fluorescence imaging by consolidating the knowledge of both surgical approaches.

We hope to accumulate experience in using MIPS at many institutions in the future to realize safe and precise hepatectomy. In the future, we expect that real-time navigation using MIPS will be expanded to areas other than liver resection, as we have already reported on the usefulness of ICG fluorescence in sentinel lymph node identification in breast surgery [6] and in lung metastasis identification in hepatoblastoma [7].

Point

We have applied projection mapping technology to ICG fluorescence imaging, enabling truly real-time navigation during open surgery.

The development of new medical devices will be accelerated by obtaining research funding and reporting the results of clinical trials.

The key to success in medical-industrial collaboration is the matching of seeds and needs.

Acknowledgement This research was supported by the Industry-Academia Collaboration Medical Innovation Creation Program (ACT-M) of the Japan Agency for Medical Research and Development (AMED).

References

1. Landsman ML, Kwant G, Mook GA, et al. Light-absorbing properties, stability, and spectral stabilization of indocyanine green. J Appl Physiol. 1976;40:575–83.
2. Aoki T, Yasuda D, Shimizu Y, et al. Image-guided liver mapping using fluorescence navigation system with indocyanine green for anatomical hepatic. World J Surg. 2008;32:1763–7.
3. Ishizawa T, Fukushima N, Shibahara J, et al. Real-time identification of liver cancers by using indocyanine green fluorescent imaging. Cancer. 2009;115:2491–504.
4. Ishizawa T, Bandai Y, Ijichi M, et al. Fluorescent cholangiography illuminating the biliary tree during laparoscopic cholecystectomy. Br J Surg. 2010;97:1369–77.
5. Nishino H, Hatano E, Seo S, et al. Real-time navigation for liver surgery using projection mapping with indocyanine green fluorescence: development of the novel Medical Imaging Projection System. Ann Surg. 2018;267:1134–40.
6. Takada M, Takeuchi M, Suzuki E, et al. Real-time navigation system for sentinel lymph node biopsy in breast cancer patients using projection mapping with indocyanine green fluorescence. Breast Cancer. 2018;25:650–5.
7. Chen-Yoshikawa TF, Hatano E, Yoshizawa A, et al. Clinical application of projection mapping technology for surgical resection of lung metastasis. Interact Cardiovasc Thorac Surg. 2017;25:1010–1.

Development of a New Operating Room That Integrates Imaging Information

32

Shunsuke Tsuzuki, Jun Okamoto, Manabu Tamura, Ken Masamune, and Yoshihiro Muragaki

Summary

- The operating room is at the potential risk of incidents and accidents due to the mixture of various old and new medical equipment.
- Attempts are being made to integrate augmented reality (AR) image processing technology with new medical technology to help with surgical planning and strategies.
- To achieve a high curative resection rate and a low complication rate, we are developing a new operating room that integrates imaging information from preoperative simulations and novel image processing techniques to reflect surgical strategies.

1 Introduction

The operating room is a "place" that provides a space to perform sterilized procedures, and specialized equipment as well as basic surgical instruments are brought in according to surgical procedures. A wide variety of surgical devices are usually stocked and used, as different surgeons may use different models even if they have the same function. In addition, new intraoperative diagnostic and therapeutic devices are introduced one after another. Therefore, we are in a situation where new and old devices are mixed together and there is a potential risk of medical incidents and accidents occurring.

S. Tsuzuki (✉)
Department of Neurosurgery, Tokyo Women's Medical University, Tokyo, Japan

Faculty of Advanced Techno Surgery (FATS), Institute of Advanced Biomedical Engineering and Science, Tokyo Women's Medical University, Tokyo, Japan
e-mail: tsuzuki.shunsuke@twmu.ac.jp

J. Okamoto · M. Tamura · K. Masamune · Y. Muragaki
Faculty of Advanced Techno Surgery (FATS), Institute of Advanced Biomedical Engineering and Science, Tokyo Women's Medical University, Tokyo, Japan

We believe that the potential risk of mixing old and new equipment is the main cause of errors. In order to reduce this risk and improve the effectiveness of surgery and procedures, we have developed SCOT® (Smart Cyber Operating Theater®) with the support of the Japan Agency for Medical Research and Development (AMED) and introduced it into clinical practice. In this treatment room, we aim to achieve a high level of safety and tumor removal rate by not only synchronizing devices that have been used as stand-alone devices but also integrating new medical technologies with image processing technologies such as simulation from various preoperative images and AR.

In the removal of malignant brain tumors, we aim to achieve both a high survival rate and prevention of surgical complications, making decisions on the extent of removal based on information measured by various medical devices. Specifically, intraoperative MRI, surgical navigation systems, neuromonitoring devices, and intraoperative rapid diagnostic devices (intraoperative flow cytometry) are used. However, the use of these stand-alone devices without data sharing makes it difficult for surgeons to integrate information for selecting appropriate surgical procedures. In order to solve this problem, we developed the "Smart Cyber Operation Theater" project in 2014 and devised OPeLiNK® enabling connection to various surgical equipment. In February 2019, Tokyo Women's Medical University Hospital installed the flagship version of Hyper SCOT®, which is the world's most advanced treatment room that aims to be linked with robotic operating tables and artificial intelligence (AI). In this paper, we describe the simulation technology to enhance the safety of surgery and the integration with the latest medical technology and detail the history and outline of the development of this operating room.

T. Ishizawa (ed.), *Fluorescence-Guided Surgery*, https://doi.org/10.1007/978-981-19-7372-7_32

2 Simulation Technology Based on Preoperative Images

Malignant glioma, which is our specialty, is a tumor of brain cells, and it grows as if it seeps into the surrounding brain parenchyma. Therefore, it is important to understand the localization of the tumor, the extent of invasion, and the positional relationship with blood vessels (arteries and veins) on the surface and in the depths of the brain before surgery and to visualize which "sulcus" to mark, which "gyrus" to enter, and which blood vessels to leave behind. In some cases, the tumor does not exist at the location assumed only by head MRI or CT, because the brain gyrus itself is swollen by the tumor or the surrounding brain gyrus is crushed by the pressure. In order to balance a high removal rate and a low complication rate, 3D simulation visualizing the special relationship between the tumor and the gyrus, sulcus, and blood vessels would be essential.

We created a 3D brain model based on head MRI information before surgery and displayed the sulci and tumors in color for preoperative simulation (Fig. 32.1). An analyzing software equipped with BrainVISA visualizes the possible location of the central sulcus in red, enabling surgeons to consider surgical procedures paying attention to the motor fibers (Fig. 32.2).

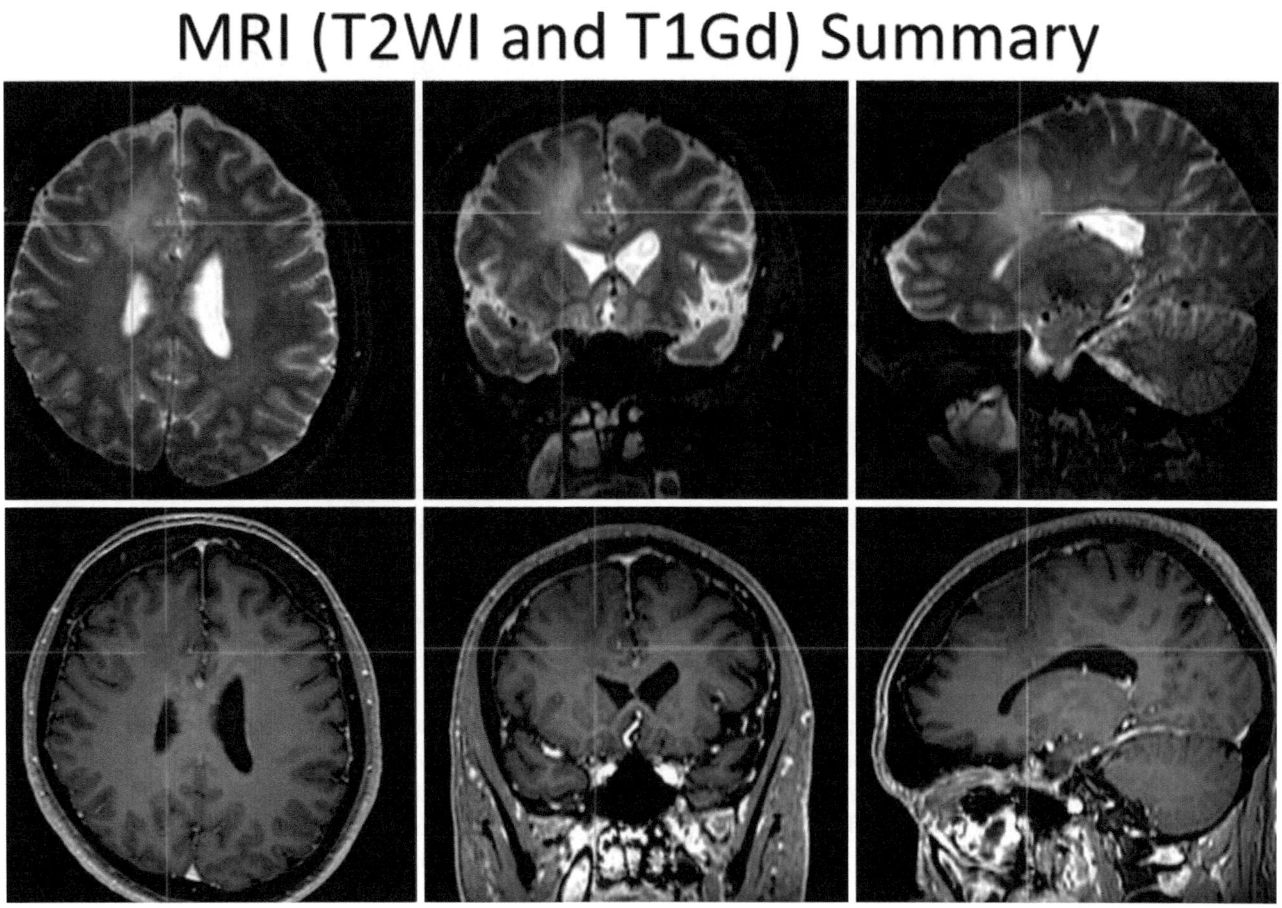

Fig. 32.1 Identification of tumor extent by MRI

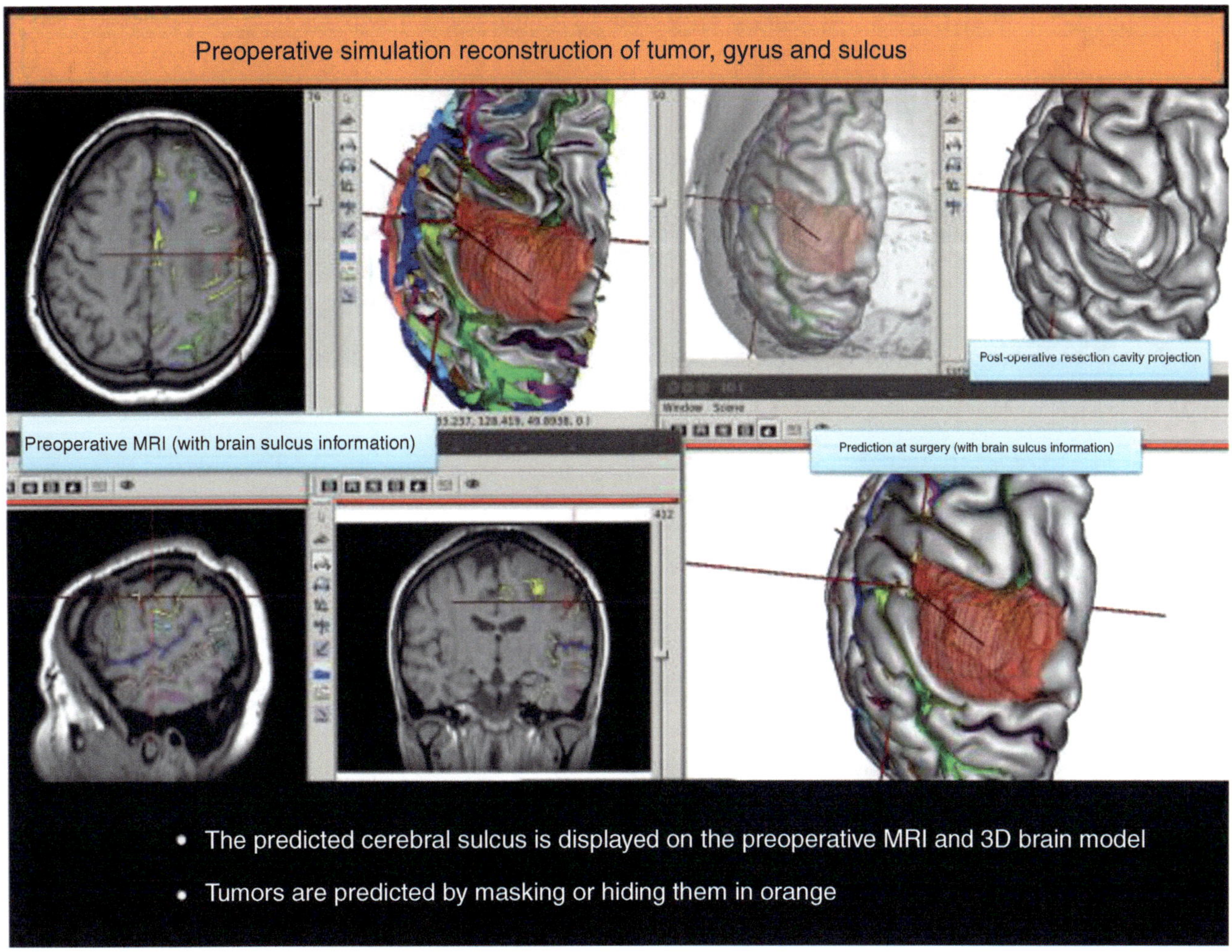

Fig. 32.2 Preoperative simulation with a 3D reconstruction of the tumor, gyrus, and sulcus. The predicted cerebral sulcus is displayed on the preoperative MRI and 3D brain model. Tumors are predicted by masking or hiding them in orange

3 How Should Intraoperative Fluorescence Imaging Be Integrated with Image Processing Technologies Such as AR?

In recent years, a variety of image display technologies have been developed and used for intraoperative support. For example, OCT (optical coherence tomography) [1], which visualizes nerve fibers, and LSCI (LASER speckle contrast imaging) [2], which quantitatively evaluates blood flow in the intestinal tract during gastrointestinal surgery, have been used. In neurosurgery, 5-ALA (aminolevulinic acid) is administered preoperatively, and intraoperative photodynamic evaluation using fluorescence imaging is performed to differentiate between the tumor and surrounding brain tissue (Fig. 32.3).

In recent years, AR has also come to be used for surgical support. The elements of therapeutic AR are basically the same as those of surgical navigation: (i) acquisition of pre-, intra-, and postoperative information and segmentation, (ii) positional registration, and (iii) visualization (presentation to the surgeon). Currently, we are continuing to develop how to utilize this treatment AR for intraoperative support. As specific examples, AR directly connected to diagnostic imaging devices such as CT and Open MRI, AR using a real 3D display, and AR using a tablet PC are being developed (Fig. 32.4).

By utilizing such AR technology, it is assumed that preoperative observation from the external surface, determination of the skin incision position, confirmation of the reoperation position, and visualization of the tumor and important structures from the skin will become possible. Intraoperative

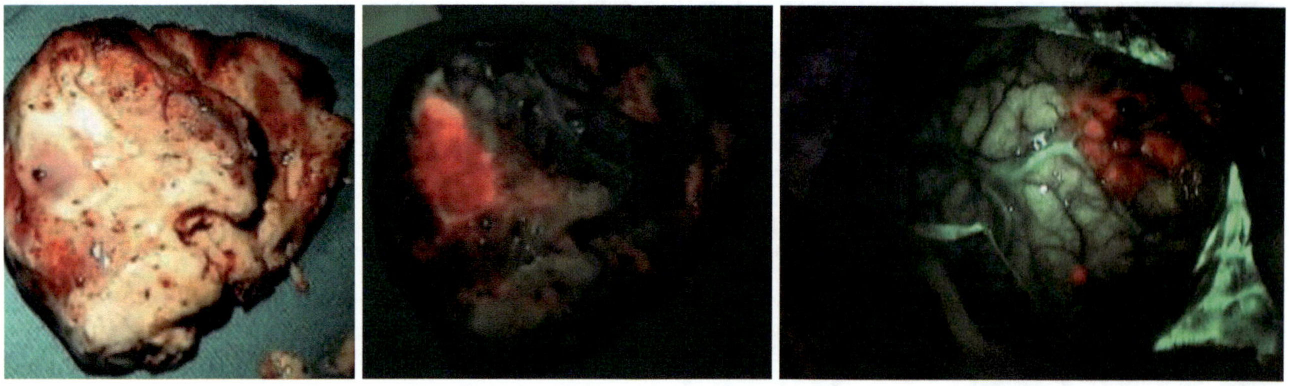

Fig. 32.3 Intraoperative fluorescence imaging using 5-ALA

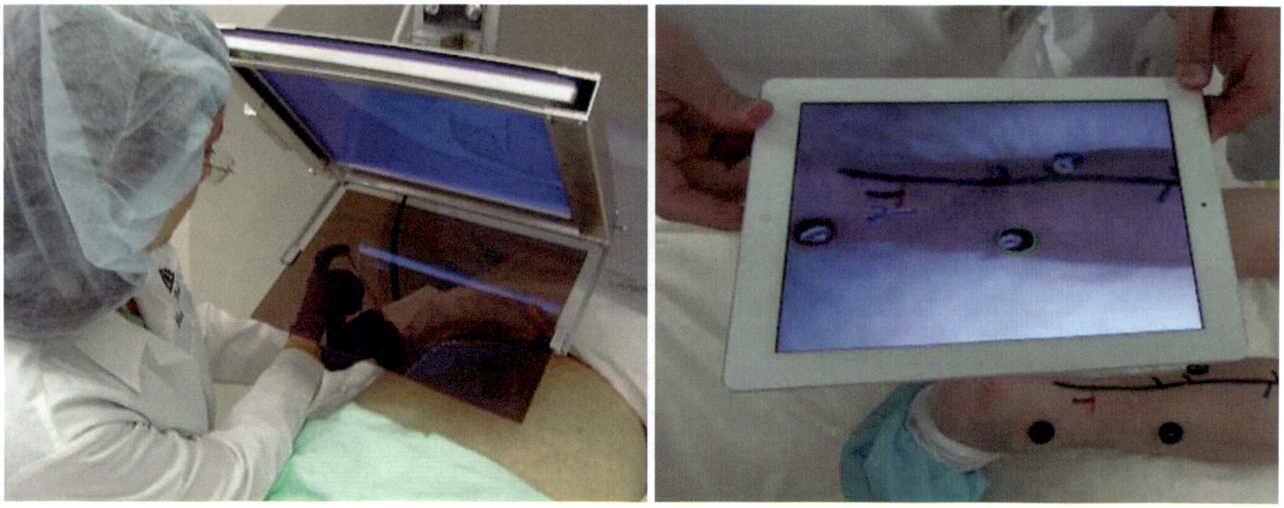

Fig. 32.4 Example of spinal nerve root block using AR-X-ray CT and varicose vein location identification using AR-Tablet PC

fluorescence imaging is a technology that enables differentiation between the tumor and surrounding tissues based on fluorescence intensity, and by combining this with image processing technology such as AR, it will be possible to accurately understand the location of the tumor and also to simulate photodynamic therapy in the excised cavity after tumor removal. In this sense, we believe that integration between AR and intraoperative fluorescence imaging can develop into a safe and effective surgical support.

4 Proposal of a Future Operating Room System That Enables Complex Image Processing and Presentation

The operation room is often filled with a lot of conventional and latest surgical devices and imaging systems according to surgical procedures and surgeons' preference. In fact, in the operation department of the Tokyo Women's Medical University, we had 747 surgical devices when investigated in 2014. In 2013, Weerakkody et al. reviewed and analyzed a paper that quantitatively assessed safety in the operating room and reported that the average number of "errors" per procedure was 15.5 [3]. Of these, 23.5% were due to malfunction or failure of the equipment or technique, 37.3% due to lack of necessary equipment or instruments, 43.4% due to combination or setting errors, and 33.5% due to failure of the equipment itself. We believe that the potential risk of mixing old and new equipment is the main cause of errors, and in order to reduce this risk and improve the effectiveness of surgery and procedures, we are developing and clinically applying SCOT® (Smart Cyber Operating Theater®) with the support of the Japan Agency for Medical Research and Development (AMED). In the following chapters, we describe the concept of the smart operating theater that can be applied to various preoperative and intraoperative diagnoses, as well as future integration with AI.

5 Smart Treatment Rooms and AI

5.1 Smart Treatment Room (SCOT®)

5.1.1 Overview and Development Requirements

Unlike conventional operating rooms that just provide a sterilized space, a smart treatment room is a "medical device" in which the entire room is a single unit offering treatment. Specifically, the necessary basic devices are selected (packaged) with the intraoperative diagnostic imaging device at the core, and the medical devices in the room, which are the components of each device, are connected to each other (networking) by ORiN (Open Resource Interface for the Network), an industrial middleware (software between OS and application). The visualized data is integrated and displayed by the network, and the necessary information for intraoperative decision-making is presented (informationization). We also aim to realize minimally invasive and highly reproducible precision medicine using the robot we have developed (robotization). In other words, packaging reduces errors and risks in the operating room, and the Internet of Things (IoT), which connects real space and cyberspace, integrates data into information. And by robotizing the equipment, we are trying to improve the effectiveness by integrating diagnosis into treatment.

In a 5-year project of development of smart operation theater since 2014, different models of SCOT systems have been installed in each institution according to a development factor. First, we installed basic SCOT at Hiroshima University in 2016, and in 2018, standard SCOT was installed at Shinshu University. In addition, a prototype of the robotized hyper SCOT was installed at Tokyo Women's Medical University in 2016, and the clinical research version will be completed in 2019.

5.1.2 A Basic Version of the Smart Treatment Room (Basic SCOT) Packaged with the Predecessor Model of the Intelligent Operating Room

The first step in developing a smart treatment room is to package the equipment, and we have experienced packaging in the "intelligent operating room," which has intraoperative MRI at its core. The intelligent operating room is an operating theater in which the presence or absence of residual tumors can be determined by intraoperative MRI in order to improve the removal rate of malignant brain tumors. For the actual installation of this system in clinical practice, it was necessary to prepare (package) MRI-compatible operating tables, anesthesia equipment, microscopes, and monitoring devices.

We have performed more than 2000 neurosurgical operations, mainly for glioma, since 2000. This intelligent operating room enables surgeons to perform resection procedures based on objective visual information (information-guided surgery) rather than a conventional judgment based on experience and intuition [4, 5], Specifically, intraoperative MRI (AIRIS-II, 0.3 Tesla, Hitachi, Ltd.) and anatomical information from a navigation device updated with intraoperative images are used as the core, and functional mapping and motor evoked potential (MEP) monitoring by awake surgery are used to confirm preservation of brain function. In addition, intraoperative rapid diagnosis and intraoperative flow cytometry provide histological information to determine whether the tumor or surrounding tissue is present. As a result, we have achieved a mean removal rate of 89% for primary glioma and 5-year survival rates of 89%, 74%, and 18% for WHO grades 2, 3, and 4, respectively.

The intelligent operating room was packaged with intraoperative MRI and MRI-compatible devices in a single-product production, but we also developed basic SCOT, which is packaged with these devices as a system, and installed it at Hiroshima University in 2016 (Fig. 32.5). The system is used not only for brain tumors but also for epilepsy surgery and bone tumor surgery in orthopedics, and its horizontal deployment has begun. In addition, the system has been installed in several places outside the project, including private hospitals.

5.1.3 Networked Standard Smart Treatment Rooms (Standard SCOT)

In conventional operating rooms, instruments are independent of each other and are not connected to the network. The data stays within each device, and the internal clocks of each device are different, making data integration extremely difficult.

In the first place, networking is impossible in an environment where a wide variety of medical devices are introduced by many companies. On the other hand, in basic SCOT, although the target devices were selected and the information of each device could be obtained, they were not networked. Therefore, we focused on ORiN, which connects many robots in a factory to a network and controls them efficiently. We thought it would be possible to connect to the network, input, and output data and control the robots without changing the inside of the devices by creating software (provider) that is similar to a device driver for PC peripherals. In this project, we also developed OPeLiNK® (Denso), a medical middleware, and have connected more than 30 devices so far. We are aiming to make OPeLiNK® a global standard. Our goal is to expand it not only to operating rooms but also to ICUs and hospital wards [6].

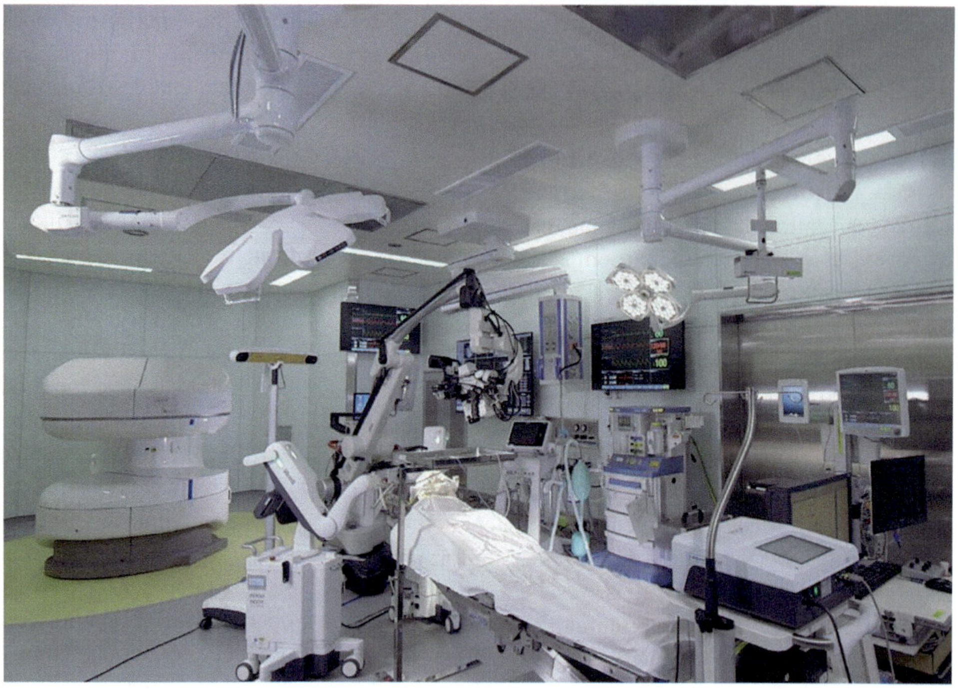

Fig. 32.5 Basic SCOT introduced at Hiroshima University

If each device can be networked, independent information can be integrated in a time-synchronized manner, and if it is combined with the location information of the navigation system, it has the advantage of providing spatial information. We developed a strategic desk system that can display both independent information and integrated information and created an application for malignant brain tumor removal. The MEP values as functional information and the intraoperative flow cytometry values as histopathological information are assigned to the manipulation sites in the navigation. Since the data between the devices are time-synchronized, the navigation system can record the operation site where the MEP value, which has a high risk of postoperative paralysis, has decreased, and the navigation system can show the site with a high percentage of cells in the proliferative phase (i.e., high malignancy) by flow cytometry. The former is an integration of functional and anatomical information, and the latter is an integration of histological and anatomical information. In 2018, standard SCOT, in which almost all devices were networked by OPeLiNK®, was established at Shinshu University. Now we are exploring the efficacy of SCOT in clinical studies.

5.1.4 A Robotized, Highly Functional Version of the Smart Treatment Room (Hyper SCOT)

Basic SCOT and standard SCOT have functions mainly for the acquisition and integration of information. In other words, these systems have been developed to aid the surgeon's eyes and brain, but in the future, robotics would replace surgical procedure and maneuver providing hands for surgeons. In Hyper SCOT, we introduced a robotic operating table and microscope, as well as a hand-held robot to support the surgeon. The robotic operating table automatically sets the patient so that the lesion is in the center of the operating room, the robot moves the microscope so that the tip of the instrument is in the center of the microscope field of view, and the hand-held robot reduces hand tremor and fatigue in the surgeon. In 2016, we developed a prototype that embodied this idea and was introduced to the media and magazines.

In addition, we are also working on the development of a new robotic therapy in Japan. Specifically, we are developing minimally invasive cancer treatment methods that combine drugs and physical force, including photodynamic therapy using lasers (light) and light-sensitive substances for superficial cancer [7] and sonodynamic therapy using focused ultrasound and acoustic-sensitive substances for deep cancer [8]. Photodynamic therapy has obtained insurance coverage in Japan for early-stage lung cancer, malignant brain tumors, and locally recurrent esophageal cancer, and sonodynamic therapy is aiming for insurance coverage for unresectable pancreatic cancer.

Currently, a hyper SCOT that can be used clinically has been installed at Tokyo Women's Medical University and has been in operation since October 2019 (Figs. 32.6 and 32.7).

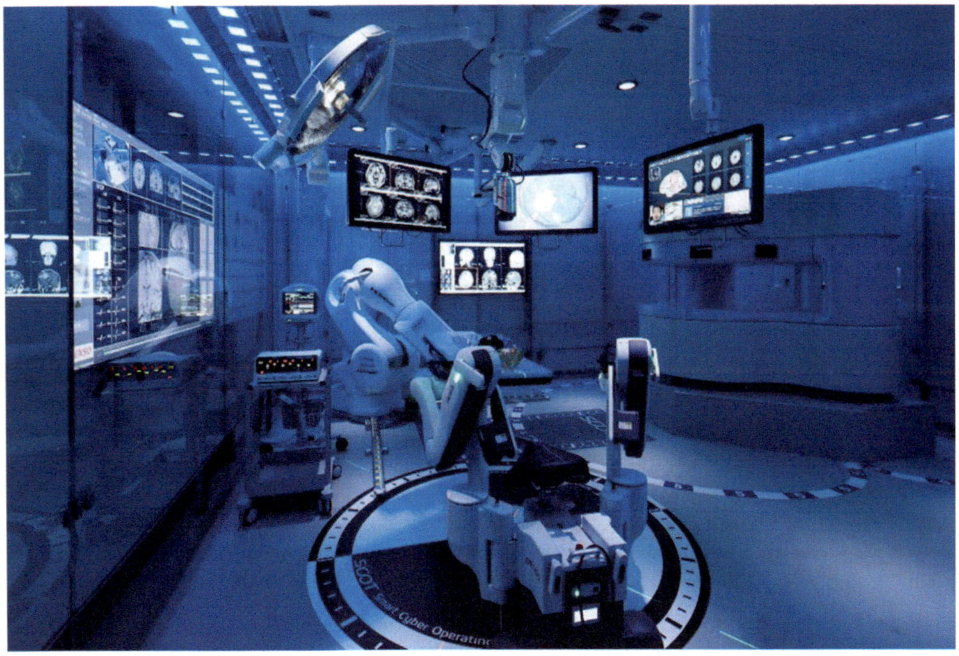

Fig. 32.6 Hyper SCOT prototype introduced to Tokyo Women's Medical University

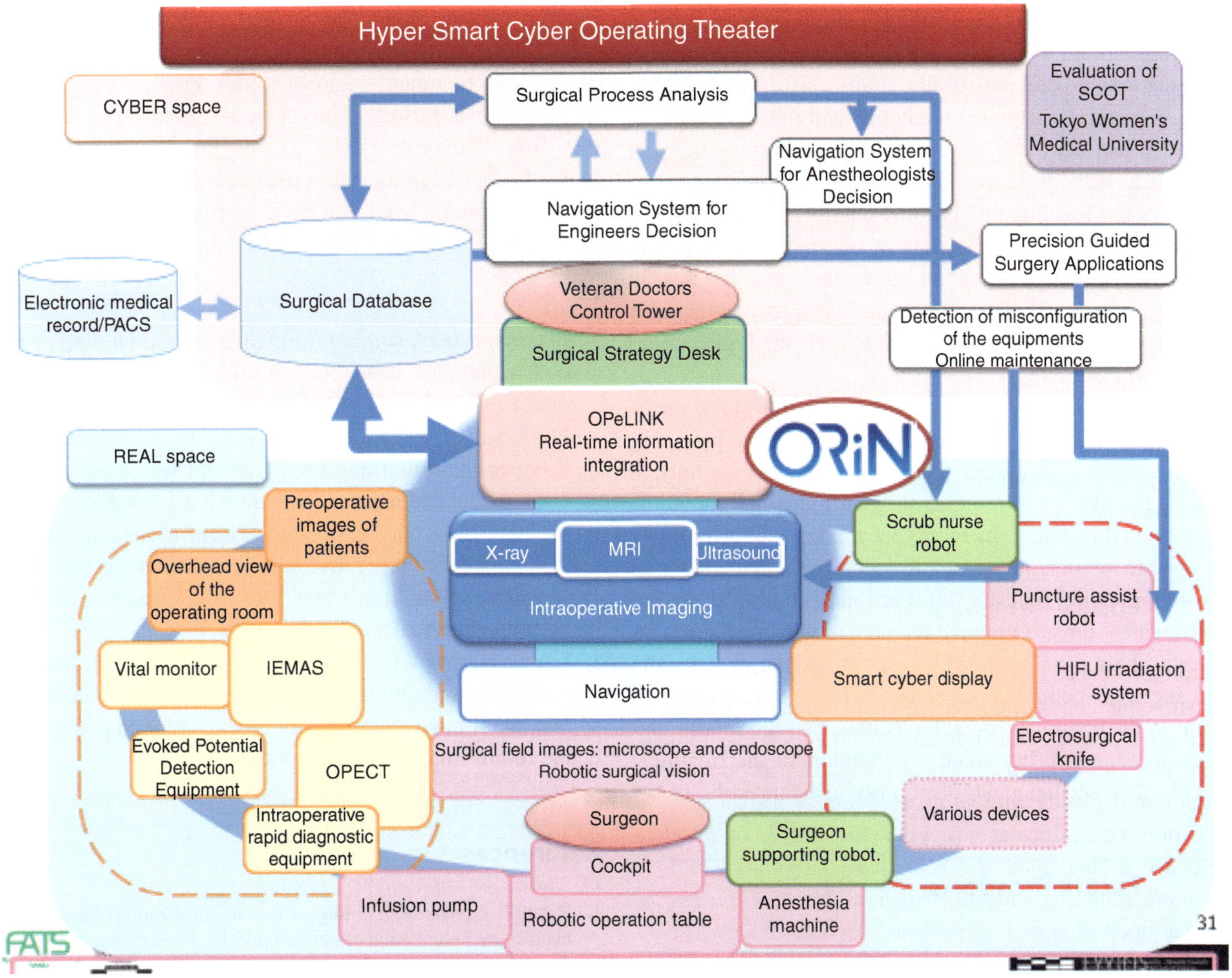

Fig. 32.7 Schematic diagram of Hyper SCOT with ORiN and OPeLiNK®

5.2 Application of IoT and AI in Smart Treatment Rooms

Standard SCOT is a system in which various objects (in this case, medical devices) are connected like the Internet (IoT), exchange information, and mutually control each other. With this system, time-synchronized information is displayed on the map (location information) to support the surgeon's decision-making, just as traffic jam information or parking lot availability information displayed on a car navigation map. However, we believe that not only intraoperative information but also prognostic information will be necessary for more advanced surgical decision-making. For example, when making a decision on whether or not to perform further removal to improve the survival rate in the final stage of removal, past data that can serve as a basis for predicting the prolongation of the survival period by improving the removal rate will be necessary.

We are currently constructing a database (data warehouse) to facilitate the analysis of the data in the electronic medical record. To predict the complications to be avoided, a risk map may be used to accumulate the records of the operation sites in the brain when the MEP is decreased and to display the locations where statistically significant decreases were more frequent. If prognosis prediction and risk map analysis progress, and if we can accumulate a large amount of structured data, decision support using AI such as machine learning and deep learning will become possible in the future. In our research project on AI, we have succeeded in predicting the time when the white blood cell count drops the most in anticancer drugs. We have also created a risk map of the parts of the body that are most likely to experience higher-order functional disorders.

In addition, hyper SCOT aims to operate the entire medical equipment by the network using IoT, just like operating home appliances remotely from a smartphone. Initially, it will be possible to turn on/off surgical lights and move the operating table, but in the future, we think it will be possible to operate robotic treatment devices. The same industrial revolution called Industry 4.0 (the fourth industrial revolution), which aims to improve the operation of real space by closely linking the real world with various sensor networks and the high computer capability of the cyberspace, will occur in the operating room. We believe that the smart treatment room will bring about a revolution in the treatment world called Medicine 4.0 (the fourth medical revolution).

6 Future Prospects

We have set precision-guided therapy as the goal of surgery in the twenty-first century. This means that various visualized information (the surgeon's new eyes) is integrated and analyzed by a strategy desk to support decision-making (the surgeon's new brain) and that minimally invasive treatment is performed with new robotic treatment devices (the surgeon's new hands). In the future, these new technologies will make it possible to perform tumor removal using fluorescence imaging accurately in real time in a smart treatment room by inputting information on tumor visualization using 5-ALA fluorescence under a surgical microscope and synchronizing time and special information obtained from a microscopic, surgical microscope, and navigation system.

Smart treatment rooms can be applied not only to intraoperative MRI-based surgery and treatment of parenchymal organs such as malignant brain tumors but also to endovascular treatments and surgeries for abdominal organs. In addition, OPeLiNK®, which can connect medical devices from different companies, has the potential to be used not only in operating rooms but also in ICUs, wards, and hospitals as a whole. The key factor would be the international standardization.

The reason we chose the term "smart treatment room" instead of "smart operating room" is that we believe that this treatment room should be a single treatment device that allows doctors to perform, or want to perform, all invasive procedures, treatments, and therapies, not just surgery. We would like to see further horizontal development so that doctors from different specialties can identify diseases that would benefit from the smart treatment room and work with us to determine what information to obtain to optimize treatment.

Point

- In our specialty, surgery for malignant brain tumors, various techniques have been introduced to achieve high tumor removal rates and low complication rates.
- Preoperative simulation indicating special relationships between the tumor and the gyrus and sulcus using a 3D brain model and augmented reality has been developed.
- We have developed a smart treatment room as a "place" that reduces intraoperative equipment problems and risks, synchronizes and integrates various information, and utilizes it in surgical strategies.
- Fluorescence imaging, which is also used in the field of malignant brain tumors, is expected to become an important factor in visualizing tumor localization in real time in the future smart treatment room.

References

1. Nakaji H, Kouyama N, Muragaki Y, et al. Localization of nerve fiber bundles by polarization-sensitive optical coherence tomography. J Neurosci Methods. 2008;174:82–90.
2. Kojima S, Sakamoto T, Nagai Y, et al. Laser speckle contrast imaging for intraoperative quantitative assessment of intestinal blood

perfusion during colorectal surgery: a prospective pilot study. Surg Innov. 2019;26:293–301.
3. Weerakkody RA, Cheshire NJ, Riga C, et al. Surgical technology and operating-room safety failures: a systematic review of quantitative studies. BMJ Qual Saf. 2013;22:710–8.
4. Muragaki Y, Iseki H, Maruyama T, et al. Information-guided surgical management of gliomas using low-field-strength intraoperative MRI. Acta Neurochir Suppl. 2011;109:67–72.
5. Muragaki Y, Iseki H, Maruyama T, et al. Usefulness of intraoperative magnetic resonance imaging for glioma surgery. Acta Neurochir Suppl. 2006;98:67–75.
6. Okamoto J, Masamune K, Iseki H, et al. Development concepts of a smart cyber operating theater (SCOT) using ORiN technology. Biomed Tech (Berl). 2018;63:31–7.
7. Muragaki Y, Akimoto J, Maruyama T, et al. Phase II clinical study on intraoperative photodynamic therapy with Talaporfin sodium and semiconductor Laser. J Neurosurg. 2013;119:845–52.
8. Maeda M, Muragaki Y, Okamoto J, et al. Sonodynamic therapy based on combined use of low dose administration of epirubicin-incorporating drug delivery system and focused ultrasound. Ultrasound Med Biol. 2017;43:2295–301.

Therapeutic Applications: Photodynamic Therapy Using Porphyrin Compounds

33

Takeomi Hamada and Atsushi Nanashima

Summary

- Photodynamic therapy (PDT) is a localized cancer therapy using a tumor-affinity photosensitive agent and low-power laser irradiation.
- PDT is safe and easy as compared with ablation or thermal coagulation using high-powered lasers.
- The incidence of photosensitivity as an adverse effect of PDT has improved with the introduction of second-generation photosensitizers.
- PDT is expected to become one of the treatment options for various types of cancer in the future.

1 Introduction

As a result of recent advances in endoscopes and optical fibers, optical engineering has been introduced to medical treatment in these years. Photodynamic therapy (PDT) is a promising treatment method that irradiates a tumor with a highly tumor-accumulating photosensitive agent and a laser with a specific wavelength and shows a cell-killing effect only on tumor cells with high drug concentration. It has already been applied clinically in the fields of pulmonary and digestive surgery, neurosurgery, ophthalmology, dermatology, and urology. As described in other chapters, photodynamic diagnosis (PDD), which is a method to diagnose the localization of tumors by irradiating biological tissues with light and detecting the fluorescence generated by photosensitive agents accumulated in tumors, is also being actively studied and applied clinically.

In Japan, PDT using the first-generation porpfimer sodium (Photofrin®, Wyeth) and the second-generation talaporphyn sodium (Laserphyrin®, Meiji Seika Pharma, Tokyo, Japan) as photosensitive agents are currently covered by the social insurance system. Compared with the first-generation porphyrin sodium, the second-generation talaporphyn sodium has the advantage of reducing the duration and degree of light shielding and is likely to be widely used in clinical practice in the future. In this chapter, we focus on PDT therapy using second-generation photosensitive substances, including future prospects.

2 What Is PDT?

PDT is a therapeutic method in which a photosensitizing tumor-affinity photosensitizer is administered into the body, and the lesion is irradiated with a laser of a specific wavelength to induce a photoreaction that kills the cancer cells. In Japan, PDT with porphyrin sodium is currently covered by insurance for early-stage lung cancer, superficial esophageal cancer, superficial early-stage gastric cancer, early-stage cervical cancer and dysplasia, age-related macular degeneration, and malignant brain tumors, while PDT with thalaborphine sodium is approved for early-stage lung cancer, recurrent esophageal cancer after chemoradiation, and primary malignant brain tumors.

3 Action Mechanism of PDT

Although the mechanism of the cell-killing effect of PDT has not been fully elucidated, it has been shown physicochemically that the PDT effect is caused by a photochemical reaction that occurs when a light-sensitive molecule with a porphyrin skeleton absorbs light energy and transfers this energy to other molecules in the presence of oxygen [1].

Two reactions, radical reaction and singlet oxygen reaction, are induced in photosensitive materials excited by light. The free radicals produced in the radical reaction react with oxygen to produce a variety of oxidants that trigger free radical chain reactions. In the singlet oxygen reaction, the photo-

T. Hamada · A. Nanashima (✉)
Division of Hepato-biliary-pancreas Surgery, Department of Surgery, Faculty of Medicine, University of Miyazaki, Miyazaki, Japan
e-mail: a_nanashima@med.miyazaki-u.ac.jp

T. Ishizawa (ed.), *Fluorescence-Guided Surgery*, https://doi.org/10.1007/978-981-19-7372-7_33

sensitive material that receives the laser light energy enters the excited singlet state and supplies energy to the oxygen in the tumor tissue, and the supplied oxygen becomes singlet oxygen. This singlet oxygen is thought to induce necrosis and apoptosis of tumor cells [2]. Recently, tumor blood vessels have also been shown to be a site of PDT injury, causing hemorrhage, hypoxia, and tumor necrosis [3–5]. Furthermore, in addition to the abovementioned direct cell-killing mechanisms by induction of apoptosis and vascular injury by reactive oxygen species, indirect therapeutic mechanisms by oxidative stress after PDT, induction of various cytokines by inflammatory changes, and activation of specific tumor immunity have also been reported [6–8].

Although the laser wavelength depends on the type of photosensitive agents, the optimal absorption wavelength range of sodium porphyrin is 630 nm and that of talaporphyn sodium is 664 nm. After administration, the light-sensitive agent is taken up by the tumor and normal tissues, and the drug concentration increases. Laser irradiation is performed 48–72 hours after intravitreal administration for porfimer sodium and 4–6 hours for talaporphyn sodium. The duration of drug retention in the body is 4–6 weeks and 2 weeks, respectively, which is shorter for talaporphyn sodium. During this time, light shielding is required, and the light shielding level is reduced from less than 300 lux to less than 500 lux with talaporphyn sodium.

4 PDT Treatment in Practice

We will introduce the current practice of PDT using talaporphyn sodium (Laserphyrin®) at our hospital for the recurrence of esophageal cancer after chemoradiotherapy.

PDT is indicated 4–6 hours after intravenous injection of 40 mg/m^2 of Laserphyrin®. For laser excitation, ZH-L5011HJP (PD Laser; Panasonic Healthcare's semiconductor laser, Tokyo Japan) was used with a wavelength set at 664 ± 2 nm. Irradiation was performed with a power density of 150 W/cm^2 and an energy density of 100 J/cm^2 for 11 minutes and 7 seconds at a time. Depending on the lesion, irradiation may be divided into one to three sessions. The PDT laser fiber used in this procedure is classified into two types: a lateral full-field irradiation type and an anterior irradiation type.

After intravenous administration of Laserphyrin®, the patient is shielded from light by a dark curtain around the patient below 500 lux for 14 days. Patients are instructed to stay out of direct sunlight for 2 weeks after discharge and to avoid skin exposure as much as possible when going out during the day.

5 PDT Using First-Generation Porphyrin Sodium

First-generation porphyrin sodium PDT is covered by insurance for early-stage lung cancer, superficial esophageal cancer, superficial early-stage gastric cancer, early-stage cervical cancer and dysplasia, and malignant brain tumors. As mentioned above, however, a light-shielding period of about 4–6 weeks is required, and the frequency of skin toxicity due to photosensitivity is as high as 20–40%. The eximer dye laser system is also very expensive and large, so there are many problems in terms of economic efficiency and simplicity for this treatment to be used widely. In addition, companies have recently announced that they will discontinue maintenance, so the number of facilities where porphyrin sodium PDT can be performed is currently limited even in Japan.

In the field of gynecology, early cervical cancer and dysplasia have been targeted for PDT treatment. Sakamoto et al. compared PDT with conical resection, a uterus-sparing treatment for early cervical cancer and dysplasia, and reported that PDT has a similar cure rate of cervical lesions but a higher possibility of fertility preservation [9].

In the treatment for early-stage gastric cancer, PDT for superficial lesions was covered by the Japanese insurance system in April 1996. While there have been some reports on the efficacy of PDT for early-stage gastric cancer [10], this strategy has not become widespread due to the development and widespread use of EMR (endoscopic mucosal resection) and ESD (endoscopic submucosal dissection).

6 PDT Using Second-Generation Thalaborphine Sodium

Second-generation thalaborphine sodium PDT is covered by insurance for early-stage lung cancer, recurrence of esophageal cancer following chemoradiotherapy, and primary malignant brain tumors in Japan. The indications for each are shown in Table 33.1.

Table 33.1 Indications of PDT using thalaborphine sodium

Indications	Recurrent esophageal cancer chemoradiotherapy	Central early-stage lung cancer	Malignant brain tumor
Selection criteria	Local remnant or recurrent esophageal cancer after CRT or RT	Stage 0 or stage I early-stage lung cancer	Primary malignant brain tumor
Adaptation	Curative treatment such as surgical resection or endoscopic treatment is not possible Wall depth does not exceed the intrinsic muscular layer The longitudinal diameter is less than 3 cm, and the circumference is less than 1/2 circumference It does not extend into the cervical esophagus No distant metastasis or lymph node metastasis	The lesion is centrally located above the area bronchus The entire lesion can be seen endoscopically in the bronchial mucosa Tumor diameter less than 10 mm No lymph nodes or distant metastases	Subtotal or greater resection of the tumor body is expected at the time of surgery The site of probable tumor cell infiltration is expected to be visible under an operating microscope and is expected to be a site where laser irradiation is feasible There are no major blood vessels in the normal cerebral circulation in the area to be irradiated with laser light
Taboo	The tumor is at the aorta T4 prior to CRT or RT Hypersensitivity to Rezafirin Current porphyria	Hypersensitivity to Rezafirin Current porphyria	Hypersensitivity to Rezafirin Current porphyria

7 PDT for Central-Type Early-Stage Lung Cancer

The results of PDT for central early-stage lung cancer were reported to be 84.3% CR rate and 84.6% CR rate in a phase II clinical trial using first-generation porfimer sodium and second-generation talaporphyn sodium, respectively [11, 12]. The response rates were 94.9% and 94.9%, respectively. In particular, CR was reported to be more than 90% when the lesion diameter was less than 10 mm, while it decreased to 50–80% in patients with lesions between 10 and 20 mm. Recently, however, the CR rate of PDT for the lesions with a larger diameter of 10–20 mm and classified as flat, early polypoid, or nodular lesions by endoscopic examinations has been improved to 90.4%, which was higher than previously reported [13]. This may be due to the improved localization diagnosis of central early-stage lung cancer lesions, enabling accurate diagnosis of the area of laser irradiation.

8 PDT for Recurrence of Esophageal Cancer after Chemoradiotherapy

A phase II study using first-generation porfimer sodium for the treatment of recurrent esophageal cancer revealed satisfactory overall outcomes: the CR rate of the primary tumor was 76%, and the 3-year survival rate was 38% [14]. In a phase II clinical trial using second-generation talaporphyn sodium, the CR rate was 88.5%, and local progression-free survival was 428 days, with no serious adverse events, suggesting a high safety profile [15]. Based on these results, PDT for recurrence of esophageal cancer after chemoradiotherapy using talaporphyn sodium was included in the Japanese insurance coverage in October 2015.

After the approval by the Japanese FDA, Amanuma et al. reported that the local complete response rate was 53.6% (60% in patients with pre-PDT depth of T1b and 37.5% in patients with T2 [16]), suggesting that PDT is an effective salvage therapy in clinical practice. In contrast, grade 3 or higher complications occurred in two cases, and grade 5 esophageal bronchopleural fistula occurred in one case, so sufficient caution is required.

9 PDT for Malignant Brain Tumors

Malignant brain tumors, especially glioblastoma, rarely have distant metastasis, and local treatment is important to improve prognosis. In recent years, there have been many reports that aggressive removal is associated with a better prognosis, and in the case of glioblastoma, removal of more than 98% of the contrast-enhancing area on MRI images improves long-term outcomes. The current consensus on the treatment of malignant brain tumors is to aim for maximum removal with minimal neurological complications. Photodynamic diagnosis (PDD) is a method to visualize tumors with obscure borders by the naked eye, and PDT is a therapeutic option offering laser eradication of invasive tumor cells that cannot be removed by surgery.

According to the results of the investigator-initiated clinical trial of talaporphyn sodium PDT started in 2009, the 12-month overall survival rate was 95.5%, and the 6-month progression-free relapse rate was 90.9% [17]. The median overall survival was 24.8 months, and based on the results in a situation where there are few prospective clinical trials with overall survival exceeding 2 years, thalaborphine sodium PDT is expected to be effective as an add-on to standard therapy and was approved by the pharmaceutical affairs bodies in Japan in 2013. Currently, post-marketing surveillance and clinical studies are underway.

10 PDT Using 5-ALA

5-ALA is metabolized into the fluorescent substance porphyrin IX after intracellular uptake and accumulates more in tumor cells, thus labeling only cells. Therefore, 5-ALA can be used for PDD, especially in the field of neurosurgery. Although PDT targeting 5-ALA has not been approved in Japan, many basic experiments of PDT using 5-ALA for brain tumors such as glioblastoma have been reported worldwide. Depending on the outcomes of current clinical trials, further development of this treatment strategy can be expected [18, 19].

11 PDT for Other Cancers (Especially with Regard to Bile Duct Cancer)

Cholangiocarcinoma arising from the epithelium of the bile duct is one of the cancer types with a poor prognosis, with a 5-year survival rate of about 10%. Surgical curative resection of cholangiocarcinoma is expected to have the best prognosis, but more than half of patients with cholangiocarcinoma are inoperable at the time of detection, and the presence or absence of local control, including bile duct stenosis, has a significant impact on the patient's prognosis. In recent years, the number of reported cases of PDT for the palliative treatment of cholangiocarcinoma has been increasing [20, 21].

Ortner et al. reported that porphyrin sodium PDT combined with biliary stent significantly prolonged survival compared with stent alone in a multicenter randomized trial for unresectable cholangiocarcinoma [22]. In a phase II clinical study, Berr et al. reported that PDT was effective in 23 patients with unresectable cholangiocarcinoma, with tumor reduction in 29–74% of patients, a 6-month survival rate of 74% after PDT, a median survival time of 11 months, no bile stasis, and improvement in QOL [23]. In a cellular-level study, Nonaka et al. reported that the combination of gemcitabine, oxaliplatin, and thalaborphine sodium PDT induced the strongest necrosis and apoptosis in cholangiocarcinoma cells [24]. The usefulness of PDT as a local treatment for cholangiocarcinoma has been almost confirmed, and it is expected that PDT will be widely used in clinical practice through large-scale clinical trials aiming at insurance coverage in the future. We have been conducting a clinical trial on the safety of PDT as a local adjuvant therapy for unresectable cholangiocarcinoma since 2017 under the approval of the hospital ethics committee. Clinical applications of PDT are highly expected also in the field of urological cancer and head and neck cancer [25, 26].

12 Future Perspectives

PDT is a minimally invasive cancer treatment method that takes function preservation into consideration, and it is one of the most promising treatments for elderly patients. In order to develop PDT as a more feasible option for cancer treatment in various fields, a new third-generation photosensitive agent with higher therapeutic efficacy, shorter light shielding period, and fewer side effects is required, as being developed actively in these years [27, 28]. Recently, a water-soluble porphyrin compound synthesized by the porphyrin complex derivatization method developed by Matsumoto et al. has been shown in basic research to have high water solubility, high biocompatibility, high quantum yield of cytotoxic singlet oxygen generation, and higher antitumor effect than Laserphyrin [29]. At our institution, we are currently confirming the efficacy of PDT using this new photosensitizer in basic research.

We hope that PDT will be one of the new methods of cancer treatment in addition to surgery, radiotherapy, and chemotherapy, leading to the improvement of prognosis in cancer patients.

Point

- PDT is a localized cancer therapy using a tumor-affinity photosensitive agent and low-power laser irradiation and has been applied clinically to the recurrence of esophageal cancer after chemoradiotherapy, central early-stage lung cancer, and malignant brain tumors.
- PDT is expected to be applied to various types of malignancies including unresectable cholangiocarcinoma.

References

1. Hilf R, Warne NW, Smail DB, et al. Photodynamic inactivation of selected intracellular enzymes by hematoporphyrin derivatives and their relationship to tumor cell viability in vitro. Cancer Lett. 1984;24:165–72.
2. Weishaupt KR, Gomer CJ, Dougherty TJ. Identification of singlet oxygen as a cytotoxic agent in the photoinactivation of a murine tumor. Cancer Res. 1976;36:2326–9.
3. Fingar VH, Siegel KA, Wieman TJ, et al. The effects of thromboxane inhibitors on the microvascular and tumor response to photodynamic therapy. Photochem Photochem Photobiol. 1993;58:393–9.
4. McMahon KS, Wieman TJ, Moore PH, et al. Effects of photodynamic therapy using mono-L-aspartyl chlorin e6 on vessel constriction, vessel leakage, and tumor response. Cancer Res. 1994;54:5374–9.

5. Gilissen MJ, van de Merbel-de Wit LE, Star WM, et al. Effect of photodynamic therapy on the endothelium-dependent relaxation of isolated rat aortas. Cancer Res. 1993;53:2548–52.
6. de Vree WJ, Essers MC, Koster JF, et al. Role of interleukin 1 and granulocyte colony-stimulating factor in Photofrin-based photodynamic therapy of rat rhabdomyosarcoma tumors. Cancer Res. 1997;57:2555–8.
7. Evans S, Matthews W, Perry R, et al. Effect of photodynamic therapy on tumor necrosis factor production by murine macrophages. J Natl Cancer Inst. 1990;82:34–9.
8. Korbelik M. Induction of tumor immunity by photodynamic therapy. J Clin Laser Med Surg. 1996;14:329–34.
9. Nanashima A, Nagayasu T. Current status of photodynamic therapy in digestive tract carcinoma in Japan. Int J Mol Sci. 2015;33:117–21.
10. Shimoyama Y, Kuribayashi S, Hosaka H, et al. Indications for gastric cancer PDT and the possibility of expanding the indications. J Res Med. 2015;16:3434–40.
11. Furuse K, Fukuoka M, Kato H, et al. A prospective phase II study on photodynamic therapy with Photofrin II for centrally located early-stage lung cancer. The Japan Lung Cancer Photodynamic Therapy Study Group. J Clin Oncol. 1993;11:1852–7.
12. Kato H, Furukawa K, Sato M, et al. Phase II clinical study of photodynamic therapy using mono-L-aspartyl chlorin e6 and diode laser for early superficial squamous cell carcinoma of the lung. Lung Cancer. 2003;42:103–11.
13. Usuda J, Ichinose S, Ishizumi T, et al. Outcome of photodynamic therapy using NPe6 for bronchogenic carcinomas in central airways >1.0 cm in diameter. Clin Cancer Res. 2010;16:2198–204.
14. Yano T, Muto M, Minashi K, et al. Photodynamic therapy as salvage treatment for local failure after chemoradiotherapy in patients with esophageal. Int J Cancer. 2012;131:1228–34.
15. Yano T, Kasai H, Horimatsu T, et al. A multicenter phase II study of salvage photodynamic therapy using talaporfin sodium (ME2906) and a diode laser (PNL6405EPG) for local failure after chemoradiotherapy or radiotherapy for esophageal cancer. Oncotarget. 2017;8:22135–44.
16. Amanuma Y, Horimatsu T, Ohashi S, et al. Association of local complete response with prognosis after salvage photodynamic therapy for esophageal squamous cell carcinoma. Dige Endosc. 2021;33:355–63.
17. Muragaki Y, Akimoto J, Maruyama T, et al. Phase II clinical study on intraoperative photodynamic therapy with talaporfin sodium and semiconductor laser. J Neurosurg. 2013;119:845–52.
18. Stepp H, Beck T, Pongratz T, et al. ALA and malignant glioma: fluorescence-guided resection and photodynamic treatment. J Environ Pathol Toxicol Oncol. 2007;26:157–64.
19. Tetard MC, Vermandel M, Mordon S, et al. Experimental use of photodynamic therapy in high grade gliomas: a review focused on 5-aminolevulinic acid. Photodiagn Photodyn Ther. 2014;11:319–30.
20. Nanashima A, Abo T, Nonaka T, et al. Photodynamic therapy using talaporfin sodium (Leserphyrin®) for bile ductcarcinoma: a preliminary clinical trial. Anticancer Res. 2012;32:4931–8.
21. Suzuki S, Inaba K, Yokoi Y, et al. Photodynamic therapy for malignant biliary obstruction: a case series. Endoscopy. 2004;36:83–7.
22. Ortner ME, Caca K, Berr F, et al. Successful photodynamic therapy for nonresectable cholangiocarcinoma: a randomized prospective study. Gastroenterology. 2003;125:1355–63.
23. Berr F, Wiedmann M, Tannapfel A, et al. Photodynamic therapy for advanced bile duct cancer: evidence for improved palliation and extended survival. Hepatology. 2000;31:291–8.
24. Nonaka Y, Nanashima A, Nonaka T, et al. Synergic effect of photodynamic therapy using talaporfin sodium with conventional anticancer chemotherapy for the. J Surg Res. 2013;181:234–41.
25. Ramsay D, Stevenson H, Jerjes W. From basic mechanisms to clinical research: photodynamic therapy applications in head and neck malignancies and vascular anomalies. J Clin Med. 2021. https://doi.org/10.3390/jcm10194404.
26. Bozzini G, Colin P, Betrouni N, et al. Photodynamic therapy in urology: what can we do now and where are we heading? Photodiagn Photodyn Ther. 2012;9:261–73.
27. Vrouenraets MB, Visser GW, Loup C, et al. Targeting of a hydrophilic photosensitizer by use of internalizing monoclonal antibodies: a new possibility for use in photodynamic therapy. Int J Cancer. 2000;88:108–14.
28. Tanaka M, Kataoka H, Yano S, et al. Antitumor effects in gastrointestinal stromal tumors using photodynamic therapy with a novel glucose-conjugated chlorin. Mol Cancer Ther. 2014;13:767–75.
29. Matsumoto J, Suzuki K, Yasuda M, et al. Photodynamic therapy of human biliary cancer cell line using a combination of phosphorus porphyrins and light emitting diode. Med Chem. 2017;25:6536–41.

Application to Therapy (2): Photoimmunotherapy Using a Near-Infrared Fluorescent Probe

34

Yutaka Tamura, Akiko Suganami, and Yoshiharu Okamoto

Summary

- Photoimmunotherapy in the near-infrared region (650 to 900 nm) has advantages over conventional photodynamic therapy in the visible light region (400 to 650 nm).
- Prof. Hisataka Kobayashi (NIH) et al. are currently conducting an international phase III clinical trial of photoimmunotherapy using an antibody-drug conjugate (IR-700: a near-infrared fluorescent dye conjugated with antibody).
- In the field of photoimmunotherapy using the enhanced permeability and retention effect (EPR effect) of indocyanine green (ICG) derivative-modified liposomes (ICG-Lipo, ICG derivative embedded on the membrane surface), clinical trials have been conducted in companion animals as translational research for human medicine.
- The development of a new imaging system using over 1000 nm in the near-infrared range (OTN-NIR: 900–2500 nm), which is called the "second and third biological optical window", is in progress.

1 Introduction

Photoimmunotherapy induces an anti-tumor immune response by cytotoxic T-lymphocytes (CTLs) activated by neoantigens released from cancer tissues by fever and singlet oxygen generation when a photosensitive agent accumulated in cancer cells is irradiated with laser light in the near-infrared fluorescent region (650–900 nm). The use of near-infrared fluorescence probes has the potential to overcome the problems of "insufficient selectivity to tumors" and "photosensitivity due to side effects" of current commercially available fluorescence probes (Photofrin, Rezafirin, Visudyne, etc.), and advantages suitable for observation and treatment deep inside the body. In fact, a phase III clinical trial of an antibody-drug conjugate using IR-700 as a near-infrared fluorescent dye (ASP-1929) is already underway for head and neck cancer, and is expected to open up a new future in cancer treatment.

In this paper, we introduce photoimmunotherapy using IR-700 binding antibody, which is currently under development by Prof. Hisataka Kobayashi (NIH) and Rakuten Medical, Inc. In addition, we will introduce our works on photoimmunotherapy using indocyanine green (ICG) derivative-modified liposome (ICG-Lipo) with ICG modifier embedded in the membrane surface, which is being studied clinically in the veterinary field through the "Consortium: Tottori University, Chiba University, Private Veterinary Hospital, ASUKA MEDICAL Inc., TOKYO IKEN CO., LTD, and TATEYAMA MACHINE CO., LTD." led by Prof. Yoshiharu Okamoto (Tottori University).

Finally, we will introduce a new approach to deep biological observation using over 1000 nm in the near-infrared range (OTN-NIR: 900–2500 nm), which is called "the second and third optical window of the living body."

2 Limitations of Conventional Photodynamic Therapy Methods

Photodynamic therapy (PDT) is defined by the Japan Photodynamic Association as "PDT, using a photochemical reaction between a photosensitive substance that accumulates in cancer and laser light irradiation, is a minimally invasive treatment method that can selectively treat cancer lesions with low energy and causes minimal damage to normal tissues." With regard to the current status of PDT in cancer treatment, the first-generation Photofrin PDT is indicated for the treatment of early-stage lung cancer, early-stage esopha-

Y. Tamura (✉) · A. Suganami
Department of Bioinformatics, Graduate School of Medicine, Chiba University, Chiba, Japan
e-mail: yutaka_tamura@faculty.chiba-u.jp

Y. Okamoto
Department of Veterinary Clinical Medicine, School of Veterinary Medicine, Tottori University, Tottori, Japan

T. Ishizawa (ed.), *Fluorescence-Guided Surgery*, https://doi.org/10.1007/978-981-19-7372-7_34

geal cancer, gastric cancer, and early-stage cervical cancer. The second-generation Rezafirin PDT is indicated for the treatment of early-stage lung cancer, primary malignant brain tumors, and locally recurrent esophageal cancer after chemoradiotherapy or radiotherapy and has shown good results. In the current rapidly aging population, PDT as a cancer therapy for the elderly is a highly promising treatment modality. In the second-generation Rezafirin PDT, photosensitivity due to photosensitive substances, which was a major issue in the first-generation Photofrin PDT, has been greatly reduced. In addition, the laser system, which was large and expensive in the first-generation Photofrin PDT, has been improved to a compact and low-cost diode laser. The indications for Photofrin PDT are expected to be further expanded in the future, and it is thought to become a pillar of a new cancer treatment strategy.

On August 15, 2017, the Japanese Ministry of Economy, Trade and Industry (METI) made a presentation titled "International Standard on Safety of Advanced Medical Devices (Photodynamic Therapy Devices) Issued - Japan-originated International Standard Issued, a Step Toward Expanding Exports of Therapy Devices" with the following contents:

> The main points of the newly published International Standard (IEC 60601-2-75) are as follows:
>
> - In order to ensure the safety of photodynamic therapy equipment, the minimum necessary technical specifications such as electrical and mechanical hazards, protection against excessive radiation, temperature and other hazards are specified.
> - We clarified the optical properties of photodynamic therapy devices related to their performance and safety, and the relationship between the properties of drugs and devices (light sources and light guides).
> - This standard was proposed by Japan to the IEC (International Electrotechnical Commission) TC62/SC62D (Medical Electronic Devices) in January 2015, and discussions were held with experts from Germany, the Netherlands, the United States, and other countries to formulate the standard. After about two and a half years of coordination with each country, the draft standard was approved by the final vote with 20 countries in favor and 0 countries against, and was officially published as an international standard.

Thus, photodynamic therapy is an ideal treatment method that can specifically kill tumor cells with low energy without causing significant damage to normal tissues by irradiation with laser light, utilizing the light-absorbing property of light-sensitive substances (absorption of light of a specific wavelength generates singlet oxygen). However, most of the light-sensitive substances (Photofrin, Rezafirin, Visudyne, etc.) currently used in clinical practice have problems related to "insufficient selectivity to tumors" and "photosensitivity due to side effects" as well as problems related to observation and treatment of deep inside the body due to light scattering in the visible light range (400–650 nm).

In the near-infrared fluorescence region (650–900 nm), which is called the "optical window of the living body," the absorption and scattering of water and in vivo substances (red blood cells, etc.) are relatively small and the tissue permeability is excellent, which is advantageous for in vivo observation.

In fact, ICG was approved by the FDA in 1957 and is used clinically (primarily for observation) in Japan for the following applications:

- Liver function tests (plasma disappearance rate, blood stagnation rate, and hepatic blood flow measurement): diagnosis of liver diseases and determination of prognosis and cure
- Circulatory function tests (measurement of cardiac output, mean circulatory time, or abnormal blood flow): diagnosis of cardiovascular disease, evaluation of blood flow in blood vessels and tissues
- Sentinel lymph node identification: breast cancer, malignant melanoma

In addition to fluorescent dyes, it has been reported that the use of fluorescent probes capable of long-wavelength excitation, such as quantum dots, enables observation of deeper regions of the body than is possible using visible light [1]. Furthermore, photoimmunotherapy using an antibody-drug conjugate using IR-700, a near-infrared fluorescent dye, has emerged as an innovative therapy that overcomes the problems of conventional photosensitive substances (such as Photofrin, Rezafirin, and Visudyne) in insufficient selectivity to tumors, photosensitivity causing side effects, and inadequate light penetration deep into the body.

In this paper, we report details of the current status of photoimmunotherapy using near-infrared fluorescent dye such as IR-700 conjugated antibody and ICG derivative-modified liposome (ICG-Lipo).

3 Photoimmunotherapy Using Near-Infrared Fluorescent Probes (ICG, IR-700, IR800dye, etc.)

3.1 Photoimmunotherapy Using IR-700 Conjugated Antibody

Photoimmunotherapy using IR-700 conjugated antibody has been studied by Prof. Hisataka Kobayashi (NIH) and became famous when it was adopted by US President Barack Obama in his 2012 State of the Union address.

IR-700 is an antibody-drug conjugate consisting of an anti-EGFR antibody (cetuximab) conjugated with IR-700 (which is activated by near-infrared (700 nm) irradiation with a laser device and physically destroys cell membranes). In overseas clinical trials, complete responses were observed in patients with recurrent squamous cell carcinoma of the head and neck. In Japan, the Japanese Ministry of Health, Labour and Welfare (MHLW) designated it as a drug subject to the "system for designated prior review," and in September 2020, it became the first drug in the world to be approved for the treatment of "unresectable locally advanced or locally recurrent head and neck cancer."

3.1.1 Photoimmunotherapy Using IR-700 Conjugated Antibody: Basics

In photoimmunotherapy using IR-700 conjugated antibody, IR-700 conjugated antibody, which is a phthalocyanine derivative IR-700 conjugated to an antibody, is used as a drug. The IR-700 conjugated antibody administered to the living body first specifically recognizes and binds to the membrane antigen of cancer cells. Subsequent irradiation with near-infrared light causes changes in the chemical structure of IR-700, as well as deformation and aggregation of the complex of membrane antigen and IR-700-binding antibody, which is thought to induce impairment of the membrane structure function of cancer cells. Then, it is thought that dendritic cells take up antigens specific to cancer cells released when cancer cells rupture and die and cytotoxic T-lymphocytes (CTLs) are induced.

EGFR (overexpressed in lung, pancreatic, and colorectal cancer), HER2 (overexpressed in breast cancer), and PSMA (overexpressed in prostate cancer) have been developed and validated as antibodies that bind IR-700 (Fig. 34.1) [2].

3.1.2 Photoimmunotherapy Using IR-700 Conjugated Antibody: Clinical Application

In the clinical trial on photoimmunotherapy for head and neck cancer started in March 2019 at the National Cancer Center Hospital East and other institutions, IR-700 conjugated antibody, ASP-1929, which is IR-700 conjugated to cetuximab (an antibody drug that specifically recognizes and binds to EGFR), is used. In this treatment, multiple optical fibers are inserted into the affected area and near-infrared irradiation is performed for about 5 minutes under local or general anesthesia.

At the 54th Annual Meeting of the American Association for Cancer Therapy, a phase IIa study of IR-700 conjugate antibody, ASP-1929, in 30 patients with recurrent head and neck cancer who had received multiple prior therapies demonstrated an overall response rate of 43% (complete response: 13%, partial response: 30%), a median progression-free survival of 5.2 months, a median overall survival of 9.3 months, and 43.3% of serious adverse events (10% of which were related to photoimmunotherapy with ASP-1929).

3.1.3 Photodynamic Therapy Using IR-700 Conjugated Antibody: Additional Information

The international phase III clinical trials of photoimmunotherapy using IR-700 conjugated antibody (LUZERA-301) are currently in progress. In this randomized controlled trial (183 patients in the photoimmunotherapy group and 92 patients in the standard chemotherapy group), progression-free survival and overall survival in patients (aged 18 years or older) with locally recurrent squamous cell carcinoma of the head and neck that have failed to respond to treatment or

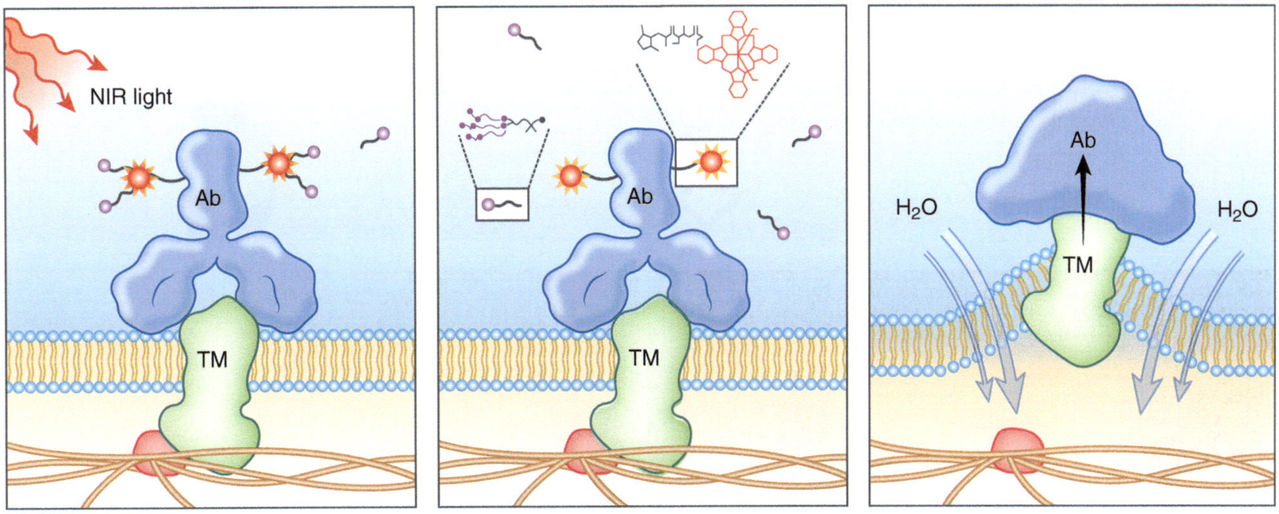

Fig. 34.1 Mechanism of rupture and death of cancer cells using IR-700 conjugated antibody. (From *ACS Cent Sci* 2018; 4: 1559–1569)

has recurred despite at least two standard systemic chemotherapies, including platinum-based agent, will be evaluated.

3.2 Photoimmunotherapy Using ICG Derivative-Modified Liposome (ICG-Lipo)

Photoimmunotherapy using ICG derivative-modified liposome (ICG-Lipo) has been clinically studied in the veterinary field by the "Industry-Academia Collaboration Consortium: Tottori University, Chiba University, Private Veterinary Hospital, ASUKA MEDICAL Inc., TOKYO IKEN CO., LTD, and TATEYAMA MACHINE CO., LTD." led by Prof. Yoshiharu Okamoto (Tottori University).

3.2.1 Photoimmunotherapy Using ICG-Lipo: Basics

In this treatment, ICG-Lipo, which is a liposome preparation containing ICG derivatives modified with alkyl chains or phospholipids on the basic skeleton of ICG, is used as a drug (Fig. 34.2) [3, 4].

ICG-Lipo administered in vivo accumulates in cancer tissues through enhanced permeability and retention effect (EPR effect, Fig. 34.3) [5, 6]. Subsequently, dendritic cells recognize neoantigens, which are released from cancer tissues by heat generation and singlet oxygen generation caused by near-infrared light irradiation, as antigens, leading to induction of anti-tumor immune responses by CTLs (Fig. 34.4) [7]. ICG-Lipo is reported to remain in cancer tissues for about 3 weeks.

3.2.2 Photoimmunotherapy Using ICG-Lipo: Application

Photoimmunotherapy using ICG-Lipo is a therapy to kill cancer cells by using ICG-Lipo, which can specifically accumulate and retain cancer cells for a long time due to its EPR effect, and irradiation with near-infrared light (810 nm).

In the clinical study in the field of veterinary medicine, one cool of the treatment requires 3 weeks (Fig. 34.5). Specifically, ICG-Lipo is administered intravenously on day 1 and then irradiated with near-infrared light (weak light: 2.5 W in the table) for approximately 20 min. Thereafter, the patient was administered with *Mycobacterium* tuberculosis thermal extract (Answer 20, solution A: 1/10 concentration, solution B: 1/100 concentration) and irradiated (5 W and 2.5 W, alternatively) with a commercially available device (ASUKA MEDICAL's DVL-15 semiconductor laser, TOKYO IKEN's Hyper5000 super riser) in an outpatient clinic. In addition, a near-infrared radiation therapy device for home use (TOKYO IKEN's Super Riser Mini) may be used for irradiation at a physician's discretion.

The ICG-Lipo used in the clinical study contains Answer 20 (solution B) and anticancer drugs (carboplatin, bleomycin, vincristine, and doxorubicin: each 1/10 of the usual dose). Therefore, the ICG-Lipo is supposed to be equipped with the following four functions.

1. ICG site: induction of acquired immunity by photodynamic hyperthermia

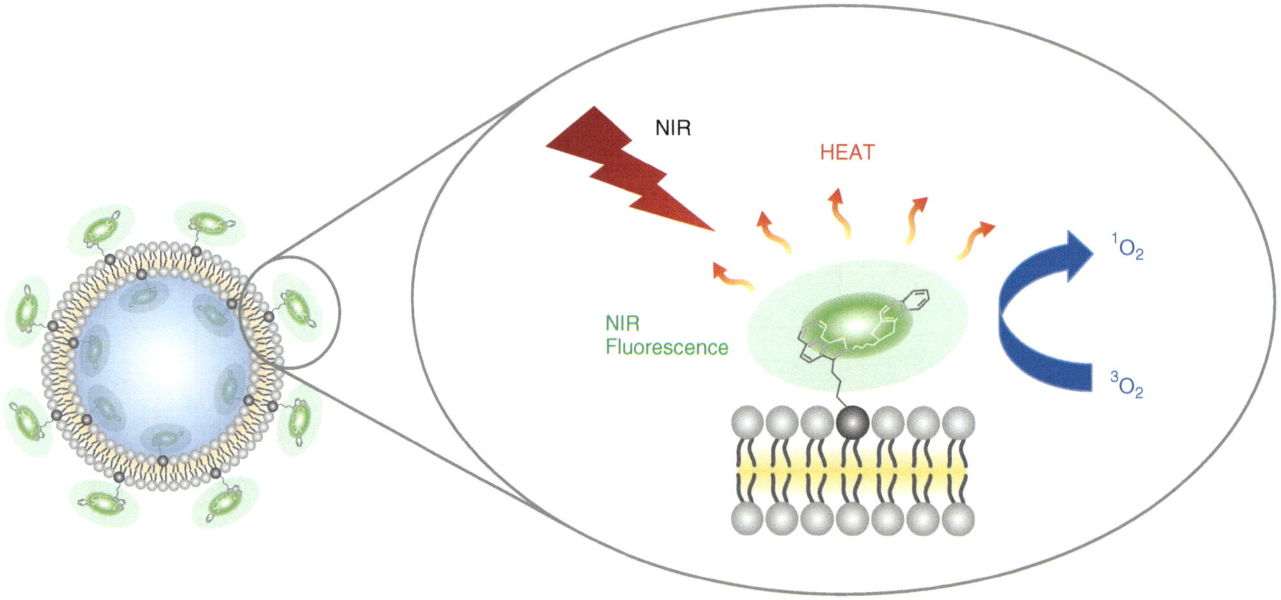

Fig. 34.2 Structure of ICG-Lipo and mechanism of luminescence, heat generation, and singlet oxygen production by near-infrared irradiation. (From *Bioorg Med Chem Lett* 2012; 22: 7481)

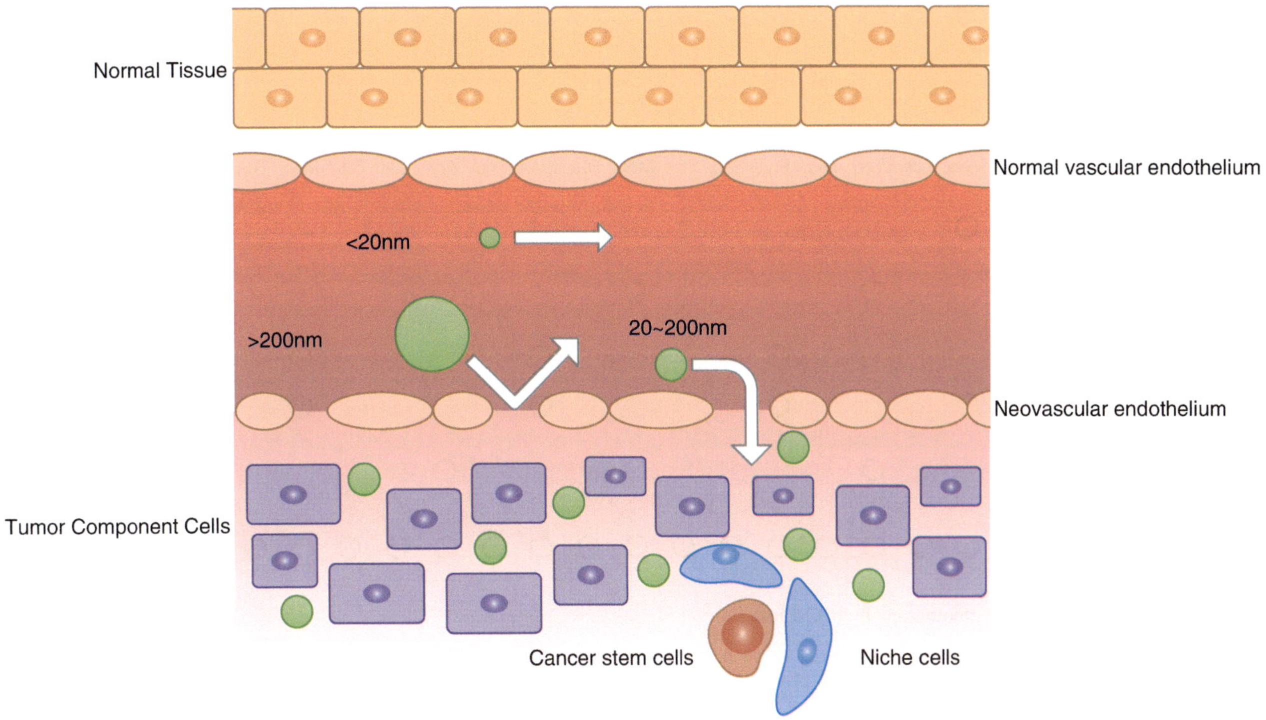

Fig. 34.3 Specific accumulation in tumor tissue by EPR effect. (Created based on *Cancer Res* 1986; 46: 6387)

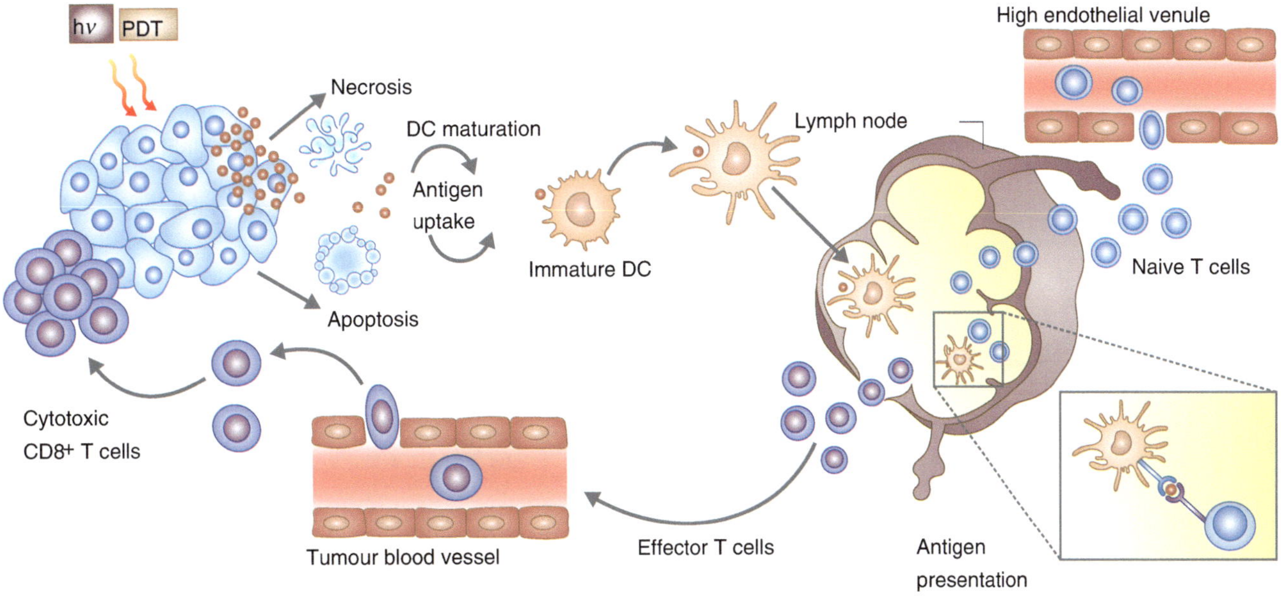

Fig. 34.4 Mechanism of action of photoimmunotherapy. (Created based on *Nat Rev Cancer* 2006; 6: 535)

2. Lipo portion: specific and long-term accumulation of cancer tissues
3. Answer 20: induction of innate immunity that was suppressed by cancer
4. Anticancer drug: inhibition of cancer cell growth

Since September 2013, an "industry-academia consortium: Tottori University, Chiba University, private veterinary hospitals, ASUKA MEDICAL Inc., TOKYO IKEN CO., LTD, and TATEYAMA MACHINE CO., LTD." led by Prof. Okamoto has been conducting veterinarian-led clinical trials for diagnosis and treatment of companion animals (Fig. 34.6). As a result of this treatment for 354 dogs and cats from October 2014 to July 2018 after the change to the four anticancer agents, 37% of cases were markedly effective and efficient, 42% were stable, and 22% were ineffective.

	1 course																				
	1	2	3	4	5	6	7	8	9	10	11	12	13	14	15	16	17	18	19	20	21
	Mon.	Tue.	Wed.	Thu.	Fry.	Sat.	Sun.	Mon.	Tue.	Wed.	Thu.	Fry.	Sat.	Sun.	Mon.	Tue.	Wed.	Thu.	Fry.	Sat.	Sun.
ICG-Lipo	O																				
Answer 20 A			O					O				O					O				
Answer 20 B					O					O					O				O		
at the hospital 2.5W	O				O					O					O				O		
5.0W			O					O				O					O				
at home Super riser mini	O	O	O	O	O	O	O	O	O	O	O	O	O	O	O	O	O	O	O	O	O

Fig. 34.5 Schedule of photoimmunotherapy by ICG-Lipo

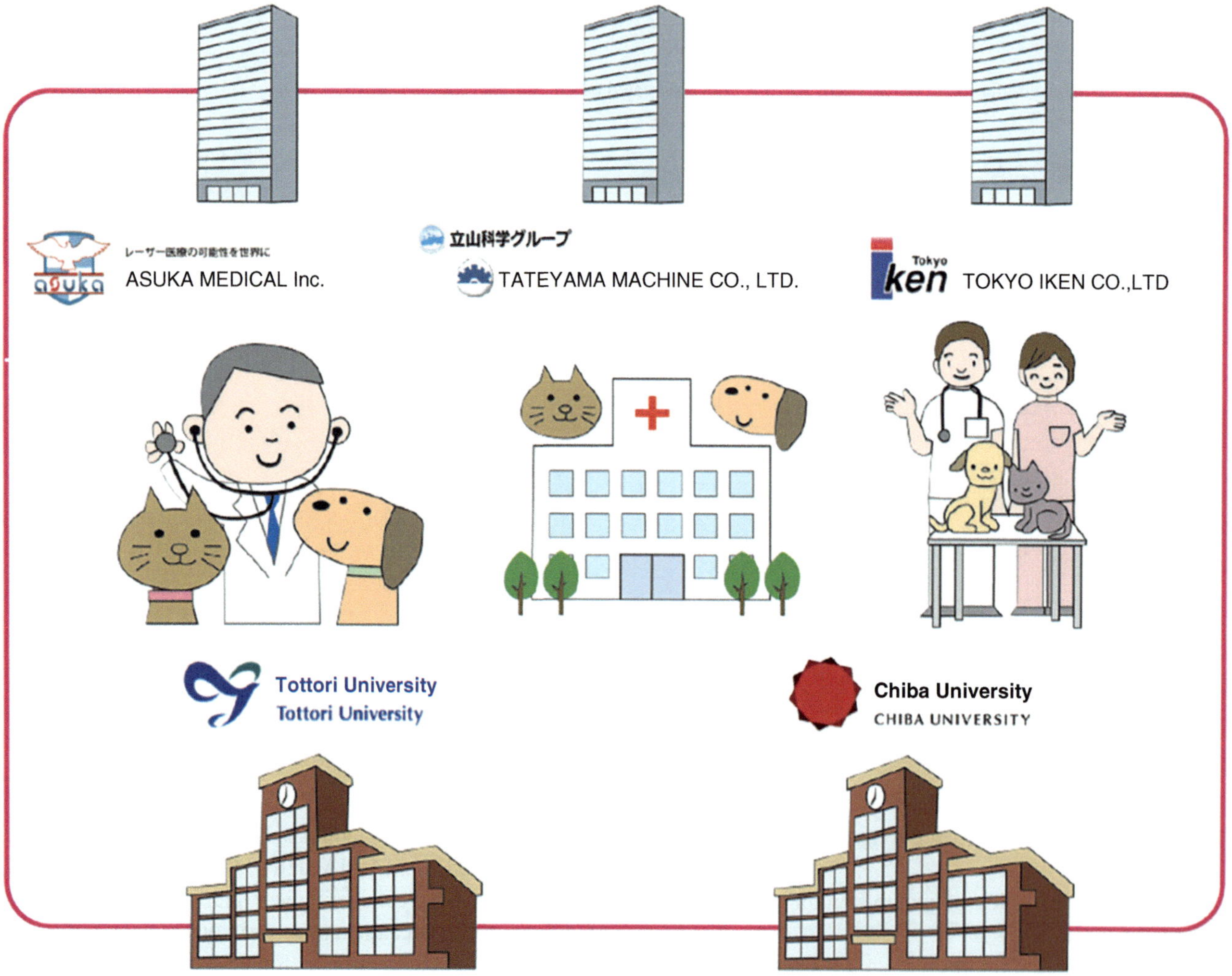

Fig. 34.6 Industry-university consortium for photoimmunotherapy using ICG-Lipo

3.2.3 Photoimmunotherapy Using ICG-Lipo: Animal Clinical Cases (Answer 20 in Combination with Anticancer Drug)

Case 1: Cat, Female, 10 Years old, 4.1 kg

Intranasal Lymphoma (Maxillary Bone Infiltration)

[Present Illness]

The patient presented to a nearby veterinary hospital with the chief complaint of sneezing and eye discharge in the right eye. Since there were no significant lesions at the first visit, antibiotics were administered, and the symptoms improved. However, 2 months later, the sneezing recurred, and the patient presented with mild epistaxis and mild swelling from the cheek to the right side of the nose. Antibiotics were started again, but the swelling of the right cheek gradually increased, and protrusion of the right eyeball and mild epistaxis was observed (Fig. 34.7). Head radiographs showed osteolysis of the maxillary region (Fig. 34.8). Based on the results of cytology, an epithelial tumor was suspected (the final diagnosis was lymphoma).

At the time of informed consent, the owner did not wish to be treated aggressively with radiation therapy, so we decided to treat the patient with ICG-Lipo.

[Treatment]

ICG-Lipo was administered intravascularly (day 0), followed by outpatient treatment (irradiation 5 times a week). A near-infrared light source (DVL-15 semiconductor laser, ASUKA MEDICAL) was used to irradiate the affected area for 20 minutes (output 5 W). As adjuvant therapy, infusions were administered when appetite decreased.

[Outcomes]

Facial swelling decreased gradually after the start of treatment, but swelling was observed again after 1 week. Thereafter, the symptoms repeatedly improved and worsened (first to fourth course) (Fig. 34.7). And finally, the facial swelling rapidly decreased from the fifth course, and the nasal embolization almost improved (98th day) (Fig. 34.7). The results of X-ray examination also showed improvement in the bone resorption image in the maxillary region (Fig. 34.8).

However, about 3 months after treatment with ICG-Lipo was discontinued (day 168), the tumor regrew, and treatment with ICG-Lipo was started again on day 173, but the patient's general condition worsened due to renal failure, and euthanasia was performed on day 359.

Case 2: Dog (Miniature Dachshund), Male, 9 Years old, 5.6 kg

Intracardiac Metastases of Cardiac Tamponade and Splenic Angiosarcoma

[Present Illness]

The patient had fainted 5 days before the visit to a nearby veterinary hospital and fainted again on the day of the visit. His condition at the time of the visit was pale mucous membranes, Ht: 23%, and unclear heart sounds.

The patient was referred to the Veterinary Medical Center, Faculty of Agriculture, Tottori University, for a thorough

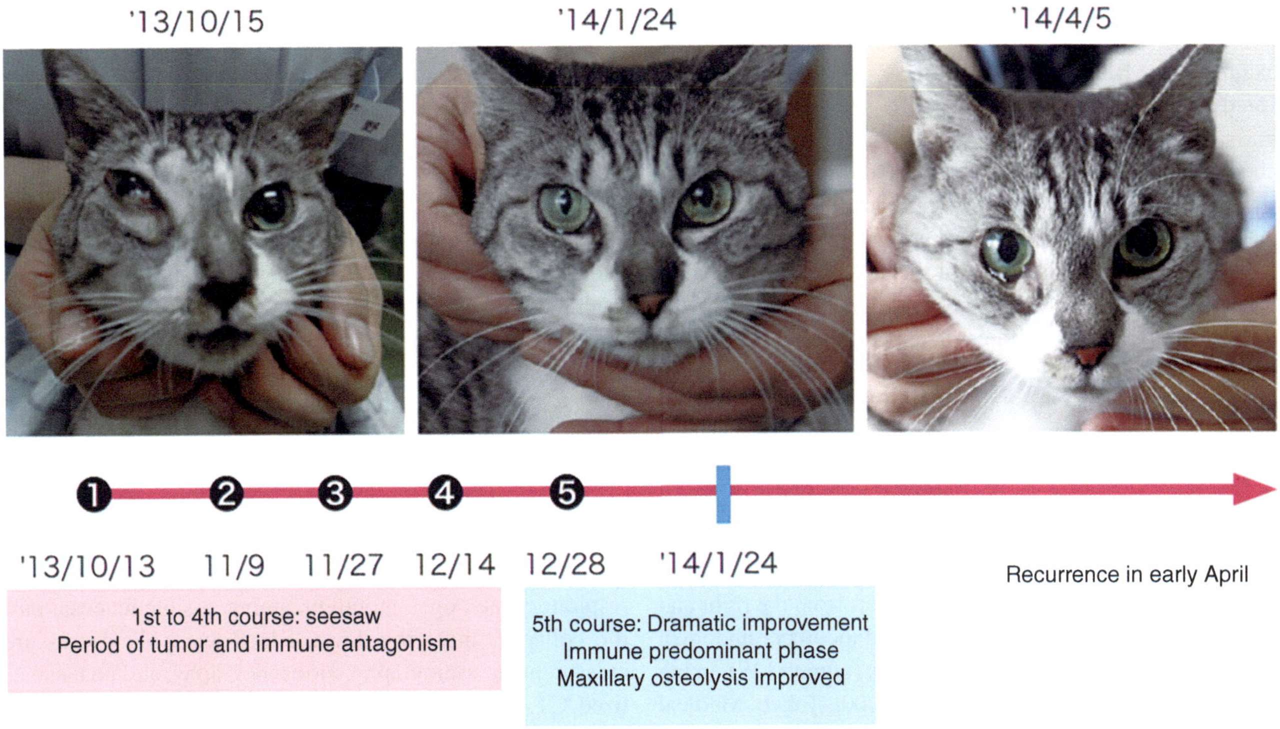

Fig. 34.7 Case 1: right eye protrusion and minor epistaxis

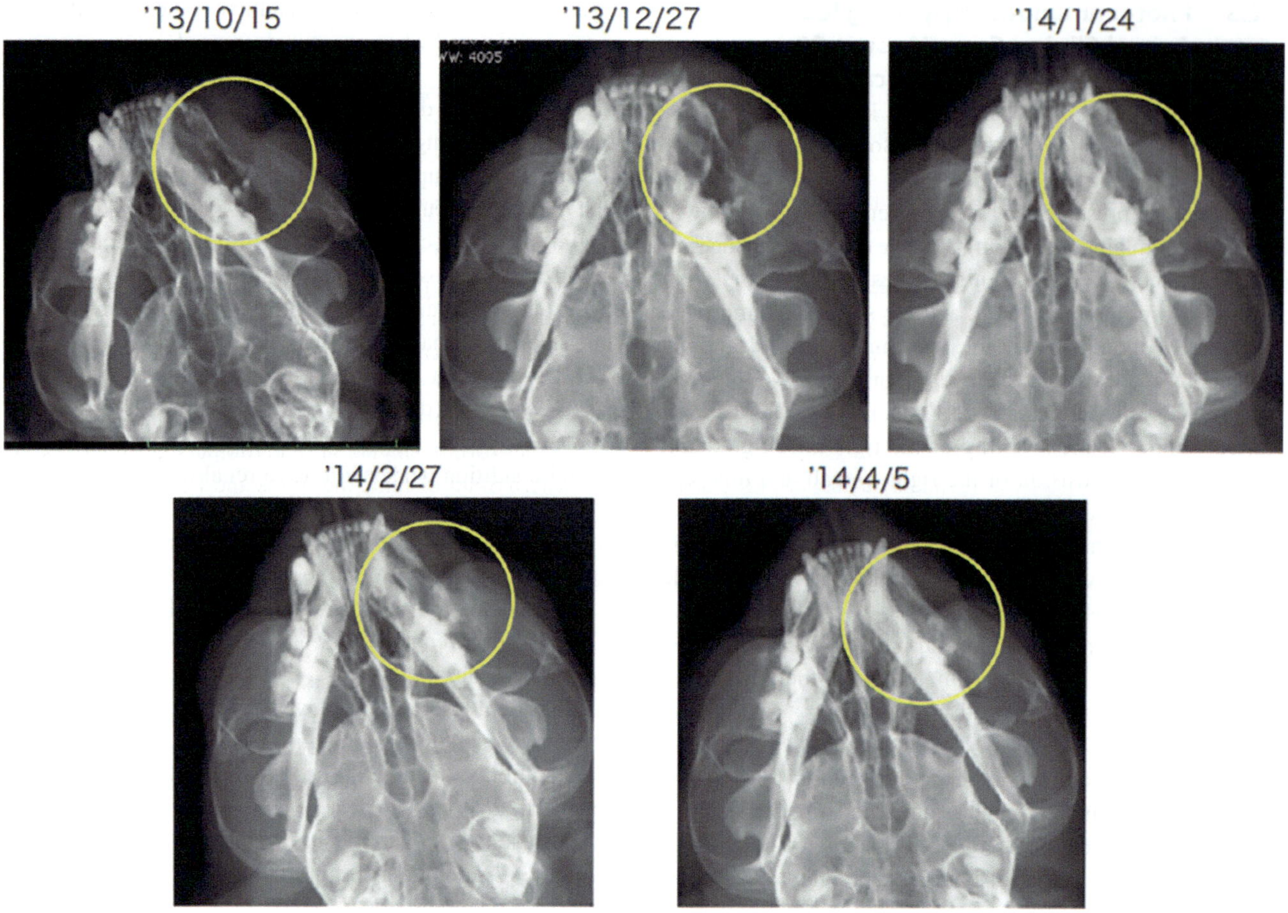

Fig. 34.8 Case 1: improvement of maxillary osteolysis

examination. Blood tests revealed severe anemia and elevated liver enzymes. Ultrasonography showed a right atrial mass, pericardial effusion, bloody ascites, mass in the abdominal wall, and splenomegaly. In addition, a chest X-ray showed cardiomegaly caused by cardiac tamponade, and the patient was tentatively diagnosed with a ruptured hemangiosarcoma of splenic origin and cardiac metastasis (Fig. 34.9).

At the time of informed consent, we explained to the owner that the prognosis of cardiac metastasis of hemangiosarcoma of splenic origin is very poor and that there is no effective treatment at this time and decided to treat the patient with ICG-Lipo. At that time, no adjuvant therapy was given.

[Treatment]

Considering the fact that the patient came from a distant place, ICG-Lipo was administered intravascularly at our hospital, followed by home treatment (daily irradiation). The owner was instructed to irradiate the heart from the right and left chest walls for 20 minutes each and the upper abdomen for 20 minutes, for a total of 60 minutes (output 5 W), using a near-infrared light source (Hyper5000, Tokyo Medical Research Institute, Inc.) loaned to the owner.

[Outcomes]

From the first week of treatment, the patient's appetite recovered and her energy increased. In the third week, the patient's general condition became very good, and blood test results showed improvement in anemia and normalization of liver enzymes. In addition, ultrasonography showed the disappearance of ascites, abdominal mass, and cardiac tamponade. Furthermore, although there was no significant change in the size of the right atrial mass, an echo-free area was observed, suggesting degeneration and necrosis inside the tumor (Fig. 34.9).

Case 3: Dog (Miniature Dachshund), Male, 13 Years old, 8.2 kg

Adrenal Medullary Tumor with Intravascular Invasion

[Present Illness]

On arrival at the hospital (day 0), the patient was in a state of shock with a body temperature of 37 °C, vomiting, rapid respiratory rate, open mouth breathing, visible mucosal pallor, collapse, and lying down. Blood tests, blood pressure measurement, radiography, ultrasonography, and unanesthetized CT scan were performed, and an adrenal medullary tumor with intravascular invasion was suspected (Fig. 34.10).

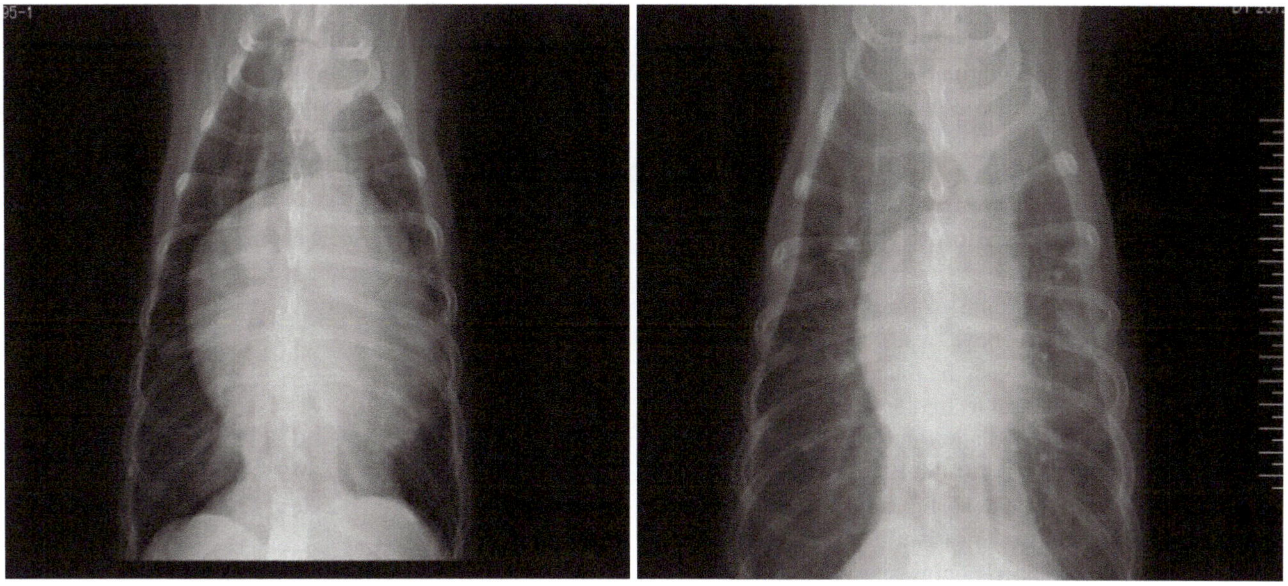

Fig. 34.9 Case 2: cardiac tamponade and intracardiac metastases of angiosarcoma

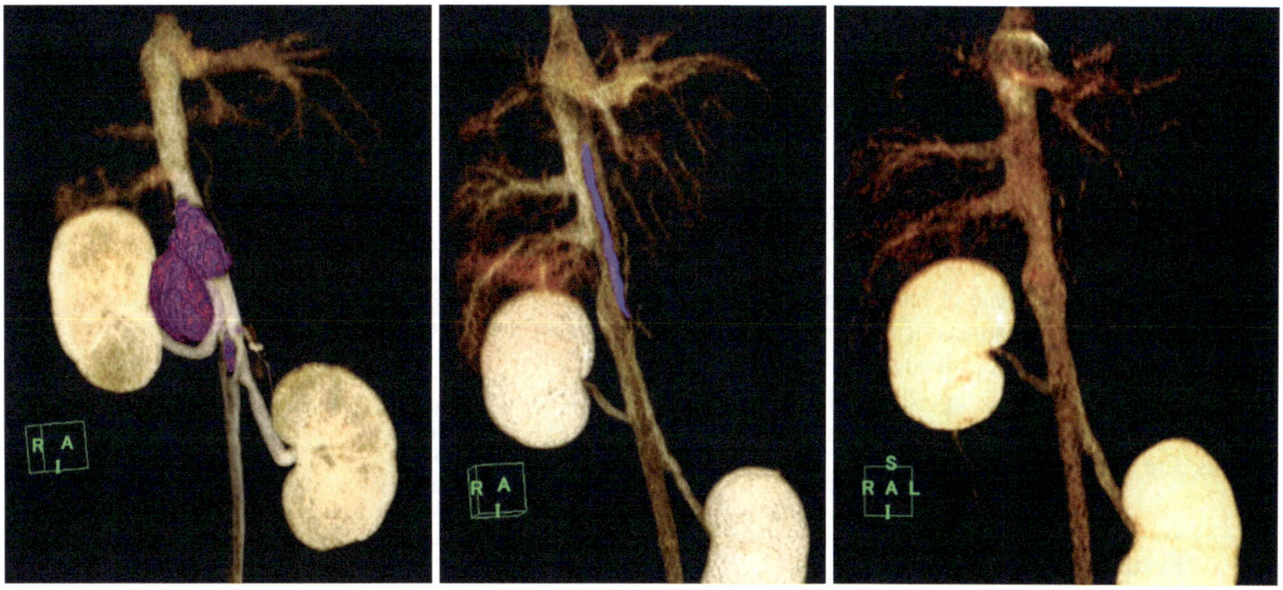

Fig. 34.10 Case 3: adrenal medullary tumor with intravascular invasion

After informed consent for the owner of the dog, it was decided to start treatment with ICG-Lipo from the 18th day of the disease.

[Treatment]

For photoimmunotherapy using ICG-Lipo, light irradiation was performed three times a week using a near-infrared light source device (Hyper5000, Tokyo Medical Research Institute), and Answer 20 (alternating between Solution A and B) was administered subcutaneously.

A total of four courses were conducted with periodic CT examinations.

[Outcomes]

The size of the tumor before treatment was 20 mm × 40 mm extravascularly, and a thrombotic tumor plug was identified in the posterior vena cava, with associated reduced hepatic venous blood flow. On the 69th day, the extravascular mass had disappeared and a linearly reduced thrombotic tumor plug measuring 2.5 mm × 57 mm was identified in the posterior vena cava. An increase in hepatic venous blood flow was also observed with the reduction of the tumor (Fig. 34.10). On the 121st day, ultrasonography, which also revealed the disappearance of the linear thrombotic tumor plug, revealed no abnormalities in the blood flow in the hepatic veins (Fig. 34.10).

3.2.4 Photoimmunotherapy Using ICG-Lipo: Potential Application to Human Clinical Care

Through the "Consortium for Industry-Academia Collaboration" described above, research and development is being conducted to establish the foundation for translating this safe and secure medical technology based on ICG-Lipo and near-infrared light irradiation into human medicine. In the future, the combined use of the ICG-Lipo drug delivery system and near-infrared diagnostic and therapeutic devices is expected to create a "noninvasive identification method" for early cancer detection, a "noninvasive treatment method" for cases where surgery is not possible, and a "high-quality palliative medicine" for terminal cancer patients.

Point

- The near-infrared fluorescence probe enables a noninvasive medical treatment called photoimmunotherapy using the optical window of the body.
- The IR-700 conjugated antibody (ASP-1929, conjugate of IR-700, a phthalocyanine derivative, and anti-EGFR antibody) is now in phase III clinical trials in patients with squamous cell carcinoma of the head and neck and is also indicated for the treatment of breast cancer and prostate cancer by converting anti-EGFR antibody to anti-HER2 antibody and anti-PSMA antibody, respectively.
- ICG derivative-modified liposomes (ICG-Lipo) were shown to be effective in clinical trials in companion animals with various cancer types.

4 Future Perspectives

Photoimmunotherapy using the IR-700 conjugated antibody, ICG-Lipo, described in this paper is a therapeutic method that utilizes the subcutaneous penetration (10–20 mm) in the "optical window of the living body (650–900 nm)." Recently, in over 1000 nm in the near-infrared range (OTN-NIR: 900–2500 nm) where autofluorescence and tissue absorption/scattering decrease, the use of the second optical window (OTN-NIR-II: 1100–1350 nm) and the third optical window (OTN-NIR-III: 1550–1800 nm) have been reported to improve the subcutaneous penetration up to 20–30 mm [8, 9]. In fact, imaging systems and OTN fluorescent probes have already been developed for in vivo imaging [10]. In the near future, novel fluorescence probes using OTN-NIR-II and OTN-NIR-III will also be used for enhancing the efficacy of photoimmunotherapy.

References

1. Chinnathambi S, Shirahata N. Recent advances on fluorescent biomarkers of near-infrared quantum dots for in vitro and in vivo imaging. Sci Technol Adv Mater. 2019;20:337–55.
2. Sato K, Ando K, Okuyama S, et al. Photoinduced ligand release from a silicon phthalocyanine dye conjugated with monoclonal antibodies: a mechanism of cancer cell cytotoxicity after near-infrared photoimmunotherapy. ACS Cent Sci. 2018;4:1559–69.
3. Suganami A, Toyota T, Okazaki S, et al. Preparation and characterization of phospholipid-conjugated indocyanine green. Bioorg Med Chem Lett. 2012;22:7481–5. [3] Suganami A, Toyota T, Okazaki S et al Preparation and characterization of phospholipid-conjugated indocyanine green as a near-infrared probe
4. Toyota T, Fujito H, Suganami A, et al. Near-infrared-fluorescence imaging of lymph nodes by using liposomally. Bioorg Med Chem. 2014;22:721–7. [4] Toyota T, Fujito H, Suganami A et al: Near-infrared-fluorescence imaging of lymph nodes by using liposomally formulated indocyanine green derivatives
5. Matsumura Y, Maeda H. A new concept for macromolecular therapeutics in cancer chemotherapy: mechanism of tumoritropic. Cancer Res. 1986;46:6387–92. [5] Matsumura Y, Maeda H. A new concept for macromolecular therapeutics in cancer chemotherapy: mechanism of tumoritropic accumulation of proteins and the antitumor agent smancs
6. Suganami A, Iwadate Y, Shibata S, et al. Liposomally formulated phospholipid- conjugated indocyanine green for intra-operative. Int J Pharm. 2015;496:401–6.
7. Castano AP, Mroz P, Hamblin MR. Photodynamic therapy and antitumour immunity. Nat Rev Cancer. 2006;6:535–45.
8. Starosolski Z, Bhavane R, Ghaghada KB, et al. Indocyanine green fluorescence in the second near-infrared (NIR-II) window. PLoS One. 2017;12:e0187563.
9. Ding F, Zhan Y, Lu X, et al. Recent advances in near-infrared II fluorophores for multifunctional biomedical imaging. Chem Sci. 2018;9:4370–80.
10. Uemura M, Soga K. Development of an in vivo imaging method using a near-infrared fluorescent probe. Bunseki. 2019;3:114–7.

Afterword

Mitsuo Kusano

This book describes the introduction, practice, and future prospects of fluorescence-guided surgery in various surgical fields. The book is very easy to understand not only for doctors who have already introduced this method in their clinical practices but also for those who are about to do so.

The *in vivo* fluorescence imaging has been used in ophthalmology and sentinel node identification of breast cancer since around 2000 and has been vigorously applied to cardiovascular surgery, neurosurgery, and digestive systems such as liver, gallbladder, and GI tracts. As you know, rapid development has been seen in less than 20 years since then.

During this period, various fluorescent agents have been used for fluorescence imaging, and ICG has been one of the triggers for the widespread use of this method because of its high safety and low cost. Now that it has been 20 years since the clinical application of fluorescence imaging has become widespread, the publication of this book, which covers all surgical fields, is very timely.

In the last few years, this method has been applied to almost all organs, including pulmonary, gynecological, urological, and other new fields, as well as robot-assisted surgery and photodynamic therapy.

All of the authors, including the editor, Dr. Ishizawa, are active in the frontline medical field in Japan and abroad and are practitioners of this method, and this book contains practical know-hows of this method that can be applied immediately from tomorrow. It should be obvious to readers who read this book that there are many things to learn from different fields of expertise through the common approach of fluorescence imaging. In addition, this book contains a wealth of essentials that can be applied to various fields, and we encourage readers to read through the descriptions of fields other than their own.

Although some applications of intraoperative fluorescence imaging, such as evaluation of blood perfusion, sentinel lymph nodes mapping during breast cancer surgery, and brain tumor identification, have been covered by the Japanese insurance system, there remains a lot of unapproved imaging methods. It is necessary to promote further practice and development to feedback to clinical practice. On the other hand, it is also expected that the investigation of fluorescence mechanics in cancer cells and specific organs leads to the development of new diagnostic and therapeutic approaches.

I would thank all contributors who have worked on the medical application of fluorescent imaging and those who have been involved in the development of imaging equipment such as fluorescent endoscopes. I would also like to take this opportunity to thank Mr. Shingo Nagasawa and his staff at Medical View and Springer for their efforts in producing this English edition. I hope that this book will be a stepping stone for the further development of this field and that more patients will benefit from the spread of this method.

July 2022

M. Kusano
Chief Clinical Director, Yoichi Hospital, Hokkaido Social Work Association, Yoichi, Japan

Honorary Director, Kushiro Rosai Hospital, Kushiro, Japan

Honorary Member, Japanese Society for Fluorescence Guided Surgery, Tokyo, Japan

T. Ishizawa (ed.), *Fluorescence-Guided Surgery*, https://doi.org/10.1007/978-981-19-7372-7

GPSR Compliance

The European Union's (EU) General Product Safety Regulation (GPSR) is a set of rules that requires consumer products to be safe and our obligations to ensure this.

If you have any concerns about our products, you can contact us on ProductSafety@springernature.com

In case Publisher is established outside the EU, the EU authorized representative is:

Springer Nature Customer Service Center GmbH
Europaplatz 3
69115 Heidelberg, Germany

Batch number: 10731528

Printed by Printforce, the Netherlands